T0192834

An Introduction to Cyber Analysis and Targeting

Jerry M. Couretas

An Introduction to Cyber Analysis and Targeting

Jerry M. Couretas
Washington, DC

ISBN 978-3-030-88561-8   ISBN 978-3-030-88559-5   (eBook)
https://doi.org/10.1007/978-3-030-88559-5

This Springer imprint is published by the registered company Springer Nature Switzerland AG
The registered company address is: Gewerbestrasse 11, 6330 Cham, Switzerland

The development of this book had many hands. Mr. Ed Waltz was key in the early discussions, feedback and support for this work. In addition, I would like to thank Vicky Pate and Pat Adrounie for reviewing each of the chapters during the writing phase.
I would like to dedicate this book to Monica, Sophie and Ella, for the time and patience they provided. In addition, I would also like to thank Jorge and Aida Carpio, for their support and mentoring. And finally, to my parents, Gus and Mary, for providing an example of persistence and faith.

Foreword

The cyber threat landscape has grown to pose risks to every facet of our lives – infrastructure, finance, communications, health, personal and social media, and even our smart homes and vehicles. As the complexity of the cybersphere has grown, even so have the threat vectors and targeting mechanisms. To defend a network, you must understand how the attacker strategizes, analyzes, and targets a network. This book uniquely describes the offensive analysis and targeting process as a set of conceptual models.

The book market is replete with books at the high, strategic level of cyber warfare and the deep, tactical level of hacking methods unique to enterprise systems. This book stands alone in providing conceptual models for addressing cyber *analysis and targeting* – the systematic analysis and prioritization of cyber entities considered for possible cyber engagement, and the planning of vectors for access.

The sophistication of the cyber-attack process has grown with the complexity of networked systems and their operations. The disciplines of cyber-Intelligence (CI) and cyber counterintelligence (CCI) conduct detailed analyses of the cybersphere and carefully select targets to exploit and conduct operations. Cyber operators, offensive or defensive, need to understand the methods to perform analysis of targeted networks and the means to select targets and then conduct cyber operations.

This book follows the traditional approach of introducing a new discipline: Grammar, Logic and Rhetoric. The grammar (unique terminology) of the cyber operational world is introduced throughout; next, the text describes the logic of how cyber analysis is conducted, how targeting selection is performed, and the means by which cyber operations are conducted. Finally, the rhetoric of cyber operations is narrated by real-world use cases that illustrate the mechanisms introduced throughout.

Jerry Couretas is uniquely equipped to introduce this subject because of his broad expertise in the fields of military and cyber operations, analysis, modeling, and simulation. Dr. Couretas has spent the last decade modeling and simulating cyber systems for network defense. As Editor-in-Chief of the *Journal of Defense Modeling and Simulation* (JDMS), Dr. Couretas produced over 20 special issues on subjects of national security importance simulating complex military operations. He has also served on the North Atlantic Treaty Organization's (NATO) Modeling and Simulation Group 117 (NMSG 117), cyber modeling and simulation (M&S). Those who build computational models of systems and operations must know the details, and Jerry has that depth of knowledge in the cyber field. His experience encompasses cyber risk mitigation, cyber ops analysis, cyber analytics, and targeting. I have enjoyed working with Jerry for over 4 years as he conceived and prepared this text. His depth of understanding and expertise in explaining this topic is evident as he introduces defensive and offensive cyber operators to the state of the practice in cyber analysis and cyber targeting.

Ed Waltz

Contents

Chapter 1
Cyber Analysis and Targeting

The goal of this book is to describe cyber analysis and targeting for defensive applications. One objective of developing a cyber analysis and targeting methodology is to add information technology (IT) considerations into traditional military operations research (OR). For example, we will include cyber threats, cyber terrain, IT architectures, and other information-related capabilities (IRCs) in a developing cyber analysis and targeting methodology, accounting for the steady ingress of cyber into military operations through IT-based improvements in weapons systems, telecommunications, and online media. In developing this cyber analysis and targeting methodology, we will leverage use cases that span from analysis to modeling and simulation. This includes a look at assessment, for resilient systems development, along with using novel modeling and simulation approaches to describe the target as a discrete event process that we will use to estimate the effects from a cyber attack.

Policy applications for cyberspace generally focus on resilience, or defensive applications, for the United States, the United Kingdom, Canada, and Australia. Similarly, the information assurance (IA) community has distilled both rule-based approaches (e.g., SANS 20) and standards (e.g., NIST: US National Institute of Standards and Technology) to guide network engineers in the development of secure cyber systems. We will therefore review this body of developing cyberspace policy and doctrine, which covers the increasing use of information-related capabilities (IRCs), in Chap. 2.

While cyberspace is still developing, in terms of policy and doctrine, conventional military engagements, missions, and campaigns have a rich history to draw on for analysis and targeting exemplars. One scenario might be to use traditional military analysis and targeting to focus on, and describe, adversary order of battle (OOB). For example, the locations and movements of Soviet divisions were thoroughly analyzed, for conventional analysis and targeting, over the course of the Cold War. Similarly, targeting for this kind of conventional engagement focused on affecting an enemy's ability to maneuver, which includes controlling the lines of communication (LOC). This might consist of destroying (e.g., denial) transport

© Springer Nature Switzerland AG 2022
J. M. Couretas, *An Introduction to Cyber Analysis and Targeting*,
https://doi.org/10.1007/978-3-030-88559-5_1

techniques (e.g., trucks, rail, associated lines of communication), reducing the ability to communicate or perform command and control (C2), or direct targeting of enemy forces and weapons. In addition, attacking enemy C2 included using information-related capabilities to target both the availability (i.e., jamming) and integrity (i.e., trust) of communication between a command headquarters and units in the field.

Much of the current military analysis is still based on Cold War era operations research approaches, a time period before microprocessors were key components in the trucks, trains, and telephones that a modern force relies on to conduct war. Cyber analysis and targeting should therefore incorporate considerations for the information technologies (IT) included in every element of a fighting force's order of battle (OOB). The ubiquitous cyber in modern fighting forces is defined as follows:

Cyber "of, relating to, or involving computers or computer networks" (Merriam Webster).

Cyber is often not considered in current conventional analysis, which includes estimating likely adversary courses of action (COAs). While communications intelligence (COMINT) is a factor in performing engagement through campaign-level planning, this is usually high level, and does not include the scale or scope of cyber effects. For example, standard tabletop exercises and wargames consider equipment effectiveness, physical terrain, and command and control (C2), among other force components, when doing force-on-force simulations. In fact, one method of bringing cyber into current, conventional, military modeling, and analysis is to turn the communications off, which only provides denial; missing the cyber effects that accrue from compromising the confidentiality of data stores, or modifying the integrity of an organization's key data sets (e.g., orders, geo location). We can therefore define cyber analysis as follows:

Cyber Analysis (1) the process of decomposing cyber information to synthesize explanation of adversary actions, networks, and cyber objects; (2) describing the physical, logical, or persona target in terms of the full spectrum of confidentiality, integrity, and availability (CIA) effects achievable via cyber means.

In defining cyber analysis, our first definition leans more toward the classical intelligence use of cyber for collection and determination of adversary intent. One of the current challenges is placing cyber alongside kinetic options when planning an operation. Definition (2) is therefore focused on the types of effects that the cyber analyst will be looking for when developing a target.

Both definitions of cyber analysis contribute to describing an order of battle, which can miss much of the cyber attack surface, or IT Achilles heel, of conventional C2 and maneuver elements. In addition, much of the targeting is based on traditional line of communication (LOC) elements, missing the more nuanced effects available via cyber. For example, cyber targeting is currently more likely to be thought of in terms of communications availability, and subsequent C2 challenges, than the longer-term effects that can result from confidentiality or integrity attacks. Chapter 3 will therefore address the scope of a cyber threat for both the individual components and the overall cyber system. This will include a use of cyber threat intelligence (CTI) to roll up the different elements of risk analysis in order to find and classify IT system vulnerabilities over the steps of a cyber attack process (Launius, 2019).

Chapter 4 transitions from the physical systems that hackers target to compromise data, to the human terrain. Information operations (IO) are used for strategic purposes to attack leadership and key personnel in order to create organizational friction via false information, potentially changing the course of their target's operations. One example was the outing of private e-mails in the 2014 Sony attack (Zetter, 2014a, b), causing Sony to cancel the release of a film that parodied the President of North Korea. In addition, we will look at the operation of cyber-based information operations (IO), including how popular news sources, including social media, were used to deliver "fake news" during the 2016 US Presidential Election, inciting dozens of riots within the United States (Mueller, 2019). For strategic effects, we will also look at how cyber operations compare to active measures from the former Soviet Union to create confusion and doubt. Similarly, for tactical effects, we will look at how cyber can be used to provide the wrong coordinates for a munition or mapping application, causing a bomb to hit the wrong target or a unit to go to the wrong location. IO therefore covers the overall spectrum described as information-related capabilities (IRCs) that govern current cyberspace operations.

As discussed in Chap. 4's use of social media to affect the news feeds and change the thinking of the general population, cyber means can also be used to collect and develop potentially compromising information on key decision makers. For example, in Chap. 5, we will look at how confidentiality attacks, where the acquiring of secure data (e.g., private keys, personnel information), can be used to unlock communication channels, determine centers of gravity (COG) (e.g., understand an organization's structure), or, in longer term, compromise key personnel with private information.

While Chap. 5 shows the detrimental effects that an intelligent adversary can have through the collection, and processing, of private information, Chap. 6 reviews security technologies that compose current cyber terrain. These end points, connections, and key nodes use security operations centers (SOCs) to secure an organization's data and key operating information. In addition to using component technologies for network protection, Chap. 6 will provide conceptual architecture techniques, including defense in depth, to layer the technologies and decrease the likelihood of attacker success. Similarly, we review implementation guidance, including denial and deception, to provide overall guidance in developing network security solutions.

While our primary focus is to use cyber analysis and targeting for defensive means, we can also look at the cyber threats in terms of well-developed attack models. One example is the Lockheed Martin Attack Cycle (Eric M. Hutchins, 2012), which provides a step-by-step map of how an attacker maneuvers from initial reconnaissance of a target to actions on objectives. This includes increasing knowledge across the steps. For example, the reconnaissance phase informs weaponization concerning the types of vulnerability evaluations that we perform in Chaps. 2 and 3, along with the intelligence, surveillance, and reconnaissance (ISR) development

(Chap. 5). In addition, we can use our understanding of defensive technologies (Chap. 6) to better understand target terrain.

Target analysis, as discussed in Chap. 7, will require information on appropriate delivery methods, and candidate techniques, for exploitation of target vulnerabilities and installation of the cyber weapon. We can therefore use the Lockheed Martin Attack Cycle to compare analysis and target intelligence development for kinetic and cyber munitions, at different stages of the attack life cycle. For example, we can look at "dumb bombs," precision-guided munitions (PGMs), unmanned autonomous systems (UAS'), and cyber, in parallel, to compare the information collection and support requirements across the spectrum of conventional, precision-guided, and cyber munitions (Fig. 1.1).

As shown in Fig. 1.1, analysis and targeting are done in advance of the mission for conventional "dumb bombs" and PGMs. With the introduction of drones (i.e., UAS'), and especially cyber, ISR is now part of the mission, a tool that operators use to refine the targeting solution both before attack cycle initiation and over the course of the attack. One advantage of UAS' and cyber is that they are more flexible for addressing dynamic, time-sensitive targets (TSTs). Conventional and "smart" munitions, on the other hand, maintain ISR as a decoupled, independent, process that occurs prior to the attack.

An additional challenge is that cyber also operates as a pseudo-kinetic actor, challenging analysis and targeting to assess effects that can take multiple forms over a range of time periods. Successful kinetic effects simply remove their targets from the battlefield. Cyber effects, however, span the compromise of information (i.e., confidentiality), the misuse of information (i.e., integrity), or the denial of information (i.e., availability). Quantifying any of these effects is still an art, with one source

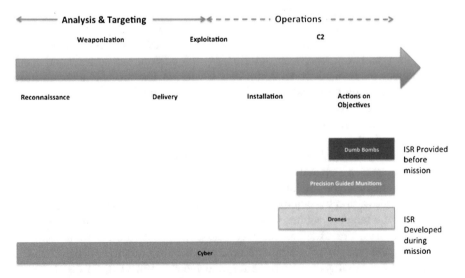

Fig. 1.1 ISR for analysis and targeting: "dumb bombs" to cyber operations

of inspiration coming from legacy joint munition effectiveness manuals (JMEMs), used to describe the performance of conventional bombs and guided munitions.

Due to the decoupling of ISR from kinetic strike, conventional munitions lend themselves to a more detached, technical assessment, with methods and techniques developed over decades. The Joint Munitions Effects Manual (JMEM) (US Army), for example, is used to document an explosive munitions' effects in engineering-level detail. A history of kinetic JMEMs defines weaponeering as follows:

> "… the process of determining the quantity of a particular type of weapon required to achieve a specific level of target damage by considering the effects of target vulnerability, warhead damage mechanism, delivery errors, damage criterion and weapon reliability, p. 1 of http://www.weaponeering.com/." (Weaponeering)

It is currently a challenge to provide this JMEM level of detailed engineering estimate for cyber munitions. For example, one of the analysis and targeting challenges for drones and cyber is that the process continues over the course of the mission, has the operator in the loop, and implicitly requires more operator assessment (Fig. 1.2).

As shown in Fig. 1.2, like conventional munitions, cyber requires a method to estimate effects. Because cyber effects usually include more than physical effects (e.g., information and psychological) and potentially cause effects at a wider scale, the description of all of these properties of cyber munitions is required before use. While kinetic effects have the Joint Munition Effectiveness Manual (JMEM), (US Army), cyber is still a challenge to describe as either an effects alternative (Mark Gallagher, 2013) (George Cybenko G. S., 2016), or an ISR complement, to conventional munitions. An additional difference is that cyber penetration and collection operations are often used for intelligence collection, as a strategic asset; with relatively few comparable effects measures.

Accounting for cyberspace operations' ill-defined effects in the application of information-related capabilities, or quantifying cyber, is a challenge that we will approach through currently known capabilities. One approach is to look at the application of cyber, either defensive or offensive, in terms of resource requirements, as an intermediate-term solution for describing cyberspace operations. For example, in Chap. 6 we describe network defense, along with component technologies and

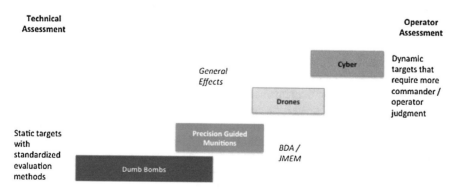

Fig. 1.2 Technical/operator assessment of munitions: dumb bombs to cyber

architecture techniques. A natural question in planning a network defense architecture is the cost to purchase and maintain a secure network. Chapter 7 provides an example attack process methodology that estimates the time and cost for an attacker to successfully attack an objective network.

The introduction to network terrain in Chap. 6 also describes the challenges that a network attacker (e.g., penetration tester) will need to analyze prior to prosecuting a target. The attack processes in Chap. 7 therefore describe how an attacker will formulate and execute his targeting process. For example, we will review the CARVER—i.e., criticality, accessibility, recuperability, vulnerability, effect, recognizability—targeting matrix to analyze key nodes during initial operational planning. Developing the targeting solution includes reviewing computer network vulnerabilities over the Lockheed Martin Attack Cycle to identify the target elements for specific effects—effects that cyber is ideally suited to provide. We can now, therefore, provide a definition for cyber targeting.

Cyber Targeting the practice of selecting targets, and pairing up the appropriate collection plan or response to them, on the basis of operational requirements.

In addition, cyber targeting is the final step in the overall scheme of cyber operations described in Fig. 1.1, which includes the defensive, operational preparation, and offensive steps of a cyber engagement.

When looking at Fig. 1.1, it seems obvious that there is currently a need for cyber analysis and targeting. In Chap. 7, we estimate the people, process, and technology costs of maintaining a defensive portfolio against hackers, terrorist, or nation state adversaries. One of the differentiators between the respective groups includes resourcing in terms of research, ISR, and operations. For example, a hacker will be expected to have no research, ISR limited to his cognitive capability, and operational capacity based on his ability to hack. At the other end of the spectrum, a nation state hacker will be able to draw from top-notch universities, access to finished intelligence products, and have state of the art tradecraft. We can therefore compare a defended network in terms of its component technologies (e.g., security baseline) and its architecture (e.g., SOC policies) for the resources required by the respective groups to access the network and compromise a target.

The targeting methods in Chap. 7 culminate in the cyber process evaluator, a "simple" means of estimating the time/cost imposed on an attacker based on the policies, processes, and technologies that compose the defensive terrain (i.e., Chap. 6). Each of the components in the defensive system can also be described by software architectures, with well-developed frameworks available to describe the terms (i.e., reference architecture), functions (i.e., solution architecture), connections (i.e., logical architecture), or implementation (i.e., physical architecture). Describing a cyber system in terms of general artifacts provides the network defender with a method for ensuring that the respective components are up to date, in terms of individual components and overall system security. In addition, architectures help with organizing the respective network architecture elements for easier management by the Security Operations Center (SOC). We review an example cyber system architecture, including its solution architecture, in Chap. 8.

The example cyber system solution architecture, from Chap. 8, provides us with a basic structure for studying the development of metrics for a cyber system. For example, we will look at key performance parameters (KPPs) to describe the solution architecture as a system. This includes system characteristics (e.g., stealth and speed). Similarly, we will use measures of performance (MOPs) to describe cyber system operator characteristics. MOPs for cyber operations are likely the limit of what can be measured for individual cyberspace targeting engagements. As we move to cyber missions and campaigns, MOPs and measures of effectiveness (MOEs) are used to show the types of unique effects that a cyber system can provide for a military commander. In Chap. 9, we will review the KPPs, MOPs, and MOEs of a cyber system, showing their merits through a compare/contrast with an unmanned aerial system (UAS), and its measurable improvements over a piloted aircraft.

Leveraging Chap. 9's cyber system metrics, Chap. 10 will include a discussion of modeling and simulation for cyber analysis and targeting. Chapter 10 will look at the inherent parallelism in cyber systems, providing conceptual models, along with example analytics, to describe how we might compute the risk associated with performing courses of action (COAs) at different portions of a generalized cyber attack cycle. These generalized operational approaches are complimented by a discussion of a cyber target as a dynamic, discrete event system, for operational effects estimation.

1.1 Key Cyber Analysis and Targeting Questions

Cyber currently provides unique value with effects that are comparable to both intelligence collection systems and munitions, occupying a strange space between the previously separable domains of spies and bombs. This presents a challenge in measuring cyber's effects, beyond relatively straightforward information collection.

Each cyber operation, whether it is used to perform intelligence collection or to provide effects, has a relatively fixed process over which it occurs. Therefore, we will use a few example questions to guide us as we progress through the development of this analytical framework:

- Which policies and doctrine are specifically written for cyber operations? (Chap. 2)
- How does cyber threat intelligence (CTI) contribute to analyzing cyber? (Chap. 3)
- How are information operations (IO) currently executed via cyber means? (Chap. 4)
- How is intelligence, surveillance, and reconnaissance (ISR) performed both through cyber and to develop cyber-specific effects? (Chap. 5)
- How do current security technologies make up cyber terrain? (Chap. 6)
- Which key targeting processes lend themselves to cyber operations? (Chap. 7)

- What are the tools and technologies that can be used for designing cyber systems? (Chap. 8)
- What are examples of cyber system metrics? (Chap. 9)
- How is modeling and simulation used for cyber analysis and targeting? (Chap. 10)
- What is a summary of the use cases in this book? (Chap. 11)

We will now look at the overall organization of this book.

1.2   Organization of This Book

This book comprises three major sections, as shown in Fig. 1.3.

Chapter 2 will provide an overview of policy, doctrine, and tactics, techniques, and procedures (TTPs). This will include a review of current policies from multiple countries, current cyber doctrine use, and TTPs applied by cyber defenders on a daily basis.

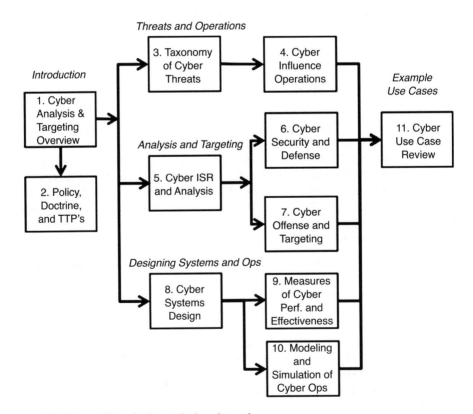

Fig. 1.3 Chapter flow of cyber analysis and targeting

Chapter 3, looking at the taxonomy of cyber threats, will discuss government industry standard cyber threat frameworks (e.g., NIST, MITRE), describing how they are used to facilitate cyber analysis and targeting.

Chapter 4 will describe influence operations, providing the reader with a general framework for analyzing information operations, providing current examples of point and area targeting in cyber influence operations.

Chapter 5 will compare current cyber to mature intelligence, surveillance, and reconnaissance (ISR) frameworks and techniques. This will include a look at how current cyber data collection and aggregation can be used for analysis and targeting.

Chapter 6 will review defensive cyber operations methodologies (e.g., DHS resilience framework, Australian eight-step approach) and evaluate current techniques, along with suggesting methods for increasing probability of detection, while decreasing false alarm rates.

Chapter 7 will expand on the targeting doctrine discussion in Chap. 2, looking at the overall people, process, and technology elements of a cyber system to guide the reader through targeting components of socio-technical stack of a cyber system.

Chapter 8 will expand on cyber systems as described by architectural products. This includes the development of an example cyber collection system via an architecture description method.

Chapter 9 will expand on effects evaluation using a comparison between cyber and autonomous systems. This will include looking at the cyber collection system architecture from Chap. 8 to develop metrics for the example cyber system.

Chapter 10 will provide a review of cyber modeling and simulation, discussing the parallel and series elements of a cyber system. This will include using a discrete event system to describe an example target process, introducing Cohen's d for effect estimates. We will also discuss the current state of constructive modeling.

Chapter 11 will provide use cases, referenced throughout the chapters, to capture key analysis and targeting insights.

Bibliography

9/11 Commission. (2004). *The 9/11 Commission Report.* Retrieved 8 4, 2019, from https://www.9-11commission.gov/report/911Report.pdf

Acton, J. M. (2017). Cyber weapons and precision guided munitions. In A. L. G. Perkovich (Ed.), *Understanding cyber conflict.* Georgetown.

Andrei Soldatov, I. B. (2015). *The red web - the struggle between Russia's digital dictators and the new online revolutionaries.* Public Affairs.

Barton Whaley, & Susan, S. A. (2007). *Textbook of political-military counterdeception: Basic principles and methods.* National Defense Intelligence College.

BBC. (2016, October 27). *18 revelations from Wikileaks' hacked Clinton emails*. Retrieved 21 Aug 2018, from BBC: https://www.bbc.com/news/world-us-canada-37639370

Ben Collins, G. R. (2018, 3 1). Leaked: Secret documents from Russia's election trolls . Retrieved 9 9, 2018, from Daily Beast: https://www.thedailybeast.com/exclusive-secret-documents-from-russias-election-trolls-leak?ref=scroll

Bennett, C. (1995). *How Yugoslavia's destroyers harnessed the media*. Retrieved 8 27, 2018, from PBS frontline: https://www.pbs.org/wgbh/pages/frontline/shows/karadzic/bosnia/media.html

Bernstein, J. (2017). *Secrecy world - inside the Panama papers investigation of illicit money networks and the global elite*. Henry Holt and Company.

Bowden, M. (2011). *Worm - the first digital world war*. Atlantic Monthly Press.

Carr, J. (2012). *Inside cyber warfare: Mapping the cyber underworld*. O'Reilly Media.

Cleary, G. (2019, 6). Twitterbots: Anatomy of a propaganda campaign. Retrieved 7 6, 2019, from Symantec: https://www.symantec.com/blogs/threat-intelligence/twitterbots-propaganda-disinformation

David Leigh, L. H. (2011). *WikiLeaks - inside Julian Assange's war on secrecy*. Public Affairs.

Department of Defense. (n.d.). Operation desert fox. Retrieved 6 24, 2020.

Diresta, R. (2018, 3 8). *How Isis and Russia won friends and manufactured crowds*. Retrieved 7 7, 2019, from wired.: https://www.wired.com/story/isis-russia-manufacture-crowds/

Doman, C. (2016, 7 6). The first cyber espionage attacks: How operation moonlight maze made history. Retrieved 8 4, 2019, from Medium: https://medium.com/@chris_doman/the-first-sophisticated-cyber-attacks-how-operation-moonlight-maze-made-history-2adb12cc43f7

Domscheit-Berg, D. (2011). *Inside Wikileaks - my time with Julian Assange at the World's Most dangerous website*. Crown.

Dustin Volz, J. F. (2016, March 24). *U.S. indicts Iranians for hacking dozens of banks*. Reuters.

Economist. (2007, 7 12). *A world wide web of terror*. Retrieved 8 11, 2019, from Economist: https://www.economist.com/briefing/2007/07/12/a-world-wide-web-of-terror

England, R. (2019, 8 13). *UN claims North Korea hacks stole $2 billion to fund its nuclear program*. Retrieved 8 18, 2019, from Engadget: https://www.engadget.com/2019/08/13/un-claims-north-korea-hacks-stole-2-billion-to-fund-its-nuclear/

Ferguson, N. (2018). *The square and the tower - networks and power, from the freemasons to Facebook*. Penguin.

FireEye. (2014). *APT28: A Window Into Russia's Cyber Espionage Operations?* Retrieved 9 9, 2018, from FireEye: https://www.fireeye.com/content/dam/fireeye-www/global/en/current-threats/pdfs/rpt-apt28.pdf

FireEye. (n.d.). *APT 29*. (FireEye, Producer). Retrieved 9 9, 2018, from APT 29: https://www.fireeye.com/current-threats/apt-groups.html#apt29

Gallagher, M. (2008). *Cyber analysis workshop. MORS*. MORS.

George Cybenko, G. S. (2016). Quantifying covertness in deceptive cyber operations. In V. S. S. Jajodia (Ed.), *Cyber deception: Building the scientific foundation*. Springer.

George Cybenko, J. S. (2007). Quantitative foundations for information operations.

George Perkovich, A. E. (2017). *Understanding cyber conflict - 14 analogies*. Georgetown.

Global Security. (2005). Global security. Retrieved from Letter from al-Zawahiri to al-Zarqawi: https://www.globalsecurity.org/security/library/report/2005/zawahiri-zarqawi-letter_9jul2005.htm

Michael R. Gordon, H. C. (2017, 4 6). *Dozens of U.S. Missiles Hit Air Base in Syria*. Retrieved 8 21, 2019, from New York Times: https://www.nytimes.com/2017/04/06/world/middleeast/us-said-to-weigh-military-responses-to-syrian-chemical-attack.html

Harris, G. (2018, 3 4). State Dept. was granted $120 million to fight Russian meddling. It has spent $0. Image. Retrieved 9 9, 2018, from New York Times: https://www.nytimes.com/2018/03/04/world/europe/state-department-russia-global-engagement-center.html

Hayden, M. V. (2016). Playing to the edge: American intelligence in the age of terror.. Peguin.

Heli Tiirmaa-Klaar, J. G.-P. (2014). *Botnets*. Springer.

Eric M. Hutchins, M. J. (2012). *Intelligence-driven computer network defense informed by analysis of adversary campaigns and intrusion kill chains*. Retrieved 8 6, 2019, from Lockheed Martin.: https://www.lockheedmartin.com/content/dam/lockheed-martin/rms/documents/cyber/LM-White-Paper-Intel-Driven-Defense.pdf

Katelyn Polantz, S. C. (2018, 7 14). 12 Russians indicted in Mueller investigation . Retrieved 9 9, 2018, from CNN: https://www.cnn.com/2018/07/13/politics/russia-investigation-indictments/index.html

Koblentz, G. D., & B. M. (2013). Viral warfare: The security implications of cyber and biological weapons. *Comparative Strategy, 32*(5), 418–434.

Koerner, B.I. (2016, October 23). Inside the cyberattack that shocked the US government. Retrieved September 7, 2018, from Wired: https://www.wired.com/2016/10/inside-cyberattack-shocked-us-government/

Lamothe, D. (2017, 12 16). How the Pentagon's cyber offensive against ISIS could shape the future for elite U.S. forces. Retrieved 8 27, 2018, from Washington Post: https://www.washingtonpost.com/news/checkpoint/wp/2017/12/16/how-the-pentagons-cyber-offensive-against-isis-could-shape-the-future-for-elite-u-s-forces/?utm_term=.8cce44e017f9

Launius, S. (2019). Evaluation of comprehensive taxonomies for information technology threats. Retrieved 3 7, 2019, from SANS: https://www.sans.org/reading-room/whitepapers/threatintelligence/evaluation-comprehensive-taxonomies-information-technology-threats-38360

Lewis, D. (2015, March 31). Heartland Payment Systems Suffers Data Breach. *Forbes*.

Lewis, J. (2012). *Significant cyber incidents since 2006*. Center for Strategic and International Studies.

MacEslin, D. (2006). Methodology for determining EW JMEM.. TECH TALK.

Mark Gallagher, M. H. (2013). Cyber joint munitions effectiveness manual (JMEM). *Modeling and Simulation Journal*.

Maurer, T. (2016, 12 17). *'Proxies' and Cyberspace*. Retrieved 8 17, 2019, from Carnegie Endowment for International Peace: https://carnegieendowment.org/2016/12/17/proxies-and-cyberspace-pub-66532

Maurer, T. (2018, 7 27). Cyber proxies and their implications for Liberal democracies. Retrieved 8 17, 2019, from WASHINGTON QUARTERLY : https://carnegieendowment.org/2018/07/27/cyber-proxies-and-their-implications-for-liberal-democracies-pub-76937

Mazetti, M. (2018, July 13). 12 Russian agents indicted in Mueller investigation.

McGee, M. K. (2017, January 10). *A New In-Depth Analysis of Anthem Breach*. Retrieved September 7, 2018, from Bank Info Security: https://www.bankinfosecurity.com/new-in-depth-analysis-anthem-breach-a-9627

McGlasson, L. (2009, 1 21). *Heartland Payment Systems, Forcht Bank Discover Data Breaches*. Retrieved 9 9, 2018, from Bank Info Security: https://www.bankinfosecurity.com/heartland-payment-systems-forcht-bank-discover-data-breaches-a-1168

Merriam Webster. (n.d.). *cyber*. Retrieved from https://www.merriam-webster.com/dictionary/cyber

Middleton, C. (2018, 6 25). Cyber attack could cost bank half of its profits, warns IMF. Retrieved 8 22, 2018, from Internet of Business: https://internetofbusiness.com/fintech-cyber-attack-could-cost-bank-half-of-its-profits-warns-imf/

Mueller, R. (2019). *Report on the investigation into Russian interference in the 2016 presidential election*. U.S. Department of Justice.

Nakashima, E. (2012, 9 12). *Iran blamed for cyberattacks on U.S. banks and companies*. Retrieved 8 18, 2019, from Washington Post: https://www.washingtonpost.com/world/national-security/iran-blamed-for-cyberattacks/2012/09/21/afbe2be4-0412-11e2-9b24-ff730c7f6312_story.html?noredirect=on

Nakashima, E. (2018, 10 23). *Pentagon launches first cyber operation to deter Russian interference in midterm elections*. Retrieved 3 12, 2019, from Washington Post: https://www.washingtonpost.com/world/national-security/pentagon-launches-first-cyber-operation-to-deter-russian-interference-in-midterm-elections/2018/10/23/12ec6e7e-d6df-11e8-83a2-d1c3da28d6b6_story.html?utm_term=.8c47d573557b

Nakashima, E. (2019, 2 27). *US disrupted Internet access of Russian troll factory on day of 2018 midterms*. Retrieved 3 12, 2019, from Washington Post.

Nancy A. Youssef, S. H. (2017, 11 25). *Why did team obama try to take down its NSA Chief?*. Retrieved 8 27, 2018, from The Daily Beast: https://www.thedailybeast.com/why-did-team-obama-try-to-take-down-its-nsa-chief

Newman, L. H. (2017, September 8). *THE EQUIFAX BREACH EXPOSES AMERICA'S IDENTITY CRISIS*. Retrieved September 7, 2018, from Wired: https://www.wired.com/story/the-equifax-breach-exposes-americas-identity-crisis/

Nichols, M. (2019, 8 5). North Korea took $2 billion in cyberattacks to fund weapons program: U.N. report. Retrieved 8 17, 2019, from Reuters: https://www.reuters.com/article/us-northkorea-cyber-un/north-korea-took-2-billion-in-cyber-attacks-to-fund-weapons-program-u-n-report-idUSKCN1UV1ZX

Parham, J. (2017, 10 18). Russians posing as black activists on FACEBOOK is more THAN fake news. Retrieved 8 22, 2018, from Wired: https://www.wired.com/story/russian-black-activist-facebook-accounts/

Paul Ducheine, Jelle van Haaster (2014). Fighting power, targeting and cyber operations. Retrieved 8 4, 2019, from CCDOE - 2014 6th International Conference on Cyber Conflict: https://www.ccdcoe.org/uploads/2018/10/d2r1s9_ducheinehaaster.pdf

Pegues, J. (2018). *Kompromat - how Russia undermined American democracy*. Prometheus.

Richard, A. C., & Robert, K. (2012). *Cyber war: The next threat to National Security and what to do about it*. Ecco.

Richard Clarke, R. K. (2011). *Cyber war: The next threat to National Security and what to do about it*. Ecco.

Riley, C. (2019, 7 9). *UK proposes another huge data fine. This time, Marriott is the target*. Retrieved 8 21, 2019, from CNN: https://www.cnn.com/2019/07/09/tech/marriott-data-breach-fine/index.html

Sanger, D. E. (2016, 4 24). U.S. cyberattacks target ISIS in a new line of combat. Retrieved 8 27, 2018, from New York Times: https://www.nytimes.com/2016/04/25/us/politics/us-directs-cyberweapons-at-isis-for-first-time.html

Sanger, D. E. (2017). Cyber, drones and secrecy. In A. E. G. Perkovich (Ed.), *Understanding cyber conflict*. Georgetown.

David E. Sanger (2018, July 15). Tracing Guccifer 2.0's many tentacles in the 2016 election. Retrieved September 9, 2018, from New York Times: https://www.nytimes.com/2018/07/15/us/politics/guccifer-russia-mueller.html

David E. Sanger, Eric Schmitt. (2017, 6 12). U.S. Cyberweapons, used against Iran and North Korea, are a disappointment against ISIS. Retrieved 8 27, 2018, from New York Times: https://www.nytimes.com/2017/06/12/world/middleeast/isis-cyber.html

Stoll, C. (2005). *The Cuckoo's egg: Tracking a spy through the maze of computer espionage*. Pocket Books.

Strategy and Tactics of Guerilla Warfare. (n.d.). Retrieved 9 9, 2018, from Wikipedia: https://en.wikipedia.org/wiki/Strategy_and_tactics_of_guerrilla_warfare

Catherine A. Theohary, John Rollins (2011, 3 8). Terrorist use of the internet: Information operations in cyberspace. Retrieved 8 22, 2018, from Congressional Research Service: https://digital.library.unt.edu/ark:/67531/metadc103142/m1/1/high_res_d/R41674_2011Mar08.pdf

U.S. Joint Forces Command. (2011, 5 20). Commander's handbook for attack the network. Retrieved 8 4, 2019, from https://www.jcs.mil/Portals/36/Documents/Doctrine/pams_hands/atn_hbk.pdf

U.S. Justice Department. (2019, March). Report on the investigation into Russian interference in the 2016 presidential election. Retrieved May 9, 2019, from U.S. Justice Department: https://www.justice.gov/storage/report.pdf

US Army. (n.d.). Joint Technical Coordinating Group for Munitions Program Office.

Weaponeering. (n.d.). History of the joint technical coordinating Group for Munitions Effectiveness. Retrieved 8 19, 2019, from Weaponeering: http://www.weaponeering.com/jtcg_me_history.htm

Wikipedia. (n.d.). Chaos computer Club. Retrieved 9 9, 2018, from Wikipedia: https://en.wikipedia.org/wiki/Chaos_Computer_Club

Zetter, K. (2014a). *Countdown to zero day - Stuxnet and the launch of the World's first digital weapon*. Crown.

Zetter, K. (2014b, 12 3). SONY GOT HACKED HARD: WHAT WE KNOW AND DON'T KNOW SO FAR. Retrieved 9 9, 2018, from Wired: https://www.wired.com/2014/12/sony-hack-what-we-know/

Chapter 2
Cyber Policy, Doctrine, and Tactics, Techniques, and Procedures (TTPs)

The purpose of this chapter is to present a general background on cyber policy, doctrine, and tactics, techniques, and procedures (TTPs) and describe their role in providing guidance for cyber analysis and targeting. This will include a listing of national cyber policies, a look at the current cyber doctrine, and a review of TTPs as both examples and frameworks that capture analysis and targeting in cyberspace operations.

Policy, Doctrine, and TTP Questions to be Addressed in Chapter 2

1. What are the definitions of policy, doctrine, and TTPs? What are cyber examples of each?
2. How is cyber policy used for both defensive and offensive cyber operations?
3. What are the key drivers for cyber doctrine?

2.1   Background

Policy, doctrine, and TTP development over the last century directly influence both the composition and operation of current cyber systems and provide a framework for cyber analysis and targeting. For example, cyber policy protections span from national infrastructure to an individual's privacy rights when using the Internet. Similarly, doctrine, in providing guidance for future cyber operations, distills the "lessons learned" from successful employment of particular TTPs. In getting started, we will review foundational definitions before going into examples of policy, doctrine, and TTPs in current usage for cyber.

- Policy: usually cyber protection
- Doctrine: leverage existing targeting documents as they apply to cyber
- Tactics, techniques, and procedures (TTPs): developing; unique to cyber due to novel/fungible maneuver space

© Springer Nature Switzerland AG 2022
J. M. Couretas, *An Introduction to Cyber Analysis and Targeting*,
https://doi.org/10.1007/978-3-030-88559-5_2

2.1.1 Policy, Doctrine, and TTP Definitions

We will start this chapter with definitions for cyber policy, doctrine, and TTPs. As shown in Fig. 2.1, policy provides overall direction for moving toward an end-state. Doctrine describes lessons learned and best practices, or teachings in the field. And TTPs provide the attack process, style, and specific steps taken during both cyber analysis and targeting.

As shown in Fig. 2.1, policy provides the overall scope for an organization's approach to an issue. We use the following definition for policy:

Policy   A course or principle of action adopted or proposed by a government, party, business, or individual. Also, a high-level, overall, plan, embracing the general goals and acceptable procedures, especially of a governmental body (Merriam-Webster)

While policies are organizational practices, or a deliberate system of principles, to guide decisions and achieve rational outcomes, doctrines are beliefs, developed through experience, that are taught as a form of institutional knowledge.

Doctrine   A principle or position or the body of principles in a branch of knowledge or system of belief (Merriam-Webster)

While doctrine provides teachings, and policies represent organizational practices, TTPs provide details concerning how, specifically, an individual goal is achieved.

Tactics, Techniques, and Procedures (TTPs): The term tactics, techniques, and procedures (TTPs) describes an approach for analyzing an actors operation or can be used as means of profiling behavior (e.g., MITRE CARET (MITRE)).

Tactics: outline the way the actor chooses to operate over a course of action (COA). For example, tactics are associated with the achievement of short−/

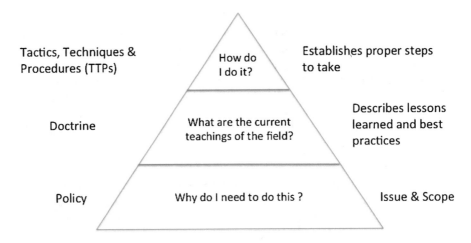

Fig. 2.1 Policy, doctrine, and TTP triangle

medium-term goal(s) via one out of many possible ways that involve human factors as the subject and means of these tactics.

Techniques: related to the target, its technicalities, and specific details, which imply or suggest a specific way to get something done; for example, installing/uninstalling a backdoor to an information system, performing a scan, etc.

Procedures: more prescriptive; procedures are the steps to get something done.

Another definition for TTPs is that tactics are means to implement strategies. Techniques are means to implement tasks. Procedures are then the standard, detailed steps that prescribe how to perform a specific task.

In tracing the development of current operational guidance and procedures, we will review examples that illustrate current national, and international, policy/doctrine /TTP implementation.

2.2 Introduction

Mass networked government, business, and personal computers emerged in the mid-1990s. The 1990s to present is therefore the time period that we have to cite examples, case studies, and lessons learned that provides the current corpus of guiding cyber policy, doctrine, and best practices for cyber analysis and targeting. Because of this relatively short history of networked computers, or cyber systems, especially for operations, the terminology for cyber analysis and targeting policy, doctrine, and TTPs are often borrowed from legacy fields (e.g., radio).

Cyber policy is used to protect information systems. Estonia's recent passing of the "Huawei Law," (Reuters, 2020) as a means of providing telecom gear reviews, is an example of using policy, via law, to protect their information infrastructure. In addition, resilience, defined as "the capacity to recover quickly from difficulties, or toughness," is the foundation for many national cyber policies. This is similar to ensuring continuity of government through any other national emergency (e.g., hurricanes, power outages).

Many National Cyber Strategies, or policy implementations, focus on the security aspects of cyber, in accord with the legacy of maintaining secure radio communications. This is often called cyber resilience, or an entity's ability to sustain operations in a cyber-contested environment. In addition to this focus on resilience, some national strategies (e.g., the United States and Canada) include commercial considerations for expanding the use of cyberspace. The US Department of Defense (DoD) is a pioneer in discussing cyber as a contestable operational domain, potentially challenging the other warfare domains (e.g., Space, Air, Sea, Land) with denied or inaccurate information transmissions.

The DoD Cyber Strategy (Department of Defense (DoD), 2018) provides policy guidance to several Joint Publications (JP) as doctrinal guidance for cyber operations; the primary documents are:

- JP 2-0, Joint Intelligence: describes the role of cyber-related intelligence collection and analysis in developing an overall intelligence picture.
- JP 2-01.3, Joint Intelligence Preparation of the Operational Environment (JIPOE): defines the role of cyber in describing the operational environment.
- JP 3-12, Cyber Operations: defines effects (e.g., denial, integrity), along with discussing the different missions over which cyber operations should be considered.
- JP 3-0, Joint Operations: the top-level publication for military operations that shows how cyber fits into the context of all military operations.
- JP 3-12: describes military operations in cyberspace.
- JP 3-13, Information Operations: where a broader view of potential cyber applications is considered for operations.
- JP 3-13.2, Military Information Support Operations (MISO): describes the role of MISO to collect, analyze, and disseminate information, including over cyber channels.
- JP 3-13.4, Military Deception: explains the role of deception principles and mechanisms that are delivered to targets over channels; cyber is but one channel used to deliver deception to military targets.
- JP 3-60, describes the integration of cyberspace operations in joint targeting.

For example, JP 3-12, Cyber Operations, defines effects (e.g., denial, integrity), along with discussing the different scenarios /TTPs over which cyber operations should be considered. JP 3-12 is complimented by JP 3-13, Information Operations (IO), where a broader view of potential cyber applications is considered for operations. This includes using both MISO for intelligence collection and the use of deception for prosecuting a military target. JP 3-60, Joint Targeting, provides the overall process for any targeting operation, including cyber.

In addition to military use of cyber operations, there are "best practices" that are derived from observing tactics, techniques, and procedures (TTPs), in the form of use cases that occur frequently enough to be categorized together. The SANS Critical Security Controls (CSCs) (SANS, 2016), 20 best practices, in order of priority, is one example of practical knowledge, distilled from known TTPs, to provide defensive cyber personnel with step-by-step approaches for securing cyber infrastructure from possible attack. In addition, the SANS CSCs are a bottoms-up view of providing cyber defenders with "doctrine" for defending our networks. Cyber policy, providing broader guidance than the use case based implementation of current cyber doctrine, benefits from a historic look at naval policy formation and implementation.

2.3 Policy

Using policy for cyber analysis and targeting provides a general view. One approach for looking at cyber policy is to compare it to a similar issue we faced almost a half a millennia ago—maritime threats (Table 2.1).

As shown in Table 2.1, cyber (in)security has similarities to maritime analogies through which we get the majority of our goods and services. In addition, the rapid growth of computer-based systems, or cyber, as a business and socialization tool, is paired with its adoption across supply chains that transparently underpin many of our day-to-day transactions. Because of the risk of cyber threats to these transactions, cyber policies have been adopted by many major governments (Table 2.2).

As shown in Table 2.2, most of the national cyber strategies focus on defense, or resilience. An additional policy introduced by the European Union is the General Data Protection Regulation (GDPR), a general law on data protection and privacy (European Union, n.d.). Slightly different than the national-level policies provided in Table 2.2, the GDPR provides more individual protections, sometimes fining organizations for unauthorized release of private data (Riley, 2019).

The US DoD, however, concerns itself with performing operations in a cyber-contested environment, which includes maintaining resilience of supporting infrastructure (e.g., critical infrastructure). In addition to the documents given in Table 2.2, multiple executive orders (EOs) (White House) and doctrine have been published to provide more specific steps to mostly defend current cyber equities (U.S. National Archives) (George Washington University).

As shown in Table 2.3, multiple executive orders (EOs) address cyber from both a personnel and technical standpoint. However, the frequency of EOs increased rapidly in the mid-2010s, and accelerated after Russian involvement in the 2016 US presidential election, with additional EOs designed to manage foreign participation in the US telecommunication sector (Trump, 2020). An additional EO was signed just before this book's publishing in response to the recent Solar Winds software supply chain attack (Trump, 2020).

While EOs add national-level emphasis to a particular thread of cyber defense, a key challenge to developing policy for cyber analysis and targeting is the ability to understand an adversary, especially with respect to attributing an attack. As shown in Table 2.1, the range of attackers spans from individuals to nation states. While civil and criminal laws are the domain for individual perpetrators, the line between crime and war is fuzzy for cyber actions. This is a challenge because a cyber

Table 2.1 Comparison between actors on the sea and in cyberspace (Egloff, 2017)

Actor type	Sea	Cyberspace
State actors	Navy (including mercenaries; e.g., blockades)	Cyber operators, intelligence analysts, contractors, tool/capability providers (e.g., denial of service botnets, specialty information operations developers)
Semi-state actors	Shipping/transportation; mercantile companies	Major telecommunications companies, technology developers, security vendors
	Privateers	Patriotic hackers, some cyber criminal elements (e.g., exfiltrations)
Non-state actors	Pirates (e.g., theft)	Hackers, cyber criminal elements (including organized crime; e.g., exfiltrations)

Table 2.2 Cyber policy documents (White House, International Partners, Department of Homeland Security [DHS], and DoD)

Title	Description
National Cyber Strategy (White House, 2018)	Leverage cyber to accomplish 4 pillars
	Pillar I Protect the American People, the Homeland and the American Way of Life
	Pillar II Promote American Prosperity
	Pillar III Preserve Peace through Strength
	Pillar IV Advance American Influence
UK National Cyber Security Strategy 2016–2021 (HM Government, 2016)	Document describing vision for 2021 is that the UK is secure and resilient to cyber threats, prosperous, and confident in the digital world.
Australian Cyber Strategy (Australian Government, 2018)	Achieve five themes of action
	1. A national cyber partnership
	2. Strong cyber defenses
	3. Global responsibility and influence
	4. Growth and innovation
	5. A cyber smart nation
Canadian National Cyber Security Strategy (Government of Canada, 2018)	Engage in the Cyber Domain via
	1. Security and Resilience
	2. Cyber Innovation
	3. Leadership and Collaboration
Singapore's Cyber Security Strategy (Singapore, 2016)	Engage in the Cyber Domain via
	1. A Resilient Infrastructure
	2. A Safer Cyberspace
	3. A Vibrant Cybersecurity Ecosystem
	4. Strong International Partnerships
Cyber Security Strategy (Department of Homeland Security (DHS), 2018)	Leverage cyber to accomplish 5 pillars
	Pillar I Risk Identification
	Pillar II Vulnerability Reduction
	Pillar III Threat Reduction
	Pillar IV Consequence Mitigation
	Pillar V Enable Cyber Scenario Outcomes
Department of Defense (DoD) Cyber Security Strategy (Department of Defense (DoD), 2018)	The Department's cyberspace objectives are
	1. Ensuring the Joint Force can achieve its missions in a contested cyberspace environment

(continued)

Table 2.2 (continued)

Title	Description
	2. Strengthening the Joint Force by conducting cyberspace operations that enhance US military advantages
	3. Defending US critical infrastructure from malicious cyber activity that alone, or as part of a campaign, could cause a significant cyber incident
	4. Securing DoD information and systems against malicious cyber activity, including DoD information on non-DoD-owned networks
	5. Expanding DoD cyber cooperation with interagency, industry, and international partners

Table 2.3 Cyber executive order (EO) examples

Executive order number	Signing date	Title
13870	5/2/19	America's Cybersecurity Workforce
13800	5/11/17	Strengthening the Cybersecurity of Federal Networks and Critical Infrastructure
13757	12/28/16	Taking Additional Steps to address the National Emergency With Respect to Significant Malicious Cyber-Enabled Activities
13718	2/1/16	Commission on Enhancing National Cybersecurity
13694	4/1/15	Blocking the Property of Certain Persons engaging in Significant Malicious Cyber-Enabled Activities
13691	2/13/15	Promoting Private Sector Cybersecurity Information Sharing
13636	2/12/13	Improving Critical Infrastructure Cybersecurity

operation, due to its potentially strategic-level effects (e.g., changing election results, turning off electric power), would be acts of war that result in escalating to nation state conflict if the same effects were achieved by a bomb; that is, kinetic scenarios. Cyber actions, therefore, require a clear set of attribution criteria.

2.3.1 Use of Force Policy for Cyber

Presidential Executive Orders provide examples of executive-level emphasis on cyber security. Additional understanding, via frameworks and manuals, however, is required to clearly understand the nature and motivation of each cyber attack to formulate a proper response. For example, the Tallinn Manual (Schmitt, 2013), stemming from the 2007 DDoS—Distributed Denial of Service—attack on Estonia (Ottis, 2007), is the first policy-level document addressing nation state response to a cyber attack, and provides an academic, non-binding, assessment of how international law applies to cyber warfare. In addition, The Tallinn Manual has been proposed to the North Atlantic Treaty Organization (NATO) and although not adopted, is currently used as a reference.

One roadblock to implementing something like the Tallinn Manual is the multiple factors that could be in play to obfuscate a cyber attack. For example, false flag attacks, where the attacker uses tools and techniques to look like someone else, are operations that are known to occur. In the 2012 Operation Ababil attack on US banks, Iranians imitated Arab jihadist web terrorists (Sect. 2.3.2.2) with misleading personas and the use of Arabic in their communications to mask their identity (Krebs, 2013). Attribution therefore requires multiple sources for actor verification. And what is the likelihood that clear identification and attribution will happen in a reasonable time frame to take effective action against an attacker?

"Effective action" hints at possible comparison between kinetic and cyber. For example, both a bomb and a denial of service attack can remove a hostile actor or asset from the battlefield. However, some situations may find this to be an acceptable course of action (COA) and others will not. While a "playbook" (Merriam-Webster), which includes process workflows, standard operating procedures, and cultural values that shape a consistent response, may help in understanding how an organization will act in specific situations, it is very challenging to frame all of the playbook scenarios into a general manual.

The goal of the playbook is to determine where the line between cyber and kinetic action is crossed and thereby determine the correct response. A simplified approach for looking where the line is crossed between non-kinetic and kinetic operations is shown in Fig. 2.2.

Figure 2.2 is a simplification of the argument provided in Joint Doctrine Note 1–19, "Competition Continuum" (Joint Chiefs of Staff, 2019). One example provided in the "Competition Continuum" document is "The Great Game," the nineteenth-century competition between the British and Russian empires in the Middle East and Asia. The Great Game consisted of the British using proxies to develop a line of defense against Russian encroachment in central Asia on the crown of the British Empire, India. In addition, the Great Game included cooperation (below the line) to active military operations (above the line):

"The Great Game was a political and diplomatic confrontation that existed for most of the nineteenth century between the British Empire and the Russian Empire over Afghanistan and neighboring territories in Central and Southern Asia. Russia was fearful of British commercial and military inroads into Central Asia, and Britain was fearful of Russia adding "the jewel in the crown," India, to the vast empire that Russia was building in Asia.

This resulted in an atmosphere of distrust and the constant threat of war between the two empires. The Great Game began in 1830 as Britain intended to gain control over the Emirate of Afghanistan and make it a protectorate, and to use the Ottoman Empire, the Persian Empire, the Khanate of Khiva, and the Emirate of Bukhara as buffer states between both empires. This would protect India and also key British sea trade routes by stopping Russia from gaining a port on the Persian Gulf or the Indian Ocean. Russia proposed Afghanistan as the neutral zone. The results included the failed First Anglo-Afghan War of 1838, the First Anglo-Sikh War of 1845, the Second Anglo-Sikh War of 1848, the Second Anglo-Afghan War of 1878, and the annexation of Khiva, Bukhara, and Kokand by Russia. The Great Game ended in 1895 when the border between Afghanistan and the Russian empire was defined, p. 17." (Joint Chiefs of Staff, 2019)

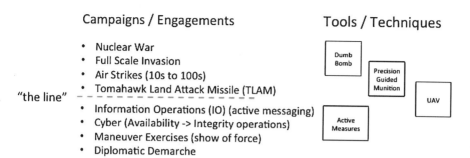

Fig. 2.2 Kinetic/non-kinetic line of hostility

As shown by the Great Game, an example of strategic competition punctuated by diplomatic agreements, nation building, and military battles, the line shown in Fig. 2.2 will continually be straddled over the course of near peer negotiation of physical assets, including geography and associated resources. The spectrum provided by Fig. 2.2, from diplomacy to military conflict, that was experienced in "The Great Game," was codified in the law of armed conflict (LOAC), which now includes cyber for collecting information, controlling processes, or compromising an adversary's message.

One study, by Dr. Marco Roscini (Roscini, 2014) analyzes the use of traditional laws for use of force to determine whether existing laws can be used in the digital age. This includes an evaluation of how cyber fits within the spectrum of armed conflict, the use of humanitarian law in cyber operations, and the legitimacy of a cyber as a military target.

While the legal framework for the incorporation of cyber into military operations continues to mature, cyber operations are ongoing. Another legal framework used to view cyber operations is the chain of command, including authorities, for the organization prosecuting missions.

2.3.2 Authorities

Military organizations are constructed with a "chain of command," or line of authority, most clearly identified by the rank on a service member's sleeve or lapel. Functional organizations also have authorities and responsibilities that they pass down from higher- to lower-ranking units. Authority to perform a function or task can also be transferred within and between different military units. Lower-ranking organizations therefore perform their operations under the authorities provided by their superiors.

The authorities question, when considering cyber, is the same kind of policy challenge faced by any commander. One new challenge, however, is that almost any operation will span multiple national boundaries (Temple-Raston, 2019), bringing the authorities question to the forefront. For example, the traversable cyber terrain for executing almost any TTP necessarily encompasses the .com, .edu, ..., and possibly the .gov, and .mil address spaces, not to mention foreign governments and businesses. Clear policy, TTPs, and doctrine that provide guidance for navigating these territories are therefore important for both immediate tactical decision-making and the possibility of pre-delegating some of the requisite authorities.

2.3.2.1 Maritime Example: Harbor Lights and World War II (Delayed Authorities)

Until 2018, US cyber policy was to operate only within US networks (Pomerleau, 2019). Going outside US networks required Presidential-level approval. This time lag between request and approval for proper authorities caused a mission delay for cyber operators, potentially missing critical opportunities to collect information or take action.

One historical policy example that relates to delaying authorities includes a lack of will, by the federal government, to apply national security level policy to state and local government, at the onset of World War (WW) II. For example, Arquilla (Arquilla, 2017) described how, due to commercial interests, the United States refused to dim harbor lights along the East Coast for over 3 months after Pearl Harbor, when German U-boats freely sailed the Atlantic:

> "For three long months, coastal cities refused even to dim their lights at night. The illumination helped U-boat skippers immensely. The eminent naval historian Samuel Eliot Morison labeled this inaction America's 'most reprehensible failure.'"
>
> "In Morison's analysis of these events, 'the massacre enjoyed by the U-boats along our Atlantic coast in 1942 was as much a national disaster as if saboteurs had destroyed half a dozen of our biggest war plants.' Indeed, during this U-boat 'happy time,' Germany sank 2.5 million tons of shipping, or about half the total losses inflicted by German submarines in the first two years of the war. Morison's unsettling bottom-line assessment is quite biting: 'Ships were sunk and seamen drowned in order that the citizenry might enjoy pleasure as usual.' And all this came at a minimal cost to the U-boat arm. Though the US Navy claimed twenty-eight kills of enemy submarines from January to March, in actuality these were all false claims made by overenthusiastic American skippers. No U-boats were sunk during this period, and only half a dozen were lost by July 1942."(Morison, 1963)

This policy was due to the lack of will, on the part of big city mayors, to implement the blackout policy, usually objecting to the "catastrophic" economic losses that would ensue. For example, Florida receives the majority of its visitors over the course of winter, and Miami's Chamber of Commerce refused to comply because it would ruin its tourist season.

Similar to the refusal to exercise national security on state and local government in the harbor lights example, any cyber operation outside US networks was once required to have executive-level approval. This changed in August of 2018, with the signing of the National Security Presidential Memorandum (NSPM) 13 (Pomerleau, 2019), with a stated goal of "persistent engagement" (Nakasone, 2019), that provides operators with a set of pre-delegated authorities to defend forward:

"When we do this, we can observe enemy techniques and procedures and their tactics as well as potentially uncover any tools or weapons that they might be utilizing."
Maj Gen Charles Moore
Director of Operations
J-3 at Cyber Command

The pre-delegated authorities in NSPM 13 help cyber operators maintain persistent engagement and reflect the dynamic nature of how cyberspace operations are developing. For example, the computer nodes and network configurations that make up the terrain in cyberspace can change in seconds, providing the cyber operator with an entirely new environment, and possibly requisite action, to move forward with her mission. Giving the cyber operator the authorities to continue in this new terrain provides for agile response; not possible when there was a requirement to request permission, wait for a response, and then move forward. The time constant, or rate of change, in cyber, therefore drives the need for a cyber operator to have the authority to act, and react, in a timely fashion to accomplish her mission. Having very short time constants, relative to the hours-to-days delay in a command/authority chain, is similar to what nuclear operators experienced a generation ago.

2.3.2.2 Pre-delegation of Authorities

Prior to the intercontinental ballistic missile (ICBM), the majority of US territory was safe from bombardment by air or sea. However, by 1958, the ICBM, and associated atomic payload, could hold any point on earth at risk by nuclear conflagration. This development required a streamlined decision structure to provide a defensive civilian (i.e., "duck and cover"), or offensive military, response, within the 1/2 hour or so it took to get from central Asia to the continental United States; cyber has the same kind of timeline. As pointed out by Feaver and Geers:

"There are three important analogues between nuclear attacks and cyber attacks: malicious code can travel across computer networks at lightning speed, successful cyber attacks are often based on novel ideas (the archetype here is the zero-day vulnerability plus exploit, which only the attacker knows about), and computer security is a complex, highly technical discipline that many decision makers do not understand. These three characteristics – speed, surprise, and specialization force national civilian leadership to give tactical military commanders a pre delegated authority to operate in cyberspace so that they are able to competently and successfully defend US computer networks." (Peter Feaver, 2017)

A key difference between ICBMs and cyber is that ballistic missiles are traceable to their launch sites with physical models—ballistic missions have "a return address." Cyber, therefore, is similar to terrorism, in that there are attribution issues that require caution when striking back. In addition, cyber actors are known to perform "false flag" operations, where they pose as another organization so as to avoid attribution. As discussed, one of the more famous attempts at mis-attribution was Iran posing as the "Izz-ad-Din al-Qassam Cyber Fighters" during Operation Ababil, where US banks were attacked with Distributed Denial of Service (DDoS) for 9 months in 2012 (Krebs, 2013).

In addition, cyber effects are usually to degrade, or manipulate, not the existential life or death threat posed by nuclear war. Therefore, it is easier for decision makers to delay their response to cyber, much like often happens for terrorism, until both the attribution and level of response, determined by cyber policy, are more thoroughly assessed. For example, once the Iranian actors in Operation Ababil were correctly identified, it would have been much easier to understand which set of authorities cyber defenders would use to operate against this malicious Iranian cyber actor. In fact, if this was a recurring issue, the authorities might be provided in advance, providing cyber defenders with the ability to maneuver against malicious cyber actors (MCAs) in real time.

Pre-delegation of authorities initially sounds risky to decision makers. However, with the cost of waiting for approval of proper authorities potentially being mission success, pre-delegation may be the smart thing to do. Attribution, a key contribution of analysis in ensuring that the target is within current authorities, is a challenge in any domain of warfare. One of the contributions of using verifiable analysis in developing a targeting solution is that the target is ensured to line up with associated authorities, and is defensible under scrutiny concerning the source of an attack. Attributing the source of a cyber attack is a key goal of Schmitt's six criteria to establish state responsibility.

2.3.3   Schmitt's Six Criteria to Establish State Responsibility

Michael Schmitt was the principal author of the Tallinn Manual and he established the Schmitt criteria (Schmitt, 2011) to determine state responsibility for an attack after the 2007 Estonian cyber attack.

According to the Schmitt criteria, "severity" evaluates the intensity of an attack, and is complemented by "measurability," or quantifying the amount of damage done, in determining the overall impact of a cyber attack. In addition, "directness" attempts to separate out the amount of harm caused by the cyber attack versus the harm caused by other, parallel, attacks. Damage directly attributable to the cyber attack weighs as an estimate of how close cyber resembles an armed attack, on a case-by-case basis. In addition, "immediacy" looks at the duration of an attack, including the time over which the effects of the cyber attacks were felt (Table 2.4).

The six different lines of questioning provided by the Schmitt criteria (Table 2.4), and the types of questions being asked, provide for an overall assessment that spans the range of complex considerations that a decision maker will address before deciding on a reasonable course of action (COA). In addition, the Schmitt criteria provide an excellent example of the broadly scoped considerations common to cyber analysis, or military assessment more generally.

Once an assessment is made, and clear attribution (e.g., from "invasiveness") is performed, the next question becomes what type of authorities are required for containing, or striking back, against a candidate attacker. Determining the type of response, and associated authorities, is an issue often worked out on a case-by-case

Table 2.4 Schmitt's six criteria for attribution (Schmitt, 2011)

Criterion	Description
Severity	Scope and intensity of an attack. Analysis under this criterion examines the number of people killed, size of the area attacked and amount of property damage done. The greater the damage, the more powerful the argument becomes for treating the cyber attack as an armed attack.
Immediacy	Duration of a cyber attack, as well as other timing factors. Analysis under this criterion examines the amount of time the cyber attack lasted and the duration of time that the effects were felt. The longer the duration and effects of an attack, the stronger the argument that it was an armed attack.
Directness	Harm caused. If the attack was the proximate cause of the harm, it strengthens the argument that the cyber attack was an armed attack. If the harm was caused in full or in part by other parallel attacks, the weaker the argument that the cyber attack was an armed attack.
Invasiveness	Locus of the attack. An invasive attack is one that physically crosses state borders, or electronically crosses borders and causes harm within the victim state. The more invasive the cyber attack, the more it looks like an armed attack.
Measurability	Tries to quantify the damage done by the cyber attack. Quantifiable harm is generally treated more seriously in the international community. The more a state can quantify the harm done, the more the cyber attack looks like an armed attack. Speculative harm generally makes a weak case that a cyber attack was an armed attack.
Presumptive legitimacy	Focuses on state practice and the accepted norms of behavior in the international community. Actions may gain legitimacy under the law when the international community accepts certain behavior as legitimate. The less a cyber attack looks like accepted state practice, the stronger the argument that it is an illegal use of force or an armed attack

basis by an incident commander's operational intent, presiding policy, and interpretation by appropriate legal counsel.

Another contribution to policy on the use of cyber by military forces includes the French publication of "Droit International Applique aux Operations dans le Cyberspace" (Schmitt, 2011). This document, still being translated at the time of this book's publication, is reportedly similar to the Tallinn Manual 2.0 in many key ways, but is more expansive in its definition of an attack in cyberspace.

The Schmitt criteria provide an excellent example of the type of framework that the cyber analyst uses to put together a case for intelligence, military, or law enforcement targeting. One example of a law enforcement targeting operation is the Coreflood Botnet, which operated for 7 years in harvesting our user names, passwords, and available credit card information, with the goal of stealing our money.

2.3.4   Policy Example: Coreflood Botnet

There are similarities between the lack of response that kept the harbor lights on during the early stages of WW II and the relatively recent delays described by the Coreflood botnet example. For example, the Coreflood botnet (Zetter, 2011a; b) was a seven-year-old botnet, comprising over two million infected computers, and was used daily to harvest usernames, passwords, and financial information. Around 190 gigabytes of data stolen from more than 400,000 victims was found in one server.

In addition, the Coreflood botnet is an example of a combined law enforcement (i.e., the Federal Bureau of Investigation [FBI]) and industry (i.e., Microsoft) cyber operation to disable a botnet and provide users with an ability to protect themselves from further infection. This active defense included the following steps:

Legal

1. The US District Court of Connecticut issued a temporary restraining order that allowed the Internet Systems Consortium (ISC) to swap out Coreflood's C2 servers for its own servers. The order also allowed the government to take over the domain names used by the botnet. When the infected machines reached out to the new C2 servers for instructions, the bots were commanded to stop. This neutralized, but did not eliminate, the malware installed on the compromised machines.
2. The FBI provided the IP—Internet protocol—addresses of infected machines to Internet service providers (ISPs) so they could notify their customers.

Industry
3. Microsoft issued an update to its Malicious Software Removal Tool so victims could get rid of the code.

As shown by the Coreflood botnet removal, active defense includes enforcing policies that set boundaries to the freedom of the Internet. While dimming harbor lights and shutting down a major botnet might be viewed as one-time actions, the frequency of current cyber attacks requires guiding doctrine to ensure our front-line cyber operators are clear in their handling of daily cyber events.

2.4 Doctrine

Doctrine helps in establishing rules via use cases that provide customary ways of applying a tool or technology. As a practical means of developing guidelines, the use of "custom" prepares one with a set of basic, proven, approaches to a defined problem set, or the use of a new technique, leaving room for flexibility when necessary. Flexibility is the key term; it is what a military organization does when faced with new tactics or technologies that disrupt not only settled doctrine but also the key assumptions underlying their current strategy and scheme of maneuver.

Military doctrine might be looked at as a portal into the thinking and experience of the organizations who write it. In addition, doctrine follows practice. Doctrine has the ability, therefore, to provide a view of current capability. This is especially useful for technical intelligence, as the doctrine should be reflective of the actual capability conferred by the "do how" of practitioners using new engineering developments.

Policy and doctrine, while closely related, are clearly distinct and designed to meet different requirements. For example, according to CJCSI 5120.02, "Joint Doctrine Development System (JDDS)," policy assigns tasks, provides direction, and recommends capabilities to ensure that the United States Armed Forces are ready to carry out their required roles. The goal of doctrine, on the other hand, is to improve operational effectiveness by providing standardized terminology and authoritative guidance on topics relevant to the employment of military forces.

2.4.1 Example US Department of Defense (DoD) Instructions, Directives, and Doctrine for Cyberspace Analysis and Targeting

Cyber analysis necessarily covers operations, described, for example, in US Department of Defense (DoD) Joint Publication (JP) 3–12—"Cyberspace Operations." Operations effects, spanning from denial to integrity, are still a challenge to account for in the same way that we do for conventional munitions. US DoD JP 3–13 helps us with evaluating information operations (IO), leveraging roles and responsibilities established in US Department of Defense Directive (DoDD) 3600.01, the principal information operations policy doctrine. Information operations, also known as influence operations, include collecting adversary tactical information as well as disseminating propaganda with the goal of achieving a competitive advantage over an opponent. US Chairman of the Joint Chiefs of Staff Instruction (CJCSI) 3210.01 uses DoDD 3600.01 as the authoritative document for joint IO doctrine (Fig. 2.3).

As shown in Fig. 2.3, JP 3–12 describes cyberspace operations in terms of IO, using JP 3–13 to leverage policy definitions from the DoDD 3600.01 technical assurance standard. Similarly, cyber targeting expands standard targeting doctrine, using operational description from JP 3–12. Each of the initial, intermediate, and final target development steps is required as a precursor to the standard six-step targeting method, employed to provide JP 3–12 effects of the battlefield. An example set of cyber, and IO, directives, instructions, and joint publications is provided in Table 2.5.

Table 2.5 provides an example flow from established IO instructions, directives, and policy, through cyberspace operations to targeting. These are examples of the broad guidelines that cyber warfighters have available to do their job.

While Table 2.5 provides the flow from information operations to targeting, many cyber activities are defensive, prescribed, and distilled from best practices of protecting computer-based systems. Critical Security Controls (CSCs) are an example of best practices distilled into rules of operation.

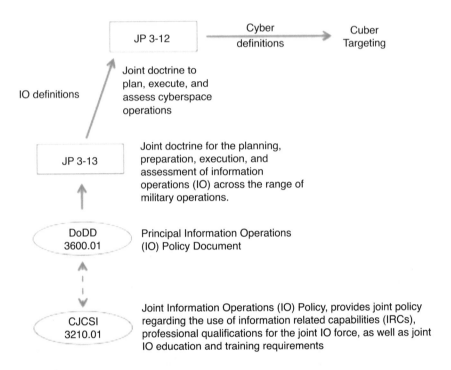

Fig. 2.3 US policy and doctrine for cyber targeting

Table 2.5 Example cyber doctrine documents (US DoD)

Title	Description
Department of Defense Directive (DoDD) 3600.01 Information Operations (IO) (DoD, 2001)	Information Operations (IO) will be the principal mechanism used during military operations to integrate, synchronize, employ, and assess a wide variety of information-related capabilities (IRCs) in concert with other lines of operations to effect adversaries' or potential adversaries' decision-making while protecting our own
DoD Joint Publication (JP) 3-13	This publication provides joint doctrine for the planning, preparation, execution, and assessment of information operations across the range of military operations
Department of Defense (DoD) Joint Publication 3-12 Cyberspace Operations	This publication provides joint doctrine to plan, execute, and assess cyberspace operations

2.4.2   Critical Security Controls (CSC)

DoD doctrine both introduces and guides cyber operators through the sometimes challenging situations in conducting cyber analysis and targeting. In addition, we also have more specific rules, or controls, based on real-world events, that provide checklist-level guidance to perform cyber defense. Cyber Security Controls (CSCs) are the result of several decades of experience. In addition, CSCs are prioritized, from most to least effective, providing the user with an added benefit of getting the highest effect with the lowest numbered CSC. Ideally, each CSC will be automated, leveraging each "rule" encoded into a CSC to provide machine intelligence based network protection.

CSCs are prioritized, based on years of experience, starting with the most important actions to be taken on the part of a cyber defender. For example, the first course of action includes providing an inventory of devices and software on a network. This is followed by ensuring that hardware and software have secure configurations for each of the endpoint devices. Along with performing an inventory and doing configuration management, continuous monitoring (e.g., vulnerability assessment/remediation), controlling passwords, and managing audit logs are common defensive practices. Similarly, maintaining firewalls, locking down network ports, and continuity of operations planning are methods to ensure both secure operations and the ability to recover from an event.

CSCs effectively provide a list of activities to provide the analyst with the TTPs for network defense. Each of the CSC steps is reflected in Australia's "Top 4" (really top 3; i.e., Application Whitelisting, Patching Systems, Restricting Administrator Privileges) to provide an estimated 85% of network security requirements. Both CSCs and the Australia Top 8 provide security controls and a cross-institutional memory that help security researchers manage key challenges in defending networks based on multiple use cases.

2.5 Tactics, Techniques, and Procedures (TTPs)

Doctrine, and its implementation (e.g., CSCs), provides general guidance for operating in cyberspace. TTPs, however, are the language and structure used to describe each cyber event. For example, TTPs are often laid out via general attack frameworks that help profile adversarial behavior, tool use, and step-by-step operations for what is being performed at each attack step. In addition, attack frameworks provide the analyst with a method to define a clear signature of a cyber attacker's TTP that includes the tools used (e.g., malware family), attack patterns, supporting infrastructure, and victim type. This signature is then used to identify a malicious cyber actor in future attacks. A malicious cyber actor (MCA) is an entity or person that is responsible for an incident or event that impacts, or threatens, the safety or security of another entity.

There are currently a variety of descriptions used for the TTPs associated with a cyber attack. For example, the analyst is often faced with the job of being a detective, pulling together specific technical indicators, or evidence, of TTPs, that include details about the specifics of an attack, with threads similar to the following:

- Hacker/team signatures: a malicious cyber group may be known to target specific individuals (e.g., business/government officials) to get access to information, money, etc.
- Compromise processes: spear phishing and watering holes are common ways of gaining initial access to a system.
- Technologies: this might be the use of a specific tool to exploit commonly used software (e.g., word processing, spreadsheets).

The analyst coordinates these different pieces of evidence for an attack to find patterns that match attack types to malicious cyber actors. Conceptual frameworks are a common method for structuring the different elements of an attack into a coherent "story" that describes what an attacker did, how she did it, and the type of infrastructure used. A good analyst therefore wants to organize each attack via a framework to understand both how an operation took place and to compare similarities and differences between operations. Frameworks exist for both defensive and offensive descriptions of cyber operations. An example set of techniques for defensive cyber description is shown in Table 2.6.

As shown in Table 2.6, there are multiple frameworks for describing a cyber attack, from a pure cyber defense viewpoint. Example frameworks for describing the steps in a cyber attack cycle are shown in Table 2.7.

As shown in Table 2.7, attack cycles are currently profiled by research (i.e., MITRE ATT@CK framework) and governmental (i.e., Cyber Threat Framework (CTF)) organizations. In addition, Table 2.7 provides current examples of frameworks that generalize on TTPs to reassemble the people, processes, and technologies behind a cyber attack.

An additional resource for documenting the people, processes, and tools used by each attacker, or group, is the Cyber Analytics Repository Exploration Tool (CARET). (MITRE) CARET profiles example advanced persistent threat (APT) actors across example attack cycles. In addition, CARET includes the tool type, by phase, of the respective APT, providing example signatures for the malicious cyber actors.

2.6 Summary

Providing a detailed structure to describe the step-by-step actions of a cyber attacker, as provided by a tool like CARET, results in understanding not only the attacker TTPs, but also possibly the guiding doctrine and policy behind the attack organizations. For example, the attack process will provide clues as to the doctrine that the attack team is following. In addition, the target type reveals the broader policy goals that the team is tasked with in using cyber for information gathering. We can now revisit the questions that we posed at the beginning of Chapter 2 in order to better understand the linkages between policy, doctrine and TTPs -

Table 2.6 Descriptive frameworks for cyber risk descriptions

Approach	Description
Risk Bow-Tie	General technique that looks like a "bow-tie," with the attacker, controls, and candidate mitigations on the left side, the event in the middle, and the remediations on the right side (Rick Nunes-Vaz, 2011)
Factor Analysis for Information Risk (FAIR)	Portfolio-based approach for understanding a given enterprise's risk; based on an International standard (Jack Freund, 2014)
Cybersecurity Framework (CSF)	NIST—US National Institute of Standards and Technology—standard on defensive cyber (Director of National Intelligence)
ISO 31000	International standard for cyber risk

Table 2.7 Descriptive frameworks for cyber attack cycles

Approach	Description
Lockheed Martin Attack Cycle	Step-by-step method for showing how an attacker gets to a target
Mandiant Attack Cycle	
Cyber Threat Framework	Layered approach that includes both the steps to get to a target (i.e., phases) and the signature that the attackers are likely to have during the attack; this is an agreed-up aggregation of several, independent, attack cycles (Director of National Intelligence)

1. What are the definitions of policy, doctrine, and TTPs? What are examples of each?

   (a) Section 2.1 provides example definitions for policy, doctrine, and TTPs.
   (b) Table 2.2 provides example national policy frameworks. In addition, Table 2.3 provides example cyber Executive Orders (EOs), issued by the President of the United States.
   (c) Table 2.5 provides example US doctrine on cyber.
   (d) Tables 2.6 and 2.7 provide descriptive frameworks for decomposing tactics, techniques, and procedures (TTPs) for a cyber attack.

2. How is cyber policy used for both defensive and offensive cyber operations?

   (a) Cyber policy drives resourcing to ensure an infrastructure resilient to cyber attack. In addition, policy frameworks (e.g., Schmitt Criteria) help policy makers determine a reciprocal response to properly attributed cyber attacks.
   (b) For offensive operations, policy determines at what level of an organization authorities reside for conducting operations. For example, the literature mentions that NSPM 13 provides cyber operators with additional latitude for conducting operations in the broader Internet, outside of organizationally owned networks.

3. What are the key drivers for cyber doctrine?

   (a) Information Operations (JP 3–13) informs cyberspace operations (JP 3–12) doctrine. Targeting leverages existing doctrine (JP 3–60), and can be used to guide cyberspace operations (Table 2.5).

Analysis and targeting in the cyber domain are guided by a hierarchy of explicit rules at three levels. Policies have established the terminology, roles, and responsibilities for cyber operations—and their relationships to other non-cyber disciplines. Doctrine is derived from policy to establish the processes and rules that guide cyber operations. At the lowest level, TTPs then establish how cyber analysts collect, analyze, and report their activities.

TTPs are used to describe both offensive and defensive operations, and benefit from the use of attack cycles to portray the progression of a cyber attack. One view of attacker TTPs is to look at the tools that they use (Table 2.7), to provide a description of what is needed for each step of an attack cycle. Tools are also included for each of the APTs, making them identifiable by attack phase via standard frameworks (e.g., CARET). How the tools are used, or the tactics and techniques used by an attacker, describe the practices of an attacker or team, providing an example of applied doctrine.

While resilience is an overriding theme for most national cyber policies, the response to a cyber attack still challenges international law and frameworks for categorizing an attack, and, thereby, determining a response. For example, the Tallinn Manual, informed by the Schmitt criteria, provides a guideline for evaluating when a cyber attack has crossed over into being an armed attack. To date, only ISIS—Islamic State in Iraq and Syria—(2015) and Hamas (2019) have had been bombed, or received a kinetic response, due to their cyber attacks. Responding to cyber with conventional weapons is uncommon due to the lack of clear categorization of cyber attacks, as compared to conventional, armed, attacks. For example, nation state cyber aggression, namely Russian cyber attacks against Estonia (2007), Georgia (2008), and Ukraine (2014), did not result in an effective military or diplomatic response.

While international policy for categorizing, and determining the appropriate response to, cyber effects is still being worked out, the United States is moving forward with clear doctrine describing targeting and operations in the cyberspace domain. The core cyber operations doctrine for the US Department of Defense (DoD) is Joint Publication (JP) 3–12—"Cyberspace Operations." JP 3–12 describes operational effects, spanning from denial to integrity. In addition, JP 3–12 leverages JP 3–13 to help us with incorporating information operations (IO) effects. A key use, by the cyber analyst, in using current joint publications (i.e., 3–12 and 3–13) is to describe cyber attacks within a common framework and using defined terms. Policy, doctrine, and TTPs are therefore focused on the actions that are performed in cyber operations (both offensive and defensive) against a generally defined cyber threat. But the threat is a broad set of activities that must be more carefully defined. In the next chapter, we introduce the structured approaches to categorize the myriad of threat types in a taxonomic structure.

Bibliography

Arquilla, J. (2017). From Pearl Harbor to the "harbor lights". In A. E. G. Perkovich (Ed.), *Understanding cyber conflict - 14 analogies*. Georgetown.

Australian Government. (2018). *Cyber security strategy*. Retrieved 3 10, 2019, from Cyber Security Strategy: https://cybersecuritystrategy.homeaffairs.gov.au/executive-summary-0

Barnes, J. E. (2019, 8 18). U.S. cyberattack hurt Iran's ability to target oil tankers, Officials Say. *New York Times*.

Biddle, S. (2004). *Military power: Explaining victory and defeat in modern Battle*. Princeton University Press.

Biden, J. (2021, 5 12). *Executive order on improving the nation's cybersecurity*. Retrieved 5 22, 2021, from White House: https://www.whitehouse.gov/briefing-room/presidential-actions/2021/05/12/executive-order-on-improving-the-nations-cybersecurity/

Carr, J. (2012). *Inside cyber warfare*. O'Reilly.

Center for Computational Analysis of Social and Organizational Systems (CASOS). (n.d.). *Center for computational analysis of social and organizational systems (CASOS)*. Retrieved 10 14, 2018, from http://www.casos.cs.cmu.edu/

Chapman, J. (1909). *Doctors of the church*. Robert Appleton Company.

Cleary, G. (2019, 6). *Twitterbots: Anatomy of a propaganda campaign*. Retrieved 7 6, 2019, from Symantec: https://www.symantec.com/blogs/threat-intelligence/twitterbots-propaganda-disinformation

Crowdstrike. (2016, 6 15). *Bears in the Midst: Intrusion into the Democratic National Committee*. Retrieved 9 21, 2019, from Crowdstrike: https://www.crowdstrike.com/blog/bears-midst-intrusion-democratic-national-committee/

CrowdStrike. (n.d.). Retrieved 11 29, 2018, from https://www.crowdstrike.com/

Department of Defense. (2012). *Joint Publication 1–13.4 Military Deception*. Department of Defense.

Department of Defense (DoD). (2018). *DoD Cybersecurity strategy*. Retrieved 3 10, 2019, from DoD Cybersecurity Strategy: https://media.defense.gov/2018/Sep/18/2002041658/-1/-1/1/CYBER_STRATEGY_SUMMARY_FINAL.PDF

Department of Homeland Security (DHS). (2018, 5 15). *DHS cybersecurity strategy*. Retrieved 3 10, 2019, from DHS Cybersecurity Strategy: https://www.dhs.gov/sites/default/files/publications/DHS-Cybersecurity-Strategy_1.pdf

Director of National Intelligence. (n.d.). Cyber threat framework. Retrieved 10 20, 2018, from https://www.dni.gov/index.php/cyber-threat-framework

Djabatey, E. (2019, 10 17). Reassessing U.S. cyber operations against Iran and the use of force. Retrieved 10 20, 2019, from Just Security: https://www.justsecurity.org/66628/reassessing-u-s-cyber-operations-against-iran-and-the-use-of-force/

DoD. (2001, 10). *DoDD 3600.01 Information operations*. Retrieved 3 10, 2019, from Homeland Security Digital Library: https://www.hsdl.org/?abstract&did=439849

Dorothy, E., & Denning, B. J. (2017). Active cyber defense - applying air defense to the cyber domain. In A. E. G. Perkovich (Ed.), *Understand cyber conflict - 14 analogies*. Georgetown.

Dragos. (n.d.). *Dragos*. Retrieved 29 2018, 11, from https://dragos.com/

Egloff, F. (2017). Cybersecurity and the age of privateering. In A. E. G. Perkovich (Ed.), *Understanding cyber conflict - 14 analogies*. Georgetown.

European Union. (n.d.). *General data protection regulation*. Retrieved 6 26, 2020, from General Data Protection Regulation: https://gdpr-info.eu/

Feldman, S. (2019, 2 25). *Russia has the fastest hackers*. Retrieved 9 3, 2019, from Statistica: https://www.statista.com/chart/17151/government-hack-speed/

FireEye. (2014). *APT28: A window into Russia's cyber espionage operations?* Retrieved 9 9, 2018, from FireEye: https://www.fireeye.com/content/dam/fireeye-www/global/en/current-threats/pdfs/rpt-apt28.pdf

FireEye. (2017). *APT 29.* (FireEye, Producer) Retrieved 9 9, 2018, from APT 29: https://www.fireeye.com/current-threats/apt-groups.html#apt29

FireEye. (n.d.). *FireEye.* Retrieved 29 2018, 11, from https://www.fireeye.com/

Frank, A. B. (2017, January). Toward computational net assessment. *Journal of Defense Modeling and Simulation, 14*(1).

George Washington University. (n.d.). *National security archive.* Retrieved 1 30, 2020, from https://nsarchive.gwu.edu/news/cyber-vault/2018-11-07/presidential-orders

Government of Canada. (2018). *National cyber security strategy.* Retrieved 3 10, 2019, from National Cyber Security Strategy: https://www.publicsafety.gc.ca/cnt/rsrcs/pblctns/ntnl-cbr-scrt-strtg/index-en.aspx

HM Government. (2016). *National security strategy 2016–2021.* Retrieved 3 10, 2019, from National Security Strategy 2016–2021: https://assets.publishing.service.gov.uk/government/uploads/system/uploads/attachment_data/file/567242/national_cyber_security_strategy_2016.pdf

IDS International. (n.d.). *SMEIR.* Retrieved 1 3, 2019, from https://www.smeir.net/

International And Operational Law Department . (2015). *Law of armed conflict (LOAC).* Retrieved 3 22, 2020, from The United States Army Judge Advocate General's Legal Center and School: https://www.loc.gov/rr/frd/Military_Law/pdf/LOAC-Deskbook-2015.pdf

Jack Freund, J. J. (2014). *Measuring and managing information risk: A FAIR approach.* Butterworth-Heinemann.

Joint Chiefs of Staff. (2019, 6 3). *Joint Doctrine Note 1–19 - Competition Continuum.* Retrieved 5 10, 2020, from Joint Chiefs of Staff.

Joint Staff. (2016). *CJCSI 3370.01B target development standards.* Joint staff, .

Joint Staff. (2018, 6 8). *Joint Publication 3–12 Cyberspace operations.* Retrieved 9 16, 2019, from Joint Publications: https://www.jcs.mil/Portals/36/Documents/Doctrine/pubs/jp3_12.pdf

Krebs, B. (2013, 6 13). *Iranian elections bring lull in bank attacks.* Retrieved 9 14, 2019, from KrebsonSecurity: https://krebsonsecurity.com/2013/06/iranian-elections-bring-lull-in-bank-attacks/

Leversage, D. (n.d.). *RiSI.* Retrieved 9 11, 2019, from RISI: https://www.risidata.com/Database

Lynn Arnhart, M. K. (2019). *Analytics, Operations Research and Strategic Decision Making in the Military.* IGI Global.

Merriam-Webster. (n.d.). *Dictionary.* Retrieved . 1 30, 2020, from www.merriam-webster.com/dictionary

MITRE. (n.d.-a). *CARET.* Retrieved 9 30, 2018, from CARET: https://car.mitre.org/caret/#/.

MITRE. (n.d.-b). Cyber analytics repository exploration tool (CARET). Retrieved 9 30, 2018, from CAR Exploration Tool: https://car.mitre.org/caret/#/.

MITRE. (n.d.-c). MITRE ATT@CK Framework. Retrieved 9 20, 2019, from MITRE: https://attack.mitre.org/

Morison, S. E. (1963). *The two-ocean war: A short history of the United States navy in the second world war.* Little, Brown.

Nakashima, E. (2018, 10 23). *Pentagon launches first cyber operation to deter Russian interference in midterm elections.* Retrieved 3 12, 2019, from Washington Post: https://www.washingtonpost.com/world/national-security/pentagon-launches-first-cyber-operation-to-deter-russian-interference-in-midterm-elections/2018/10/23/12ec6e7e-d6df-11e8-83a2-d1c3da28d6b6_story.html?utm_term=.8c47d573557b

Nakashima, E. (2019, 2 27). *US disrupted Internet access of Russian troll factory on day of 2018 midterms.* Retrieved 3 12, 2019, from Washington Post.

Nakasone, P. M. (2019). A cyber force for persistent operations. *Joint Forces Quarterly, 92*(1), 10–15.

National Cyber Security Center. (2018, 10 11). *Joint report on publicly available hacking tools.* Retrieved 3 10, 2019, from National Cyber Security Center: https://www.ncsc.gov.uk/joint-report

National Research Council. (2010). *Committee on deterring cyberattacks: Informing strategies and developing options for U.S. policy.* NAP.

Nicole Perlroth, Q. H. (2013, 1 8). Bank hacking was the work of Iranians, officials say. Retrieved 9 21, 2019, from New York Times: https://www.nytimes.com/2013/01/09/technology/online-banking-attacks-were-work-of-iran-us-officials-say.html

O'Flaherty, K. (2019, 5 6). *Israel retaliates to a cyber-attack with immediate physical action in a world first.* Retrieved 5 15, 2019, from https://www.forbes.com/sites/kateofla-hertyuk/2019/05/06/israel-retaliates-to-a-cyber-attack-with-immediate-physical-action-in-a-world-first/#1319bce9f895

Office of the Director of National Intelligence. (n.d.). Cyber threat framework. (DNI, Producer) Retrieved 9 20, 2019, from DNI: https://www.dni.gov/index.php/cyber-threat-framework

Office of the National Manager for NSS. (n.d.). *DoDCAR.* Retrieved 3 8, 2019, from https://csrc. nist.gov/CSRC/media/Presentations/DODCAR-no-class-markings-Pat-Arvidson/images-media/DODCAR_-no%20class%20markings%20-%20Pat%20Arvidson.pdf

Ottis, R. (2007). *Analysis of the 2007 cyber attacks against Estonia from the information warfare perspective.* Retrieved 9 14, 2019, from Cooperative Cyber Defence Centre of Excellence: https://ccdcoe.org/uploads/2018/10/Ottis2008_AnalysisOf2007FromTheInformationWarfarePerspective.pdf

Peter Feaver, K. G. (2017). "When the urgency of time and circumstance clearly does not permit ..." - pre-delegation in nuclear and cyber scenarios. In A. E. G. Perkovich (Ed.), *Understanding cyber conflict - 14 analogies.* Georgetown.

Pomerleau, M. (2019, 5 8). New authorities mean lots of new missions at cyber command. (F. Domain, Producer) Retrieved 9 14, 2019, from Fifth Domain: https://www.fifthdomain.com/dod/cybercom/2019/05/08/new-authorities-mean-lots-of-new-missions-at-cyber-command/

Rapid7. (n.d.). *Metasploit.* (R. 7, Producer) Retrieved 9 11, 2019, from Metasploit: https://www. metasploit.com/

Reuters. (2020, 5 12). *Estonia passes 'Huawei Law' for telecom security reviews.* Retrieved 5 23, 2020, from U.S. News and World Report: https://www.usnews.com/news/technology/articles/2020-05-12/estonia-passes-huawei-law-for-telecom-security-reviews

Rick Nunes-Vaz, S. L. (2011). A more rigorous framework for security-in-depth. *Journal of Applied Security Research, 23.*

Rid, T. (2013). *Cyber war will not take place.* Oxford University Press.

Riley, C. (2019, 7 9). *UK proposes another huge data fine. This time, Marriott is the target.* Retrieved 8 21, 2019, from CNN: https://www.cnn.com/2019/07/09/tech/marriott-data-breach-fine/index.html

Robert M. Lee, Michael J (2014). *German steel mill cyber attack.* SANS ICS CP/PE (Cyber-to-Physical or Process Effects) case study paper. SANS.

Roguski, P. (2019, 9 27). *An overview of international humanitarian law in France's new cyber document .* Retrieved 6 5, 2020, from Just Security: https://www.justsecurity.org/66318/an-overview-of-international-humanitarian-law-in-frances-new-cyber-document/

Roscini, M. (2014). Cyber operations and the use of force in international law. .

David E. Sanger, N. P. (2015, 4 15). *Iran is raising sophistication and frequency of cyberattacks, study says.* Retrieved 10 14, 2018, from New York Times: https://www.nytimes.com/2015/04/16/world/middleeast/iran-is-raising-sophistication-and-frequency-of-cyberattacks-study-says.html

SANS. (2016). *Critical security controls.* (SANS, Producer) Retrieved 9 24, 2019, from SANS: https://www.sans.org/media/critical-security-controls/critical-controls-poster-2016.pdf

SANS. (n.d.). *SANS Institute.* Retrieved 9 26, 2019, from https://www.sans.org/

Schmitt, M. (2011, 4 3). Cyber operations and the Jus in Bello: Key Issues. *Naval War College International Law Studies, 2019*(9), 10.

Schmitt, M. N. (2013). *Tallinn manual on the international law applicable to cyber warfare.* Cambridge University Press.

Singapore. (2016). *Singapore's cyber security strategy* . Retrieved 3 10, 2018, from Singapore's Cyber Security Strategy : https://www.csa.gov.sg/~/media/csa/documents/publications/singaporecybersecuritystrategy.pdf

Singer, P. W., & Brooking, E. T. (2018). *LikeWar - the weaponization of social media*. Houghton Mifflin.

Slowik, J. (2019, 8 15). *CRASHOVERRIDE: Reassessing the 2016 ukraine electric power event as a protection-focused attack* . Retrieved 9 24, 2019, from DRAGOS: https://dragos.com/wp-content/uploads/CRASHOVERRIDE.pdf

Symantec. (n.d.). *Symantec*. Retrieved 29 2018, 11, from https://www.symantec.com/

Temple-Raston, D. (2016, 9 12). Cyber bombs reshape U.S. Battle against terrorism. Retrieved 9 3, 2019, from NPR: https://www.npr.org/2016/09/12/493654985/cyber-bombs-reshape-u-s-battle-against-terrorism

Temple-Raston, D. (2019a, 1 11). Hacks are getting so common that companies are turning to 'Cyber insurance'. Retrieved 9 3, 2019, from NPR: https://www.npr.org/2019/01/11/684610280/hacks-are-getting-so-common-that-companies-are-turning-to-cyber-insurance

Temple-Raston, D. (2019b, 9 26). *National public radio (NPR)*. (NPR, Producer) Retrieved 9 29, 2019, from How The U.S. Hacked ISIS: https://www.npr.org/2019/09/26/763545811/how-the-u-s-hacked-isis

Temple-Raston, D. (2019c, 8 14). Task force takes on Russian election interference. Retrieved 9 3, 2019, from NPR: https://www.npr.org/2019/08/14/751048230/new-nsa-task-force-takes-on-russian-election-interference

Timur Chabuk, A. J. (2018, 9 1). Understanding Russian information operations. Retrieved 9 9, 2018, from Signal (AFCEA): https://www.afcea.org/content/understanding-russian-information-operations

Trump, D. (2020, 4 4). Executive order on establishing the committee for the assessment of foreign participation in the United States telecommunications services sector. Retrieved 6 4, 2020, from https://www.whitehouse.gov/presidential-actions/executive-order-establishing-committee-assessment-foreign-participation-united-states-telecommunications-services-sector/

TTP vs Indicator: A simple usage overview. (2018). Retrieved 3 11, 2019, from STIX Project: https://stixproject.github.io/documentation/concepts/ttp-vs-indicator/

U.S. National Archives. (n.d.). *Federal Register*. Retrieved 5 19, 2019, from Federal Register: https://www.federalregister.gov/presidential-documents/executive-orders

United Nations. (n.d.). *Charter of the United Nations*. Retrieved 10 20, 2019, from http://legal.un.org/repertory/art1.shtml

US Department of Defense. (2008). *Cyberspace operations*. US Department of Defense, Joint Staff.

Verizon. (2018). *2018 Data breach investigations report*. Retrieved 10 17, 2018, from Verizon: http://www.verizonenterprise.com/industry/public_sector/docs/2018_dbir_public_sector.pdf

Warner, M. (2017). Intelligence in cyber - and cyber in intelligence. In A. E. G. Perkovich (Ed.), *Understanding cyber conflict - 14 analogies*. Georgetown University Press.

Whaley, B. (2007). *STRATAGEM - deception and surprise in war*. Artech House.

White House. (2018, 9). National cyber security strategy of the United States of America. Retrieved 3 10, 2019, from National Cyber Security Strategy of the United States of America: https://www.whitehouse.gov/wp-content/uploads/2018/09/National-Cyber-Strategy.pdf

White House. (n.d.). Cyber executive orders. Retrieved 3 10, 2019, from Homeland Security Digital Library: https://www.hsdl.org/?collection&id=2724

Zetter, K. (2011a, April 26). FBI vs. Corefloot Botnet: Round 1 Goes to the Feds. *Wired*.

Zetter, K. (2011b, April 11). With Court Order, FBI Hijacks 'Coreflood' Botnet, Sends Kill Signal. *Wired*.

Zetter, K. (2014). *Countdown to zero day - stuxnet and the launch of the World's first digital weapon*. Crown.

Chapter 3
Taxonomy of Cyber Threats

A taxonomy of cyber threats includes accounting for the people, policy, processes, and technologies associated with both the threat and the system to be defended. Both governments and industry provide policy and process, respectively, for threat identification and mitigation. In addition, standard systems engineering provides hardware/software verification methodology for analysis and security validation.

Key questions for cyber threat analysis include:

1. Who are the responsible organizations that provide the policies and processes that are used to describe and mitigate cyber threats?
2. What are examples of popular cyber threat taxonomies?
3. What is the process for developing cyber threat descriptions of a system of interest?
4. What are the current data standards used to describe and share cyber threat data?
5. How do cyber protection tools use collected vulnerability data to defend computer-based systems?
6. What are some example analysis and targeting uses of cyber threat/vulnerability data?

We will therefore start this chapter with a review of cyber threat taxonomy approaches that have matured into policy and process standards used by both American and International communities. Understanding the current state of play with defensive cyber threat taxonomies, we will explore cyber system risk evaluations in order to see how cyber threats are developed and described for current security evaluations.

The purpose of this chapter is to present a general background on the taxonomy of cyber threats, how these threats are used in cyber systems analysis, and the type of analysis and targeting information that current cyber threats provide. A key item in this discussion is a review of how the cyber attack life cycle is used to describe the scope of a cyber threat in the current cyber threat intelligence (CTI) framework. In addition, we will use vulnerability evaluation frameworks to better understand

© Springer Nature Switzerland AG 2022
J. M. Couretas, *An Introduction to Cyber Analysis and Targeting*,
https://doi.org/10.1007/978-3-030-88559-5_3

both the location and potential severity of a cyber threat in the overall attack life cycle.

3.1 Background

Cyber threat intelligence (CTI) rolls up the different elements of risk analysis in terms of the cyber attack process:

> "CTI was born from the application of military intelligence doctrine to data analysis of cyberattacks. The (Department of Defense) DoD describes the intelligence process as a cycle of phases: direction, collection, processing, analysis, dissemination, and feedback (JP 2-0, 2013). While represented as a cycle, the steps may happen concurrently or may be skipped entirely depending on the situation. The intelligence cycle prescribes the process for collecting threat data and transforming it into threat intelligence, p. 5." (Launius, 2019)

In the above, Launius describes how CTI is intrinsically tied to the intelligence collection process. I believe that this process understanding, for the collection and development of CTI, serves us well in providing a method for looking at a taxonomy of cyber threats. Fig. 3.1 provides an overall cyber threat diagram, combining the potential weaknesses and vulnerabilities of the systems that we are working to protect from the possible effects of an attacker.

Looking at cyber from a purely technical standpoint, we use Fig. 3.1 to provide an overall picture of the processes used to determine the type of threat exposed by one of the cyber risk evaluation frameworks and to categorize the estimated effect from current Joint Publication 3-12 doctrine. Definitions that will help in understanding Fig. 3.1 include the following:

Cyber Risk Evaluation Framework methodical approach for evaluating an information technology system's weaknesses and vulnerabilities. These frameworks are usually associated with improving system resilience (Table 2.2).

Cyber Threat Intelligence (CTI) Standards:

Internal cyber threat intelligence information that is already within an organization. This might be lessons learned from past cyber events or information gathered from a central log management or Security Incident and Event Management (SIEM) system used to protect the organization's network.

External cyber threat intelligence there are multiple companies that provide cyber threat intelligence that cover general information including threat actors, and the malware that they employ, to more specific information about the industry-type or system components (e.g., software composing websites, operating systems, key databases) to be defended.

Cyber Effects Joint Publications (JP) 3-12 defines cyber effects in two main categories, (1) manipulation and (2) denial.

Using JP 3-12 cyber effects (e.g., deny, manipulate) highlights that the risk evaluation methods more commonly used for cyber security can also be used to perform

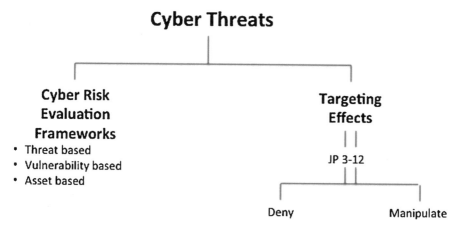

Fig. 3.1 Cyber weaknesses, vulnerabilities, and potential effects

systems analysis for targeting. Each of the JP 3-12 denial and manipulation effects will include elements of the cyber security confidentiality, integrity, and availability (CIA) model that is used by governments and industry when implementing cyber security approaches.

3.2 NIST Cyber Taxonomy Examples

The US National Institute of Standards and Technology (NIST) Cybersecurity Framework (CSF) provides computer security guidance, including a policy framework for private sector organizations to assess and improve their ability to

- Prevent
- Detect
- Respond

to cyber attacks. Similar to earlier risk approaches (Fig. 3.3), the NIST framework includes a structured approach for assessing cyber security risks and providing course of action analysis Source is https://www.nist.gov/system/files/documents/cyberframe-work/cybersecurity-framework-021214.pdf (NIST Cyber Security Framework (CSF)) NIST CSF Version 1.0, for example, published in 2014, was originally focused on critical infrastructure operators with Table 3.1's five basic steps.

Table 3.1 shows NIST's standardization of common practices shown in the system risk bow-tie (Fig. 3.2).

As shown in Fig. 3.2, the defender's view of an attack life cycle is standardized via the NIST CSF, providing both US and international cyber security professionals with standard policy and process guidelines to protect their enterprise. In addition, the NIST CSF is currently being used by a wide range of businesses and organizations, shifting them to be proactive about risk management.

The NIST Framework organizes cyber security capabilities around the five functions: identify, protect, detect, respond, and recover. Using the NIST Framework is usually performed in conjunction with NIST special publications that include the following:

- 800-39: Managing Information Security Risk: Organization, Mission, and Information System View
- 800-37: Risk Management Framework to Federal Information Systems.

In conjunction with other NIST standards and guidelines, NIST policies and processes provide users with detailed methods for making clear, risk-based decisions, and managing cyber security risks across their respective organizations. In addition to NIST, the US Office of the Director for National Intelligence (ODNI) developed the Cyber Threat Framework (CTF) for understanding both the attack process and the types of indictors to look for in a cyber attack. This includes accounting for the stages, objectives, actions, and indicators for the threat (Fig. 3.3).

As shown in Fig. 3.3, internal cyber threat intelligence is used to build a picture of the threat actor's objectives and current stage in meeting this objective. Using the threat assessment in Fig. 3.3 to understand the stage of a threat, as described in Fig. 3.2, includes having a clear understanding of the defended system. Cyber system threats are communicated via cyber security data standards and compared to standardized system descriptions in order to provide the system defender with both a qualitative and quantitative picture of her current security posture. This includes using language to describe each element of a threat and applying scoring measures to estimate the severity of each threat, and their combinations.

Table 3.1 Protect, detect, respond, and recover steps of the NIST Cyber Security Framework

Step	Description
Identify	Develop the organizational understanding to manage cybersecurity risk to systems, assets, data, and capabilities
Protect	Develop and implement the appropriate safeguards to ensure delivery of critical infrastructure services
Detect	Develop and implement the appropriate activities to identify the occurrence of a cybersecurity event
Respond	Develop and implement the appropriate activities to take action regarding a detected cybersecurity event
Recover	Develop and implement the appropriate activities to maintain plans for resilience and to restore any capabilities or services that were impaired due to a cybersecurity event

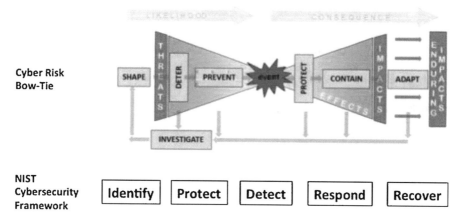

Fig. 3.2   Cyber risk bow-tie and the NIST Cybersecurity Framework

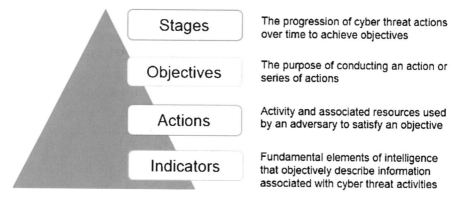

Fig. 3.3   ODNI Cyber Threat Framework (CTF): high-level view

3.3   Cyber System Threats: Risk Evaluation and Cyber Threat Understanding

Standards are used to describe a cyber system's security. This includes using policy to guide the application of proven methods to describe issues in a system with standard terms. Developing and maintaining this cyber security community of interest includes several organizations, community sites, and approved methodologies/tools. We will therefore start with a review of cyber security standards, models, and scoring approaches; wrapping up this section with a look at an end-to-end systematic approach for evaluating a computer-based system of interest.

3.3.1 Cyber Security Data Standards

A lot of effort has gone into ensuring clear communication of cyber threat information. Some common examples include the following languages and measures from the NIST Security Content Automation Protocol (SCAP) (NIST) that include:

- Languages:
  - Asset Reporting Format (ARF): used to communicate information about system assets and their reporting.
  - Open Checklist Interactive Language (OCIL): series of checklists with guidance for interpreting the feedback.

- Measures:
  - Common Vulnerability Scoring System (CVSS): used to evaluate the potential severity of reported vulnerabilities for a system of interest.
  - Common Configuration Scoring System (CCSS): derived from the CVSS, the CCSS provides a set of measures concerned with configuration issues.
  - Common Misuse Scoring System (CMSS): used to estimate the effects of how the misuse of a system's methods or functions might impair the security of the system.

In addition, Security Content Automation Protocol (SCAP) affiliations include:

- Languages: XSCCDF, OVAL
- Enumerations:
  - Common Configuration Enumeration (CCE)
  - Common Platform Enumeration (CPE)
  - Common Vulnerabilities and Exposures (CVE)
  - Common Attack Pattern Enumerations and Classifications (CAPEC)
- Measures: (CVSS)

Within the cyber community, standards that result in common reporting and sharing cyber tool data are provided by organizations like OASIS (OASIS) to address communication issues. Data standards are used to provide a common means for comparing security assessments of different systems, moving the cyber security community toward quantitative evaluations.

3.3.2 DREAD, STRIDE, and CVSS

Some of the early cyber threat models include DREAD—damage/reproducibility/exploitability/affected users/discoverability—STRIDE—spoofing/tampering/repudiation/information disclosure/denial of service/elevation of privilege—and CVSS—Common Vulnerability Scoring System—for identifying and scoring system vulnerabilities. DREAD and STRIDE are mnemonics developed at Microsoft as methods for modeling a computer system to assess cyber threats. CVSS, still widely used, was developed by Carnegie Mellon University (CMU) to account for organizational priorities, how vulnerabilities change over time, and use this information to provide base metrics for a system of interest (Table 3.2).

These initial efforts at threat modeling, guided by using security requirements engineering to help with the choosing of appropriate countermeasures, are effectively elicitation techniques that are used to discover system vulnerabilities. For example, DREAD (Schostack, 2008), an initial cyber security effort by Microsoft, was developed as a method for assessing the risk of computer security threats. Similarly, STRIDE is an approach for analyzing the security of an individual application, and is used as a model to understand the threats to a system. In addition to DREAD and STRIDE, the Common Vulnerability Scoring System (CVSS) is a more rigorous approach to vulnerability evaluation. Provided by the National Institute of Standards (NIST), CVSS is described as:

> "The Common Vulnerability Scoring System (CVSS) provides an open framework for communicating the characteristics and impacts of IT vulnerabilities. CVSS consists of three groups: Base, Temporal and Environmental. Each group produces a numeric score ranging from 0 to 10, and a Vector, a compressed textual representation that reflects the values used to derive the score. The Base group represents the intrinsic qualities of a vulnerability. The Temporal group reflects the characteristics of a vulnerability that change over time. The Environmental group represents the characteristics of a vulnerability that are unique to any user's environment. CVSS enables IT managers, vulnerability bulletin providers, security vendors, application vendors and researchers to al benefit by adopting this common language of scoring IT vulnerabilities," (Peter Mell, 2007)

While several other frameworks have been developed, the more famous DREAD, STRIDE, and CVSS provide some of the initial approaches for classifying and estimating potential vulnerability damages. Finding the vulnerabilities remains a challenge. One approach, the process for attack simulation and threat analysis (PASTA), provides a knowledge-based approach for finding vulnerabilities and estimating level of threat that they actually pose.

Table 3.2 DREAD, STRIDE, and CVSS: cyber security models

DREAD	STRIDE	Common Vulnerability Scoring System (CVSS)
Damage	Spoofing	
Reproducibility	Tampering	Environmental factors—accounts for organizational priorities
Exploitability	Repudiation	
Affected users	Information disclosure	Temporal characteristicsvulnerability changes over time
Discoverability		
	Denial of service time	
	Elevation of privilege	Base metrics—initially weighted on a scale of 0 through 10 and then modified based on temporal and environmental metrics

3.3.3 *Process for Attack Simulation and Threat Analysis (PASTA)*

The process for attack simulation and threat analysis (PASTA) is a good example of a rigorous approach used to evaluate a system for vulnerabilities and the level of threat that these vulnerabilities actually pose to the system. This includes an in-depth approach in analyzing the technologies that compose a system, their popularity/availability, and the types of transactions that provide an external attack surface to bad actors. For example, PASTA™ (Velez & M. M., 2015) provides an overall methodology for threat evaluation for a cyber system (Table 3.4).

As shown in Table 3.3, PASTA is an example of a high-level analysis approach for developing future baseline scenarios and subsequent courses of action (COAs). In addition, factors that influence how a system is modeled are provided in Table 3.4 (Velez & M. M., 2015).

As shown in Table 3.4, PASTA includes cyber threat modeling considerations for a system of interest. A key goal for the cyber security community is to scale up with systematic approaches, like PASTA, so that the collected threat data can be shared

Table 3.3 Stages of process for attack simulation and threat analysis (PASTA) methodology

Define objective	Identify business objectives
	Identify security and compliance requirements
	Technical/business impact analysis
Define technical scope	Define assets
	Understand scope of required technologies
	Dependencies: Network/software (COTS)/service
	Third-party infrastructures (Cloud, SaaS, Application Service Provider (ASP) Models)
Application decomposition	Use cases/abuse (misuse) cases/define app entry points
	Actors/assets/services/roles/data sources
	Data flow diagramming (DFDs)/trust boundaries
Threat Analyses	Probabilistic attack scenarios
	Regression analysis on security events
	Threat intelligence correlation and analytics
Vulnerability and weakness mapping	Vulnerability database or library management (CVE)
	Identifying vulnerability and abuse case tree nodes
	Design flaws and weaknesses (CWE)
	Scoring (CVSS/CWSS)/likelihood of exploitation analytics
Attack modeling	Attack tree development/attack library management
	Attack node mapping to vulnerability nodes
	Exploit to vulnerability match making
Risk and impact analysis	Qualify and quantify business impact
	Residual risk analysis
	Identify risk mitigation strategies/develop countermeasures

Table 3.4 Factors affecting time requirements for threat modeling—PASTA

Factor	Description	Example
Number of use cases	The number of actions that an application can perform as a result of a client request, scheduled job, or Application Programming Interface (API)	Actions that include buying items online, paying bills, exchanging content between entities, or managing accounts
Popularity of technology	The notoriety of a platform or software technology will provide attackers with the ability to have a sophisticated level of understanding on how to better exploit the software or platform	Any distributed servers, both open source and commercial
Availability of technology	The rarity of technology will affect probability levels of malicious users obtaining a copy of similar technologies to study its vulnerabilities for exploitation	Legacy software or proprietary software
Accessibility to technology	Cost of technology is not only a deterrent for legitimate, law abiding companies, but also for those organizations that subsidize cybercrimes	Proprietary developed systems, kernels, or software
Level of expertise	Given that exploit scenarios move beyond the theoretical in application threat modeling, the appropriate level of expertise is needed to exploit vulnerabilities and take advantage of attack vectors. Depending on the level of expertise, a threat modeler or team of security professionals may have varying levels of time constraints in trying to exploit a given vulnerability. This is very common and would require the security expert to be well versed in multiple talents to exploit vulnerable systems (J.H. Cho & H.J., 2016) (Noam Ben-Asher, 2015) (Jones, 2015)	Experience with rare software/platforms

across the community. Sharing standardized threat data therefore helps analysts to develop a common operating picture of current cyberspace events.

3.4   Data-Sharing Models

Cyber threat data models take different forms in order to provide sharing: hub and spoke, broadcast to all, point-to-point, and hybrid models. Similarly, several mechanisms make possible the dissemination of cyber threat data across national, organization, and technical team (e.g., product/service) boundaries. Federal law enforcement organizations include the Federal Bureau of Investigation's (FBI's) iGuardian, the US Department of Homeland Security (DHS) Cyber Risk and Collaboration Partnership (Department of Homeland Security), US Department of Defense's (DoD's) DIB Enhanced Cyber Security Services (DECS), and the office of Cybersecurity and Communications at the US Department of Homeland Security's Trusted Automated eXchange of Indicator Information (TAXII).

State organizations include the state and the Financial Services Information Sharing and Analysis Center (FS-ISAC), Multi-State Information Sharing and Assurance (CS/IA) Program, and the Center for Internet Security's Multi-State Information Sharing and Analysis Center (MS-ISAC).

Some of these collaborations define trust agreements and governance, while others simply enable an organization to share information among partners (Heckman & F. J., 2015). Each of these tools and methods is supported by organizations that span from non-profits to commercial corporations (Fig. 3.4).

As shown in Fig. 3.4, multiple organizations support cyber threat data sharing through communities of interest that include sharing sites and tools. One of the key benefits of data sharing is the incorporation of known threats into commercial anti-virus systems. Commercial threat providers are therefore key players in the discovery and management of cyber vulnerabilities.

3.4.1 Cyber Threat Data Providers

As discussed in the introduction to Sect. 3.4, the US government and multiple non-profits and private corporations provide data to cyber security analysts for sharing. Table 3.5 provides a small sample of cyber threat data providers.

The goal of the example cyber threat data-sharing providers in Table 3.5 is to increase security by providing actionable intelligence on adversaries for cyber analysts.

3.4.2 Cyber Threat Data and System Defense

Computer security controls (CSCs) are one of the key uses of published vulnerabilities. For example, the SANS "top 20" is a well-known, prioritized, list of computer security controls used for any information technology system. In addition, each of the CSCs is designed to be automated. Therefore, leveraging industry available tools, published vulnerabilities can be used to keep computer defenses up to date in protecting computer systems and networks.

A more elaborate architecture for automating the use of CSCs for network protection is found in Canada's Automated Computer Network Defense—ARMOUR (DRDC (Canada)., 2013) (DRDC (Canada)., 2014a, 2014b)). ARMOUR provided an early example of expert system like ability to use a rule-based system for designing and maintaining and intelligent network defense.

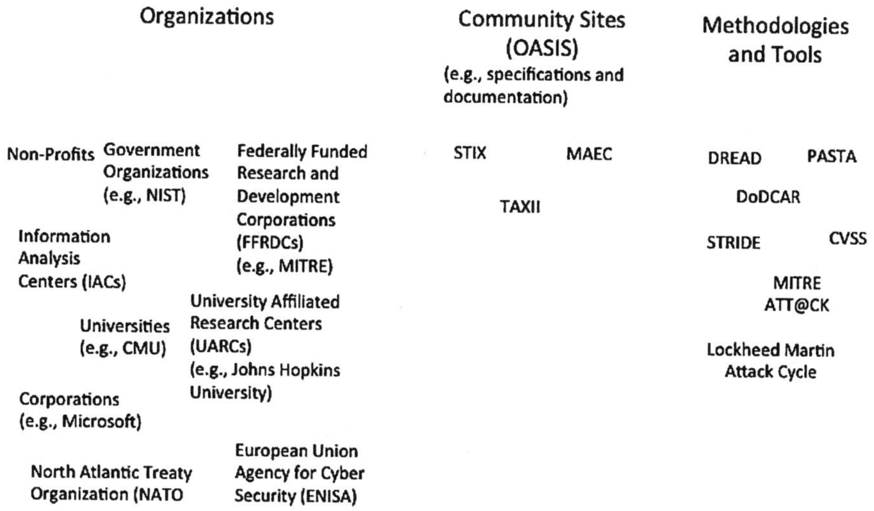

Organizations			Community Sites (OASIS) (e.g., specifications and documentation)		Methodologies and Tools	
Non-Profits	Government Organizations (e.g., NIST)	Federally Funded Research and Development Corporations (FFRDCs) (e.g., MITRE)	STIX	MAEC	DREAD	PASTA
Information Analysis Centers (IACs)			TAXII		DoDCAR	
					STRIDE	CVSS
	Universities (e.g., CMU)	University Affiliated Research Centers (UARCs) (e.g., Johns Hopkins University)			MITRE ATT@CK	
Corporations (e.g., Microsoft)					Lockheed Martin Attack Cycle	
	North Atlantic Treaty Organization (NATO	European Union Agency for Cyber Security (ENISA)				

Fig. 3.4 Organizations and their support for cyber threat management

Table 3.5 Open-source cyber threat reports: organizations and missions

Name	Mission
DHS Automated Indicator Sharing	Private companies report cyber threat indicators with the DHS, which are then distributed through the Automated Indicator Sharing website. This database reduces the effects of simple attacks through exposing malicious IP addresses, email senders, and more
FBI InfraGard	The FBI's InfraGard Portal collects information with the goal of protecting critical infrastructure (e.g., utilities, power generation, etc.). In addition, the FBI also provides information on cyber attacks and threats that they are tracking
SANS Internet Storm Center	The Internet Storm Center uses a sensor network that collects over 20 million intrusion detection log entries per day to generate alerts regarding security threats

3.5 System Engineering and Vulnerability Evaluation

The final state of a vulnerability model provides both how to look for system threats and the use of security controls to monitor for these threats and react to them before they breach system defenses. Finding the vulnerabilities, estimating their risk, and applying countermeasures can be performed via a systematic approach to document both the system/subsystem/component risks and their associated remediation (Fig. 3.5).

While Fig. 3.5 provides a general overview for system risk identification and remediation, several individual processes have been developed to methodically go through a system of interest for its associated risks. For example, the risk bow-tie (Fig. 3.2) and multiple other approaches provided in Sect. 3.3 are component-level

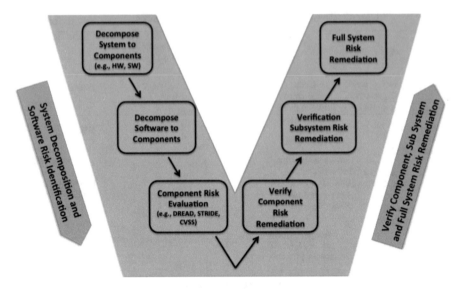

Fig. 3.5 System risk identification and remediation (V Curve)

risk evaluation approaches that can also help in overall system evaluations. Additions to these early models included the use of attack trees to both find and organize threats to a system of interest. More recently, system engineering approaches have been developed to work collaboratively with traditional gaming constructs (e.g., red and blue teams) to iteratively narrow down threats to a system of interest (Bryan Carter, 2019).

3.5.1 DoD Cyber Security Analysis Approaches and Tools

A key goal in using systems engineering, for cyber, is to use existing tools to describe the system, understand its vulnerabilities, and provide a clear assessment (e.g., quantitative scoring, qualitative reporting) that provides a roadmap that guides us to the desired system security state with cost and level of effort well characterized (Fig. 3.6).

As shown in Fig. 3.6, the DoD Architecture Framework (DoDAF) provides the artifacts for describing the concept of operations (CONOPs), or what a system does. This is complemented by the system engineering descriptions provided by the system modeling language (SysML), describing how a system does what it is supposed to do. With this basic system engineering "model," the cyber threat analyst can then do risk assessments via the cyber threat framework (CTF) or MITRE Cyber Analytics Repository Exploration Tool (CARET), to estimate how an attacker might exploit system vulnerabilities. The risk assessment is then used, in combination with a scoring approach (e.g., CVSS), to develop system resilience priorities, or

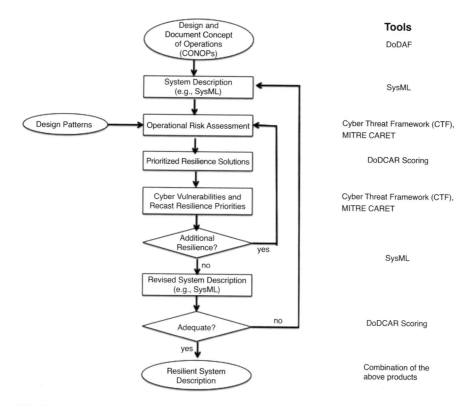

Fig. 3.6 Cyber risk assessment and system engineering tool examples

changes to the system design that take the forms of engineering modifications via the systems SysML diagram. This process continues until the system resilience scoring meets the overall security objectives.

One example of end-to-end approach for cyber system analysis is the DoD Cyber Analytics Repository (DoDCAR) (Office of the National Manager for NSS), which provides a technical cyber threat framework, including an attack life cycle and DoD Architecture Framework (DoDAF) artifacts, for acquisition professionals to describe cyber-specific system elements (Fig. 3.7).

As shown in Fig. 3.7, DoDCAR uses multiple architecture artifacts to describe a system of interest, helping security engineers understand the types of cyber capabilities needed for their system. The use of DoDCAR, especially when paired with more traditional architecture tools, provides a method for communicating architecture, diagrams, and projects. This might be used for technical roadmap development and planning, providing system-level consistency in terminology and measured data for technical reviews and in-progress reviews (IPRs). Follow-on benefits from a systematic approach include increased agility in responding to new requirements and clear understanding of current funding for reviews and future justifications.

DoDCAR Process Artifacts	Material Solution Analysis			Alignment with DAU Acquisition Phase and Associated Information										
				Technology Maturation and Risk Reduction			Engineering and Manufacturing Development			Production and Deployment			Operations and Sustainment		
	ICD	AoA	Draft CDD and TEMP	TEMP	RFP	CDD / TRA	PDR	CDR	CPD	LD	PRR	PPP	PIR	ECPs	EOL
Threat Models OV-5a OV-5b	Apply Framework (OV-5a) to proposed System's Threat Environment with Baseline of Mitigation Performance (i.e. MOE/KPP)			Identification of Test Environment, Test Cases, and Technical Performance Measures (TPMs)			Detailed system specific threat actions (OV-5b) for severity weighted most probable threat impact actions			Inform Red/Blue Team scenarios to most likely Threat Actions and Campaigns. CCORI and CCRI threat scenarios			Support 'tuning' of appliances configurations/rules/analytics as adversary behaviors change		
Architecture Models	Cybersecurity performance and affordability parameters. System Functions (SV-1, SV-10, CV-2) for Threat Mitigation			System Tradeoffs and weighted Cybersecurity performance and affordability parameters.			Detailed system specific threat mitigation functions for corresponding threat actions (updated OV-5b)			Specific Network deployment models, Ports and Protocols			Threat-based ECP/Tech Refresh designs and functions		
Scoring Model CV-6	Possible combinations of cyber system capabilities for TMRR (initial CV-6)			Cyber Effectiveness scores for capabilities in CV-6 for Trade-off Analyses			Updated scores from detailed design reviews to supplement selecting solutions			Establish capability measure of effectiveness (MOE) feedback loop for deployed systems			Adjusted scoring for threat-based ECP/Tech Refresh		

Fig. 3.7 Example DoDCAR architecture artifacts for cyber system security evaluation (Office of the National Manager for NSS)

Along these lines, Hibbs provides a thoughtful brief on the challenges that systems engineering currently has in being applied to cyber systems (Hibbs, 2019).

Much has been written on the system engineering tools referenced above. This introduction is simply to show that end-to-end systems engineering tools currently exist and are used today. In addition, the analysis performed with these frameworks, methodologies, and tools, while used to describe the system to be defended, can also be used to develop overall systems for the defense, or possibly targeting, of cyber systems of interest.

3.5.2   Analysis and Targeting Use of Cyber Threat Data Examples

Both open-source (e.g., Table 3.5) and individually collected threat information help inform cyber system defense. This includes the use of Security Incident and Event Management (SIEM) systems. This is an example application of rule-based, or intelligent, systems, in providing the computer security controls (CSCs) used for evaluating a defended system of interest. In addition, this same logic can be used to analyze a system for possible compromise. This might be a penetration test, where clear rules of engagement are used by ethical hackers to identify system vulnerabilities and report them to the system owner. This process might also include using automated tools to identify and exploit target system vulnerabilities.

3.5.2.1 Use of Vulnerabilities/Exploits for Cyber System Defense

As discussed in Sect. 3.5.1, known external threats are commonly used to support system-level defense. In one example, Defense Research and Development Canada (DRDC) developed the Automated Computer Network Defense (ARMOUR) framework (DRDC (Canada)., 2014a, 2014b)) (DRDC (Canada)., 2013) as an entire technical architecture, a knowledge-based system, that incorporates external threat understanding in developing a system-level defense. An automated technical architecture, like ARMOUR, will both leverage open-source data (Table 3.5) and collect real-time data to maintain security for a system of interest.

In addition to the use of external threat information, computer protection systems use directed collection on their own systems to develop the understanding required for protecting the system from external threat. Fig. 3.8 provides the collection, processing, analysis, and reporting steps commonly used for developing intelligence on an information technology system.

As shown in Fig. 3.8, this example computer network intelligence system leverages both open source (Table 3.5) and directed collection as input sources. Directed collection includes penetration testing to identify system vulnerabilities. One example of automated identification and possible exploitation of system vulnerabilities is provided by the tool Metasploit (Rapid7), which leverages open-source data to continually improve Metasploit's ability to find and exploit network vulnerabilities. The use of Metasploit was highlighted in a report by FireEye (FireEye, 2020) on a group called Advanced Persistent Threat (APT) 41, where APT 41 used Cisco router exploits from Metasploit in their financially motivated campaign—global campaign against commercial businesses.

Leveraging the same tools as hackers improves a defender's ability to ensure against similar types of attacks. In addition, processing external and collected threat data is done by collecting them into a threat database and cross-referencing them with existing controls across the organization's known assets and technologies. Risk analysis is then used to provide the organization with performance and security reports, which include current maturity assessment via alerts, dashboard feedback, and more formal reporting.

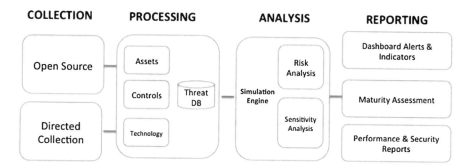

Fig. 3.8 Collection, processing, analysis, and reporting for computer network intelligence

In addition to using system vulnerabilities to assess current threats, tools like Metasploit can be used to develop offensive attack tools. While the "white hat" use of this tool helps information security officers understand their current security posture, these same tools can be used by network attackers to analyze target networks of interest for penetration, maneuver, and exploitation.

3.5.2.2   Use of Vulnerabilities/Exploits for Cyber System Attack

While ethical hackers use vulnerabilities, and their exploits, to identify weaknesses in order to protect systems of interest, malicious hackers use these same tools to attack systems to steal information and funds. And a growing vulnerability market is key to providing these hackers with the information that they need to develop tools to exploit their target systems.

Dusan and Repel (Dusan Repel, 2015) describe how the vulnerability market is key to providing the raw material required for the developing cyber weapons market. Converting software vulnerabilities to malware, or attack tools, uses many of the same cyber threat descriptions that were provided in Fig. 3.6. For example, software vulnerabilities (CVE, CWE) might be scored (CVSS, CWSS) and then developed into malware (e.g., Metasploit discussion) via a development process (e.g., agile) to produce malware (Fig. 3.7).

As shown in Fig. 3.9, each of the steps used to facilitate communicating cyber threats might also be used for attack tool development, and these steps are already built into "white hat" hacker tools (e.g., Metasploit).

Analysis and targeting use of cyber threat data can therefore be used for both designing system-level defenses and for designing attack tools. Each of the data sources, modeling approaches, and assessment techniques can be used for either defensive or offensive development of cyber system tools.

3.6   Summary

Using cyber threat intelligence data includes the standard collection, processing, and development of reporting for either defensive or offensive applications. We began this chapter with a series of questions to ensure coverage of the topic of cyber threat. And we will now revisit these questions.

1. Who are the responsible organizations for providing the policies and processes for describing and handling cyber threats?

    (a) National Institute of Standards (NIST) provides the Cybersecurity Framework.
    (b) The National Infrastructure Advisory Council (NIAC) developed the Common Vulnerability Scoring System (CVSS), and it is maintained by First.org, a US-based non-profit organization.

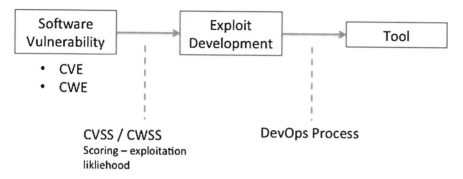

Fig. 3.9 Software vulnerability to exploit development—attack tool process

    (c) US Department of Defense provides the DoD Cyber Analytics Repository (DoDCAR).

2. What are examples of popular cyber threat taxonomies?

    (a) Structured Threat Information Expression (STIX)—maintained by OASIS Cyber Threat Intelligence (CTI) Technical Committee
    (b) ENISA (Europe)
    (c) ATT@ACK Framework (MITRE)

3. What is the process for developing cyber threat descriptions for a system of interest?

    (a) Cyber Threat Intelligence (CTI) process
    (b) DREAD/STRIDE
    (c) PASTA

4. What are the current data standards for describing and sharing cyber threat data?

    (a) STIX
    (b) TAXII
    (c) MAEC—Malware Attribute Enumeration and Characterization

5. How do cyber tools use collected vulnerability data to further defend systems?

    Using end-to-end processes for system evaluation can be supported by system decompositions for follow-on assessment. For example, the ARMOUR Framework's scenario-based framework provides the system engineer with an ability to design system security based on prescribed technical and operational specification in support of policy security objectives.

    Similarly, SANS 20 Controls are designed with rule-based implementation in mind:

"…protect critical assets, infrastructure, and information by strengthening your organization's defensive posture through continuous, automated protection and monitoring of your sensitive information technology infrastructure to reduce compromises, minimize the need for recovery efforts, and lower associated costs".

The Australian Signal Directorate's "Essential 8" preventive techniques, designed with the same end in mind as the CSCs, are said to prevent 85% of cyber attacks.

6. What are some example analysis and targeting uses of cyber threat/vulnerability data?

Systems engineering approaches (Sect. 3.5) provide an overall approach for analyzing a system of interest. In addition, this system description can be used for tailoring rules in an SIEM system. Or, for offensive purposes, a systems-level analysis might be used to design compromises; similar to what the Metasploit tool is used to do.

Cyber threat, and its developing taxonomy, therefore consists of a loosely organized set of data, models, best practices, and assessment approaches that are supported by organizations that span from non-profits to corporations (Fig. 3.4). A key goal of understanding cyber threat is primarily to help cyber defenders understand their systems with enough detail so that their assessment of current attacker capabilities can be quickly translated into technical fixes that keep the systems safe. Achieving this goal is currently being performed through processes that include using the multiple tools available for overall system evaluation (Fig. 3.6).

Bibliography

Barth, A, C. J. (2008). Robust defenses for cross-site request forgery. *Proceedings of the 15th ACM conference on computer and communications security (CCS '08)*.

Bernier, M. (2015). Cyber effects categorization - the MACE taxonomy. DRDC Center for Operational Research and Analysis. TTCP JSA TP3 Cyber Analysis .

Bryan Carter, S. A. (2019). A preliminary design-phase security methodology for cyber-physical systems. *Systems, 7*(21).

Cho, J. H., & H. C. (2016). Effect of personality traits on trust and risk to phishing vulnerability: Modeling and analysis. In *Cognitive methods in situation awareness and decision support (CogSIMA), IEEE international multi-disciplinary conference*. IEEE.

ClearSky Research Team. (2017, March 17). (C. C. Security, Producer) Retrieved February 10, 2018, from http://www.clearskysec.com/iec/#att123

Couretas, J. M. (2018). *Introduction to cyber modeling and simulation*. John Wiley and Sons.

Defense, A. G. (n.d.) *Strategies to mitigate cyber security incidents*. Retrieved November 4, 2016, from https://www.asd.gov.au/infosec/mitigationstrategies.htm

Department of Defense. (2012). *Joint Publication 1-13.4 Military Deception*. Department of Defense.

Department of Homeland Security. (n.d.). *Cyber information sharing and collaboration program*. Retrieved 3 7, 2019, from Cyber Information Sharing and Collaboration Program: https://www.dhs.gov/cisa/cyber-information-sharing-and-collaboration-program-ciscp

Department of Homeland Security (DHS). (n.d.). *Cybersecurity Framework*. Retrieved 5 20, 2019, from https://www.nist.gov/cyberframework

Director of National Intelligence. (n.d.). *Cyber threat framework*. Retrieved 10 20, 2018, from https://www.dni.gov/index.php/cyber-threat-framework

DRDC (Canada). (2013). *System technical specification for the ARMOUR TD, v2.1*. DRDC.

DRDC (Canada). (2014a). *Architectural design document for the automated computer network Defence (ARMOUR) technology demonstrator (TD) contract*. DRDC.

DRDC (Canada). (2014b). *System concept of operations (CONOPS) for the automated computer network Defence (ARMOUR) technology demonstration (TD) contract.* DRDC.

Dusan Repel, S. H. (2015). The ingredients of cyber weapons. *10th International Conference on Cyber Warfare and Security (ICCWS15)*, (pf 1-10).

FireEye. (2017, 6 15). *M-Trends 2017.* Retrieved 3 28, 2018, from M-Trends: https://www.fireeye.com/ppc/m-trends-2017.html?utm_source=google&utm_medium=cpc&utm_content=paid-search&gclid=Cj0KCQjw-uzVBRDkARIsALkZAdniLMfO9X-z1aSqYzJsuRVHLVFjroaLajoLjFaTV15jnzjdyyWEvNMaAt5sEALw_wcB

FireEye. (2020, 3 25). *This is not a test: APT41 initiates global intrusion campaign using multiple exploits.* Retrieved 3 25, 2020, from Threat Research: https://www.fireeye.com/blog/threat-research/2020/03/apt41-initiates-global-intrusion-campaign-using-multiple-exploits.html

FireEye. (2017). *M-TRENDS 2017 - A view from the front lines.* FireEye.

Heckman, K. E., & F. J. (2015). Cyber counterdeception: how to detect denial & deception (D&D). In P. S. S. Jajodia (Ed.), *Cyber Wrfare: Building the scientific foundation.* Springer.

Hibbs, E. (2019, May 15). *Systems engineering in a cyber world - connecting frameworks for program decisions.* Retrieved March 8, 2020, from DISA: https://www.disa.mil/-/media/Files/DISA/News/Events/Symposium-2019/1%2D%2D-Hibbs_Systems-Engineering-in-a-Cyber-World_approved-Final.ashx+&cd=10&hl=en&ct=clnk&gl=us

Joint Chiefs of Staff. (2017, 6 17). *Joint planning.* Retrieved 2 15, 2019, from Joint Publication 5-0.: https://www.jcs.mil/Portals/36/Documents/Doctrine/pubs/jp5_0_20171606.pdf

Joint Staff. (2017, 6 16). *Joint planning.* (J. Staff, Producer) Retrieved 2 11, 2019, from www.jcs.mil/Doctrine/Joint-Doctrine-Pubs/5-0-Planning-Series

Jones, R. (2015). Modeling and integrating cognitive agents within the emerging cyber domain. In *Proceedings of the Interservice/industry training, simulation and education conference (I/ITSEC).* NDIA.

Kick, J. (2014). *Cyber exercise playbook.* MITRE.

Kime, B. P. (2016). *Threat intelligence - planning and direction.* Retrieved 3 7, 2019, from SANS: https://www.sans.org/reading-room/whitepapers/threats/threat-intelligence-planningdirection-36857

Koerner, B. I. (2016, 10 23). *Inside the cyberattack that shocked the US Government.* Retrieved from Wired: https://www.wired.com/2016/10/inside-cyberattack-shocked-us-government/

Launius, S. (2019). *Evaluation of comprehensive taxonomies for information technology threats.* Retrieved 3 7, 2019, from SANS: https://www.sans.org/reading-room/whitepapers/threatintelligence/evaluation-comprehensive-taxonomies-information-technology-threats-38360

Robert M. Lee, M. J. (2014). *German steel mill cyber attack.* SANS ICS CP/PE (Cyber-to-Physical or Process Effects) case study paper . SANS.

Leversage, D. (2007). Comparing electronic battlefields: Using mean time-to-compromise as a comparative security metric. In I. K. V. Gordodetsky (Ed.), *Computer network security.* Springer.

Malware Attribute Enumeration and Characterization (MAEC). (n.d.). Malware attribute enumeration and characterization (MAEC). Retrieved 3 8, 2019, from https://maecproject.github.io/about-maec/

Popular Mechanics. (2018, March 13). How long does it take hackers to pull off a massive job like Equifax? Retrieved from Popular Mechanics: https://www.popularmechanics.com/technology/security/a18930168/equifax-hack-time/

MITRE. (n.d.). *CARET.* Retrieved 9 30, 2018, from CARET: https://car.mitre.org/caret/#/.

National Information Exchange Model (NIEM). (n.d.). National information exchange model (NIEM). Retrieved 3 8, 2019, from https://www.niem.gov/

National Insititute of Standards (NIST). (n.d.). *NIST SP 800.* Retrieved 3 8, 2019, from https://csrc.nist.gov/publications/sp800

NEWSWEEK. (2016, 3 30). *Alleged dam hacking raises fears of cyber threats to infrastructure.* Retrieved 2 10, 2018, from http://www.newsweek.com/cyber-attack-rye-dam-iran-441940

NIST. (n.d.). Security content automation protocol. Retrieved 3 6, 2020, from SCAP: https://csrc.nist.gov/Projects/Security-Content-Automation-Protocol/

Noam Ben-Asher, A. O. (2015). Ontology based adaptive systems of cyber defense. In *Proceedings of the 10th Internatioanl conference on semantic Technology for Intelligence, Defense, and security (STIDS)*.

O.W.A.S.P. (OWASP). (2013). *OWASP Top 10*. Retrieved 5 26, 2018, from http://owasptop10.googlecode.com/files/OWASPTop10-2013.pdf

OASIS. (n.d.). *OASIS*. Retrieved from https://www.oasis-open.org/standards

Office of the National Manager for NSS. (n.d.). *DoDCAR*. Retrieved 3 8, 2019, from https://csrc.nist.gov/CSRC/media/Presentations/DODCAR-no-class-markings-Pat-Arvidson/images-media/DODCAR_-no%20class%20markings%20-%20Pat%20Arvidson.pdf

Peter Mell, K. S. (2007, June). CVSS a complete guide to the common vulnerability scoring system version 2.0. Retrieved February 23, 2020, from National Institute of Standards (NIST): https://tsapps.nist.gov/publication/get_pdf.cfm?pub_id=51198

Ponemon Institute. (2018). Cost of a data breach. Retrieved 3 8, 2019, from https://securityintelligence.com/series/ponemon-institute-cost-of-a-data-breach-2018/

Rapid7. (n.d.). *Metasploit*. Retrieved 3 5, 2020, from https://www.rapid7.com/products/metasploit/

Rick Nunes-Vaz, S. L. (2011). A more rigorous framework for security-in-depth. *Journal of Applied Security Researh, 23*.

Rick Nunes-Vaz, S. L. (2014). From strategic security risks to national capability priorities. *Security Challenges, 10*(3), 23–49.

Schostack, A. (2008). Experiences threat modeling at Microsoft. Retrieved 2 19, 2020.

Segura, V. (2009, June 25). Modeling the economic incentives of DDoS attacks. Retrieved from Semantic Scholar: https://pdfs.semanticscholar.org/afdf/d974bc68dc05c48020e-f07a558a61ab94f8a.pdf

Suresh Damodaran, J. C. (2015). Cyber modeling & simulation for cyber-range events. In *SummerSim (p. 8)*. SCS.

Symantec. (2019). Internet security threat report. Retrieved 3 8, 2019, from Symantec: https://www.symantec.com/security-center/threat-report

Symantec. (2014). Dragonfly: Cyberespionage attacks against energy suppliers.

Velez, T. U., & M. M. (2015). *Risk centric threat modeling: Process for attack simulation and threat analysis*. John Wiley & Sons, Inc.

Verizon. (2018). *2018 data breach investigations report (DBIR)*. Retrieved 3 8, 2019, from Verizon: https://enterprise.verizon.com/resources/reports/DBIR_2018_Report_execsummary.pdf

Vogt, P., & F. N. (2007). Cross-site scripting prevention with dynamic data tainting and static analysis. In *The 2007 network and distributed system security symposium (NDSS '07)*.

Chapter 4
Cyber Influence Operations

The purpose of this chapter is to present a general background on cyber influence operations. This will include a discussion of information operations (IO), associated policy, key implementation techniques, and their uses to affect a target audience (TA). In addition, we will review example influence operations spanning from the Cold War to current cyber-based IO events.

Key Questions of the Chapter

1. What are the mechanisms used in current cyber influence operations?
2. How have influence operations evolved with changes in technology?
3. How do influence operations work in commercial, government, and military applications?
4. How did the Soviet Union use information operations during the Cold War? What are examples of recent Russian cyber-based influence operations?
5. How are cyber influence operations used in conjunction with kinetic operations?
6. What are the dynamics, or moving parts, of influence operations? How are these dynamics changed by cyber technologies (e.g., social media)?
7. How does targeting occur in cyber influence operations? For example, do ideas like area targeting vs. point targeting apply in cyber influence operations? How?

4.1 Cyber Influence Operations Background

Cyber analysis and targeting practitioners include information warfare (IW) when developing cyber-based influence operations. Influence operations are then used to link offensive and defensive missions to tactical objectives via the tactics, techniques, and procedures (TTPs) described in Chaps. 2 and 3. Information Operations (IO) draw on several types of information and take multiple forms (Table 4.1).

As shown in Table 4.1, IO examples span from what we would normally call advertising (i.e., propaganda) to deliberate falsification of information (i.e.,

© Springer Nature Switzerland AG 2022
J. M. Couretas, *An Introduction to Cyber Analysis and Targeting*,
https://doi.org/10.1007/978-3-030-88559-5_4

Table 4.1 Information Operations (IO) examples

IO example	Description
Propaganda	A term made famous by Bernays (Bernays, 1918), propaganda is used to project a narrative in order to influence a population as a center of gravity in a messaging campaign (Lasswell, 1927). Corporations, governments, and private individuals might use propaganda to influence a population concerning a product, program, or political issue, respectively. The end-state for a propaganda campaign is to steer opinion to the desired position by the entity generating the message.
Misinformation	Misinformation is incorrect information that may be spread due to plausible rumors, or partial truths, that obfuscate actual facts. For example, conspiracy theories spread via social media make it a challenge to know which trending messages are true.
Disinformation	Disinformation is the intentional spreading of false information, often with a seemingly legitimate source, to create chaos and confusion, at minimum, or slander a person or position at its most extreme.

disinformation), with the goal of creating destabilizing effects on social processes (e.g., elections) or individuals (i.e., slander). All of these activities occur in an information environment that includes the people, processes, and technical systems that collect, process, and disseminate information (Fig. 4.1).

As shown in Fig. 4.1, the information environment includes the physical, informational, and cognitive layers. Therefore, one of the challenges, for cyber analysis and targeting, is to decompose the scope of a desired operation and its proposed effects to the correct layer. For example, influencing people (i.e., cognitive) might be achieved simply with misinformation or disinformation (Table 4.1). However, the planning of this influence operation may require data from the informational layer, to make the message seem legitimate. And data from the informational layer could easily require access via the physical layer. Therefore, each of the layers might be involved to provide information, or access to information, to achieve the IO effect. In addition to the system layers, there are several system-level functions that compose Information Operations (IO) capabilities.

4.1.1 Information Operations (IO) Background

One way of describing IO is through purely military terms. For example, an older version of the US Department of Defense's (DoD) Joint Publication (JP) 3–13 (U.S. Department of Defense, 2012) and the Information Operations Roadmap (U.S. Department of Defense (DoD), 2003), which was declassified in 2006, shows IO being described in terms of five pillars (Table 4.2).

We will now look at example elements of Table 4.2 in terms of the standard information security, or the cyber security confidentiality, integrity, and availability (CIA) triad. For example, one of the most common CIA issues that people discern is when the availability of a system is denied via jamming. Another advanced

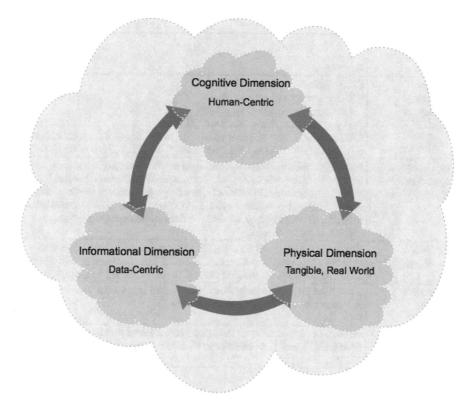

Fig. 4.1 The information environment (Joint Chiefs of Staff, 2012)

Electronic Warfare (EW) approach is spoofing a navigation system, which is called an integrity attack. An integrity attack reduces the effectiveness of a computer-based (i.e., cyber) system and requires operators to rely on non-cyber-based methods (e.g., maps for navigation) to overcome the IO effect. Therefore, a key goal of operations security (OPSEC) is to ensure that these key cyber systems are not compromised.

MILDEC, as shown in Table 4.2, might be used to achieve OPSEC by intentionally deceiving a rival. For example, in *"The Cuckoo's Egg"* (Stoll, 2005), a real-life example of how Clifford Stoll, a Lawrence Berkeley Lab systems administrator who tracked down a group of West German hackers working on behalf of the Soviet KGB—*Komitet Gosudarstvennoy Bezopasnosti* (Committee for State Security)—to extract defense secrets related to the 1980s Star Wars missile defense system, labeled files so that they may seem to have more importance to the network attacker. These documents, called "honey files," were one of the first uses of cyber deception that helped a cyber defender direct the movements of an attacker. This use of deception also kept the attackers busy with files that are unimportant to the OPSEC of the system.

In addition, PSYOP, similar to MILDEC, uses false information to target audiences in order to affect observations and decision-making of those organizations, groups, and individuals. Each of the elements in Table 4.2 is included in the cyber operations

Table 4.2 Information Operations (IO) elements

Information Operations Elements	Description
Computer network operations (CNO)	CNO originally consisted of: Computer network attack (CNA) Computer network defense (CND) Computer network exploitation (CNE) CNO later became cyberspace operations, offensive and defensive, with its own separate doctrine in Joint Publication 3–12 (Joint Chiefs of Staff, 2013)
Psychological operations (PSYOP)	PSYOP includes the planning and use of information (propaganda) to influence and shape behaviors of governments/organizations/groups/individuals. Strategic PSYOP includes using information to influence target audiences. At the operational level, PSYOP is used for mission support. The term PSYOP was changed to military information support operations (MISO) in recent doctrine
Electronic warfare (EW)	Electronic warfare involves using the electromagnetic spectrum as a means to direct attacks on the enemy; e.g., jamming (e.g., decreasing availability) communications or challenging navigation (e.g., spoofing integrity)
Operations security (OPSEC)	Operations security includes identifying critical information and ensuring procedures to keep that information secure
Military deception (MILDEC)	Military Deception includes taking action to mislead an adversary, elements of which will lead to mission accomplishment (Department of Defense, 2012)

analysis and targeting support publications, field manuals, and guiding documents previously discussed in Chap. 2. An updated version of Fig. 2.3, that includes additional guiding joint concepts for cyberspace operations, is provided in Fig. 4.2.

As shown in Fig. 4.2, in 2009 the Strategic Communications Joint Integrating Concept (Joint Chiefs of Staff, 2009) provided an overarching view of how communications are used throughout a military campaign cycle. More recently, the 2018 Joint Concept for Operating in the Information Environment addressed the incorporation of information considerations into all aspects of operations (Joint Chiefs of Staff, 2018). The overlap in communications, IO, and cyber is shown in the dependency relationships of the cyberspace targeting documents of Fig. 4.2. In addition, these documents provide a way to decompose the different elements of the information environment (Fig. 4.1), and the respective tools and techniques being used (Table 4.2) when attempting to disambiguate complicated real-world operations.

4.1.2   Influence Operations, Advertising, and Propaganda

While "fake news" and media manipulation got a lot of attention in the 2016 US presidential election, these techniques started their development with the advent of nineteenth-century mass communication (e.g., telegraph) and media distribution

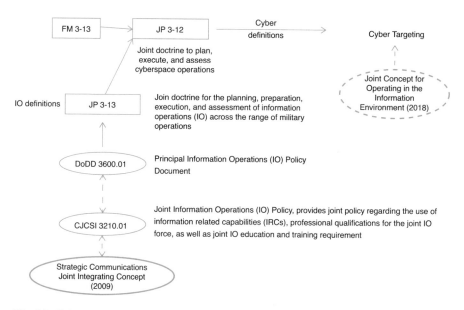

Fig. 4.2 Cyberspace targeting documents: addition of joint concepts for the communications and information environments

(e.g., newspapers). Real-time communication and editorials became political tools with early twentieth-century propaganda. These techniques were refined through the European mass socialist movements (e.g., Russia's Bolsheviks [later Soviets], Germany's National Socialists [i.e., Nazis]) anchoring messages that reinforced their cause. After World War II, radical political operatives, whose goals varied from better working conditions for labor unions to political change, used similar communication approaches. For example, Saul Alinsky (Alinsky, 1989), one of the more famous radicals, adopted many of the early propagandists' communications tenets for anchoring a message through repetition. These techniques, of course, were also used in commercial settings for advertising and marketing.

Advertising, propaganda, and disinformation are all names of influence operations used in venues that range from the commercial to the strategic (e.g., mobilizing a population for war). One of the founders of public relations, working in the early twentieth century, was Edward Bernays, a nephew of the famous psychiatrist Sigmund Freud. Mr. Bernays did many of the original mass marketing campaigns through his public relations, or influence, techniques (Bernays, 1923; Edward Bernays, 2004). In addition, Bernays' teachings set the groundwork for influence operations that became a focus for both sides during World War II. For example, the Allies countered Nazi propaganda, made famous by Joseph Goebbels, through "morale operations" (Office of Strategic Services, 1943). The subsequent Cold War between the Soviet Union and the non-communist world included several influence campaigns that only became clear with the publishing of documented disinformation operations after the Cold War ended (Christopher Andrew, 2015; Ion Mihai Pacepa, 2013).

Fig. 4.3 Washing
falsehoods to propagate
disinformation

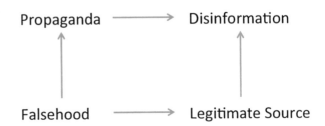

4.1.3 Influence Operations and Disinformation

According to Romanian General Ion Pacepa (Ion Mihai Pacepa, 2013; Joseph Farah, Ion Pacepa, & Moore, 2017), a key player in Soviet Bloc information operations during the Cold War, disinformation has a history that dates back to the Russian Count Potemkin's false-front villages that the Count used to impress Catherine the Great during a late eighteenth-century tour of Crimea. This included using false building fronts to make it appear that there were many more buildings, and associated economic growth, than actually existed. Count Potemkin's "fake news" had the appearance of legitimacy via the false building fronts that "proved" there were more villages; an increase in economic activity.

Transforming any falsehood into disinformation relies on the use of seemingly legitimate sources to validate a message that is continuously communicated to anchor a message into the public's mind as a truth (Fig. 4.3).

As shown in Fig. 4.3, disinformation comes from the repetition of a falsehood through propaganda or from a seemingly legitimate source. One approach, used in the coffee shops of Lisbon during World War II, was to use a known gossip as a source to spread a rumor (Peiss, 2020). Legitimate sources also include newspapers, television personalities, or other sources with a reputation for trustworthiness. Compromising these legitimate sources is a key part of disinformation activities.

4.1.4 Cold War Examples of Soviet Disinformation: Development and Dissemination

Count Potemkin's example of "fake news" preceded the modern era of fungible reporting via the Internet. In 1922, the new Soviet Union started an Office of Disinformation (Popken, 2018), which coordinated efforts to develop stories and disseminate them for the new regime's benefit. Reporting on these disinformation activities mostly occurred after the end of the Cold War. For example, Oleg Kalugin (Kalugin, 2009), an active Soviet Russian spy from the 1960s to the 1980s, described the widespread use of disinformation operations during the Cold War:

> Under Andropov (Intelligence Chief of the KGB) the disinformation branch of the KGB
> flourished. For both domestic and external consumption, it concocted stories to deceive,
> confuse, and influence targeted audiences. It conducted operations to weaken Soviet adver-

saries and to undermine the internal stability and foreign policies of the Western world in order to facilitate favorable conditions for the eventual triumph of Communism.

Kalugin notes that policy guidance for the KGB to accomplish these goals included the following:

- Manipulating mass organizations, pickets, protest groups; creation and financing of "committees of solidarity," public tribunals.
- Planting stories and rumors; giving speeches with known disinformation; forging documents.
- Disseminating radio propaganda, including clandestine broadcasting of leaflets; public scandals; strikes; underground cells; governments in exile; economic subversion; sabotage; and power supply difficulties.
- Manipulating and controlling the media.

Executing Soviet disinformation policy guidance included producing falsehoods for public consumption, for which Kalugin provided the following examples:

- The CIA:
  - Ousted President Nixon.
  - Arranged the 1978 mass suicide and murder of more than 900 people of the Jonestown cult (Guyana).
  - Coordinated the assassinations of:
    - Olof Palme in Sweden.
    - Indira Gandhi in India.
    - Aldo Moro in Italy.
  - Planned the attempt on the Pope John Paul II's life (1981).
- The United States was developing an ethnic weapon that would kill blacks and spare whites.
- The US Army scientists developed the AIDS—Acquired Immune Deficiency Syndrome—virus (i.e., Operation INFEKTION (Boghardt, 2009)).

In addition, Kalugin adds that, at the peak of the Cold War, in 1981, the KGB, according to its report to the Communist Party (CP) Central Committee, carried out the following:

- Funded or sponsored 70 books and brochures.
- Developed 4865 articles in foreign and Soviet newspapers and magazines.
- Produced 66 feature and documentary films.
- Sponsored 1500 radio and TV programs.
- Organized 3000 conferences and exhibitions.
- Published 170,000 reports to the public.

As a tactic to reduce the credibility of the US government at home and abroad, the KGB used influence operations as a key part of both shaping perception and adapting news to fit into pro-Soviet narratives. The rapid adaption of news into

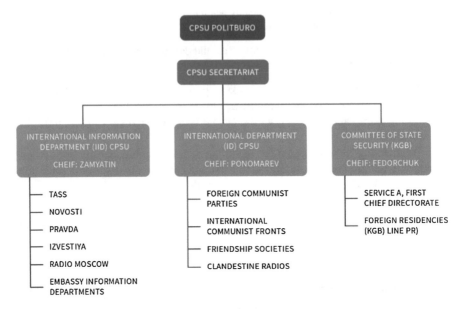

Fig. 4.4 Soviet Union's structure for active measure development (Jones, 2018)

Soviet information operations was enabled by a centralized infrastructure of news agencies, front organizations, and diplomatic operatives (Fig. 4.4).

As shown in Fig. 4.4, the Soviet Union maintained a well-oiled machine of journalists, community organizers, and diplomats to generate pro-Soviet propaganda during the Cold War. Funded at an estimated $ 3 billion per year in the early 1980s (Jones, 2018), Kalugin's data show that the machine was also managed with factory-like precision in keeping counts of the published articles, books, and protests generated.

We are fortunate to have Soviet examples of both methodology and metrics for their influence operation efforts during the Cold War. Most of this propaganda was issue based, designed to plant doubt and distrust in the minds of citizens and allies about the motives and actions of the US government. This might be called area targeting in JP 3–60 terms. Cyber, however, adds the individual in both the falsehood creation and its targeting through social media. For example, using applications like Twitter, a person established as an authority figure can be interpreted by some as a legitimate source; making Fig. 4.3's conceptual model for disinformation development applicable to current-day cyber operations.

4.2 Mechanisms of Influence

Tactical implementation of information operations for political purposes became more common in the twentieth century due to both progressive political movements and the introduction of mass communication technologies: radio, television, and

Table 4.3 White, gray, and black propaganda for cyber influence operations

Propaganda type	Description
White	Openly revealing its source, white propaganda relies on persuasion and public relations techniques (e.g., RT News, Sputnik (Rutenberg, 2017))
Black	Misinformation that claims to be coming from one side but is actually produced by the opposing side (e.g., "False Flag" operations (Krebs, 2013))
Gray	Propaganda where the source is never identified

now the Internet. One example of using communications as a weapon dates to 1943, when the US Morale Operations Branch opened under the Office of Strategic Services (OSS), the precursor to the CIA. These early political/military influence campaigns were called morale operations, and included white, black, and gray propaganda (Office of Strategic Services, 1943); each of which is currently being employed in cyber operations to achieve influence effects (Table 4.3).

Black and gray propaganda shown in Table 4.3 were the primary KGB products produced in Sect. 4.1.4's examples of Soviet propaganda. In addition, black propaganda, traced to Russia's GRU—*Glavnoje Razvedyvatel'noje Upravlenije* (Main Intelligence Directorate)—was used extensively during the 2016 US presidential election, with Russian operators using personas that included Tennessee Republicans, Guccifer 2.0, etc., to develop influential online voices during the campaign (Mueller, 2019). White propaganda, however, is more often used in American smear campaigns to directly attribute inappropriate words/actions to a person.

4.2.1 Propaganda

Each type of propaganda in Table 4.3 relies on repetition to anchor a message firmly in the target's brain (Attkisson, 2017):

1. The bigger the lie, the more people will believe it.
2. If you repeat a lie often enough, it becomes the truth.
3. An attempt to convince must confine itself to a few points and repeat them over and over. Persistence is the first and most important requirement for success.

And the Internet, through false users and the ability to repeat "memes," provides an ideal platform for the repetition that a propagandist needs to succeed in validating a message. In addition, Saul Alinsky, considered the founder of modern community organizing, provided four basic steps in his "*Rules for Radicals*" (Alinsky, 1989):

- Ridicule is man's most potent weapon.
- Keep the pressure on, with different tactics and actions.
- Develop operations that will maintain a constant pressure upon the opposition.
- The threat is usually more terrifying than the thing itself.

Many of the information operation (IO) approaches, from Bernays to Alinsky, are currently used in attempting to achieve a group's strategic objectives. In addition, using cyber to deliver these IO effects, especially through social media, can cause a message to "go viral." However, even with proven techniques and scalable technology, additional, kinetic, measures may be required to achieve policy objectives.

4.2.2 Influence Operations and Cyber Kinetic Fusion

Expansion of influence operations picked up with the Internet due to the easy placement of editorial content in social media. Influence "battles" therefore occur across the political spectrum, from local elections (Welch, 2018) to diplomatic influence (Kaplan, 2019).

Influence operations, as related to targeting, follow the JP 5.0 (Joint Chiefs of Staff, 2017) steps, from the shaping of operations (Phase 0) through dominating the battlefield (Phase 3) (Fig. 4.5).

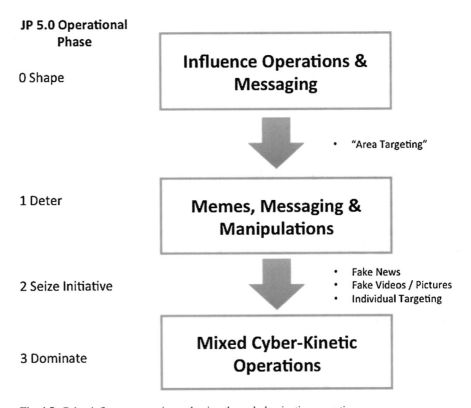

Fig. 4.5 Cyber influence operations: shaping through domination operations

Table 4.4 Media type and targeting resolution

Message receivers	Mainstream media	Social media	Group blogs	Point-to-point (individual) communications
General public	***	***	*	*
Interest groups	*	***	***	*
Individuals	*	*	***	***

***Primary focus
*Ancillary focus

Figure 4.5 is an example of the use of influence operations across the campaign phases. In addition, the "flow" of Fig. 4.5 describes the persistent messaging, or "area targeting," that precedes the more specific "point targeting" of the Phase 3 domination operations. This point targeting is facilitated by the introduction of social media to focus on smaller groups, or even individuals.

While print media and radio, "mass communication," were the primary tools at the time of the early twentieth-century propaganda operations, the early twenty-first century added the use of social media (e.g., Facebook, Twitter) to narrow down the targeted audience to specific interest groups, or even key individuals (Table 4.4).

As shown in Table 4.4, Internet-based media, always a factor in shaping broad operations, is now complimented by social media as a targeting tool for individuals. The Internet has therefore narrowed the "area" targeting of mass communications to the "point" targeting available via social media and point-to-point communications (e.g., e-mail).

4.3 People: Power Laws, Persuasiveness, and Influence

The media's mindshare, via traditional print, radio, and television, is now complimented by highly interactive, in some cases addictive, social media. A new twist to online sources is that an individual, sometimes anonymous, can quickly become a "social media star" (e.g., The Famous People (The famous people, n.d.; Vaynerchuk, n.d.)) with limited previous media or content footprint. In addition, one person can multiply their presence via the use of online technologies. For example, as described by Atkisson (Attkisson, 2017):

> In December 2015, the Department of Justice announces the shocking arrest of nineteen-year old Jalil Ibn Ameer Aziz. The teen allegedly used Twitter accounts and fake personas to operate a network that spewed violent, Islamic extremist rhetoric. All from the comfort and safety of his parents' Pennsylvania home. The feds say Aziz employed a manifold of Twitter accounts to promulgate his vile doctrine and magnify its impact under fifty-seven names, including @KolonelSham, @WiseHaqq, @AnsarUmmah, and @MuslimBroh(). During the raid of his home, police found an alarming cache of high-powered ammunition. Aziz was charged with advocating violence against U.S. troops, chatting on-line about buying a seventeen-year-old female slave, and recruiting for the terrorist group ISIS. (Attkisson, 2017).

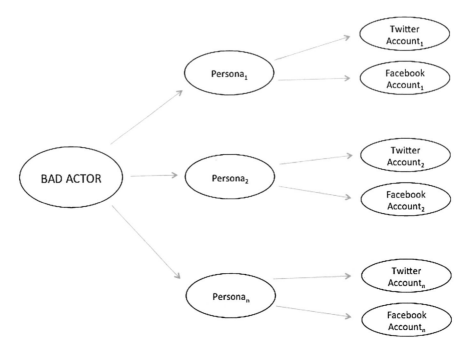

Fig. 4.6 Cyber actors, personas, and leveraging multiple accounts

Through obfuscation and multiple online identities, or personas, Aziz effectively generated gray/black propaganda (Table 4.3) from the comfort of his home. This easy production of falsehoods, via the Internet, is a big change from the thousands of people required for disinformation operations during the Cold War (Sect. 4.1.4; from Kalugin). An example of Mr. Aziz's net presence is shown in Fig. 4.6.

As shown in Fig. 4.6, through the use of multiple accounts, a single teenager in Pennsylvania can develop an online presence that looks like a community. This is sometimes called "Astroturf," the use of bots to develop a virtual following for a candidate, issue, or product/service in online media promotions. Astroturf uses the anonymity of the Internet to magnify a position (Fig. 4.7).

Figure 4.7 is an example of Pacepa's (Ion Mihai Pacepa, 2013) conceptual model for disinformation (i.e., from Fig. 4.3). In addition, Fig. 4.7 provides an example of the magnification possible for a small group of influencers, via the Internet, using multiple accounts (e.g., Fig. 4.6) (Twitter, 2020). As online followers forward his content, there is a snowball effect, often described by "power laws."

4.3.1 Power Laws

Seeing what a single teenager can do, in terms of multiplying his presence on the web, provides the kind of force multiplier effects available via social media. As described by P.W. Singer:

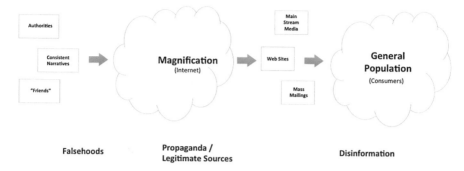

Fig. 4.7 Internet magnification of messaging

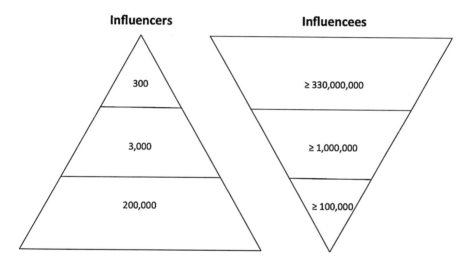

Fig. 4.8 The power law relationship between influencers and influencees

...In Internet studies, this is knows as a 'power law.' It tells us that, rather than a free-for-all among millions of people, the battle for attention is actually dominated by a handful of key nodes in the network. ... This even happens in the relatively controlled parts of the web. A study of 330 million Chinese Weibo users, for instance, found a wild skew in influence: fewer than 200,000 users had more than 100,000 followers; only about 3000 accounts had more than a million. When researchers looked more closely at how conversations started, they found that the opinions of these hundreds of millions of voices were guided by a mere 300 accounts. (Singer & E. T., 2018).

An example of the skewed influence for web-based influence operations is shown in Fig. 4.8.

This power law is not unique to cyber warfare, or influence operations. For example, Fig. 4.8 describes what a small group of determined individuals can do, and provides exemplars for the kind of numbers that we might expect to see in troll organizations planning messaging campaigns for elections or smear campaigns. A graphical version of the power law relationship, also called the Pareto or long-tailed distribution, between influencers and influencees is shown in Fig. 4.9 (Tauberg, 2018a).

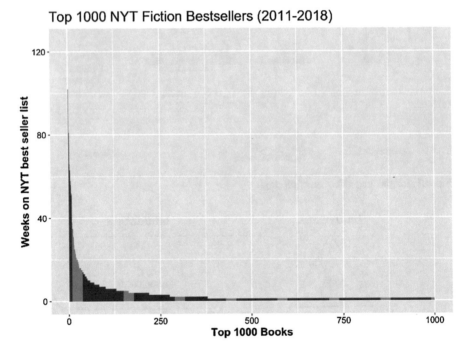

Fig. 4.9 Power law relationship: *New York Times* top 1000 fiction bestsellers (Tauberg, 2018b)

As shown in Fig. 4.9, a few key influencers usually dominate a given market space. This is expounded upon in strategic communications, where "key influencers" are targeted for shifting public opinion (DoD Joint Staff, 2009). A natural next question would be what kind of techniques a few key individuals use to attract such a large proportion of the followers. Once answer comes from the well-studied field of marketing, where creating an attractive brand is one of the principle goals of the practitioners.

4.3.2 Persuasiveness

We discussed propaganda in Sect. 4.1. Edward Bernays worked to anchor messages into future customer's minds through repetition, making the marketed product the natural choice when it came time to make a buying decision. In the late twentieth century, Robert Cialdini wrote one of the more famous books mixing human psychology with examples of how we are persuaded. In "*Influence*," Dr. Cialdini (Cialdini, 2007) provided a framework of six broad approaches used by persuaders for influence (Table 4.5).

As shown in Table 4.5, each of Cialdini's six influence approaches was, and currently is, being used in Internet information operations. For example, during the

Table 4.5 Cialdini's six influence approaches

Approach	Cyber examples
Reciprocity	It is human nature to pay back when others have given us something
Commitment	Once we make a decision, we are pressured by ourselves and others to stay consistent with that decision—We are committed to what we have decided
Consensus	When people are unsure how to act in certain circumstances, they tend to see how other people respond and act accordingly
Likability	Likability includes several traits. For example, we like people similar to us, due to familiarity. Similarly, we like people who compliment us, making us feel better
Authority	People have a tendency to obey people in authority. This includes political figures, law enforcement, medical professionals, and people with perceived expertise in other fields. To influence people, known authorities can be used to endorse, and validate, information as being legitimate
Scarcity	Things get more valuable when there seems to be less of them. This is common in the art world, as there are a finite number of works by a given painter, sculptor, etc.

2016 US presidential election, foreign actors used existing information (e.g., retweets from authority figures) to polarize target audience(s) via "Sockpuppets," which mimic authorities in one of three ways (Singer & E. T., 2018):

- Organizer of a Trusted Group (e.g., @Ten_GOP, unofficial Twitter account of Tennessee Republicans) had 136,000 followers, 10 times the number of actual Tennessee Republicans. @Ten_GOP's 3107 messages were retweeted 1,213,506 times, often amplified by Trump Campaign figures such as Donald Trump Jr., Kellyanne Conway, and Michael Flynn. Flynn followed at least 5 Russian accounts, sharing this propaganda with his 100,000 followers at least 25 times.
- Pose as a Trusted News Source: examples included the Denver Guardian, Twitter account @tparty news, which was retweeted by Trump advisor Sebastian Gorka.
- Trustworthy Individuals: grandmothers, Midwestern blue collar workers, decorated veterans, etc.

While Russian operatives established "authority" (i.e., legitimacy in Fig. 4.3) via retweets by members of the Trump campaign, they also used "likability." In addition, common personas were used by Russian actors to do social media op-eds. Guccifer 2.0, for example, was supposed to be an amiable (All World Report, 2018) Romanian commenting on the 2016 US presidential election when, in fact, it was a Russian cell actively working to sow discord. Similarly, by the use of "consensus," masquerading as a legitimate movement, Russian operatives fomented discord by actual Americans living in the United States—sometimes even organizing extremists to attack each other at a rally (ALLBRIGHT, 2017; McCarthy, 2017; Mueller, 2019; NPR, 2017). Using "authority," "likability," and "consensus" are all elements of successful information operations in the cyber domain. Using these marketing approaches is especially powerful with the individual targeting tools available with current social media.

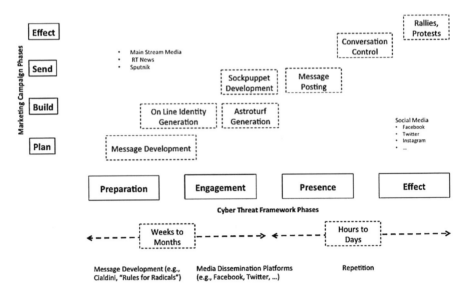

Fig. 4.10 Life cycle of an influence operation

4.3.3   Influence Campaigns and Cyber

While social engineering (Hadnagy, 2018) is a well-known part of any cyber campaign, Table 4.8 connects the respective elements of a traditional marketing (Kotler, 2017), or influence, campaign, and its cyber components.

As shown in Fig. 4.10, there is a crossover between a traditional marketing campaign's sequence of Plan/Build/Send/Effect and the Cyber Threat Framework's (CTF) (Director of National Intelligence, n.d.) Preparation/Engagement/Presence/Effect. A real example of "fake news" from the 2016 election included the Denver Guardian, a completely fictitious site, that attracted millions of followers (Sydell, 2016a, b):

Fake News Example - Denver Guardian
  "FBI Agent Suspected in Hillary Email Leaks Found Dead in Apparent Murder-Suicide."
  Denver Guardian

- Resulted in 500,000 Facebook "shares".
- Published 3 days before the 2016 US Presidential Election (November 5, 2016) where Hillary Clinton lost.
- "Denver Guardian" a completely fake source with fake news (Sydell,s 2016a, b).

Using Fig. 4.3's conceptual model, we have each of the steps for creating disinformation (Table 4.6).

As shown in Table 4.6, with a pseudo-legitimate source, creating disinformation can be as easy as putting the falsehoods out there. The implied credibility will result in targeted dissemination to millions of culpable consumers.

Table 4.6 Disinformation model example: Denver Guardian

Falsehood	"Legitimate source"	Disinformation
FBI agent suspected in Hillary email leaks found dead in apparent murder-suicide	Denver guardian	FBI agent suspected in Hillary email leaks found dead in apparent murder-suicide

4.4 The Disinformation Process: Hot Topics, Reporters, and Shades of Media

While Russian active measures date back over a century, recent cyber espionage attacks are rooted in classic Soviet KGB techniques used by "Service A" during the 1950s (Pegues, 2018). Service A was a special KGB misinformation unit, composed of mad scientists and puppeteers, who concocted nonviolent and bizarre schemes to destabilize other countries (US Information Agency, 1992). Service A distributed racially charged misinformation during the 1970s and 1980s, throughout the United States, that included creating pamphlets tailored to motivate targeted groups to action (Pegues, 2018). Some examples of groups singled out for radical messaging include:

- Militant African American groups.
- Jews—urging reprisal for the attacks on Jewish shops and Jewish people.
- White Supremacists (e.g., Ku Klux Klan).

More recently, a Russian hacker group called "Secured Borders" launched a series of Facebook and Twitter posts intended to stir up hatred. They tried to provoke an anti-immigrant rally in Twin Falls, Idaho, on August 27, 2016, by using Facebook's event and initiation tool. Secured Borders called the city a "center of refugee resettlement responsible for a huge upsurge of violence toward American Citizens." (Ben Collins, 2017). Similarly, when taking a historical accounting of Russian active measures against the United States, Jeff Pegues (Pegues, 2018) argues that:

> … there were the more recent incarnations of Russian-government-led attempts to use our vulnerabilities against us. In the 1960s, Russian-led operative units tried to exploit the country's Achilles' heel – racism – and Jim Crow laws (Ioffe, 2017). For example, they tried to portray Dr. Martin Luther King Jr. as an Uncle Tom who had been bribed by President Lyndon Johnson to quell the fire of the movement and keep black people in subordinate roles (Bringuier, 2013; Casey, 2017).

Finding social weaknesses, and the interest groups on each side of key issues, is therefore a time-tested technique in misinformation/disinformation operations. One of the key tools in developing these operations is reporters and journalists who have access to decision makers and the ability to disseminate seemingly authoritative information.

Table 4.7 Books and periodicals with KGB influence (Kalugin, 2009)

Title	Publication type	Method of compromise
"A minority of one"	Erudite American Liberal periodical	KGB funding and publishing of several articles prepared by the KGB
"The liberator"	African American periodical	Completely funded by the KGB; stories on the racist regime in South Africa and how the Soviet Union had solved its race problems were used to highlight race problems in the United States
"Nazi war criminals in our midst"	Book	KGB provided "documentary material" to bolster Charles Allen's book project on Nazi rocket scientists working in the United States
"The Russian voice"	New York based Russian-language periodical	Completely funded by the KGB

4.4.1   Journalists, Venues, and Operations Examples

As shown in Fig. 4.3, a falsehood has two primary paths to acceptance:

1. Repetition of propaganda through a variety of sources:

    (a) Pay journalists to work with fabricated information (I.F. Stone (Harvey Klehr, 2009)).
    (b) Pay for the news outlet (Table 4.7).

2. Dissemination through legitimate sources: leverage journalists and editors of accepted news outlets.

The Cold War's disinformation campaigns were broadcast through the media, with newspapers and journalists becoming the "front line" for shaping the narrative. For example, according to former Soviet KGB General Oleg Kalugin, who himself spent years spying in the United States as a journalist, his tasks included (Harvey Klehr, 2009):

- Recruiting agents in American left-leaning magazines and newspapers.
- Financing newspapers and magazines, and then planting stories to reflect the Soviet line for dissemination.

In his book, Kalugin goes on to describe the role of journalists:

The KGB recruited journalists in part for their access to inside information and sources on politics and policy, insights into personalities, and confidential and non public information that never made it into published stories. Certain journalistic working habits also lent themselves to intelligence tasks. By profession, journalists ask questions and probe; what might seem intrusive or suspect if done by anyone else is their normal modus operandi. Consequently, the KGB often used journalists as talent spotters for persons who did have access to sensitive information, and made use of them to gather background information that would help in evaluating candidates for recruitment .... There was also much less risk that a journalist having contact with a government official or engineer would attract the

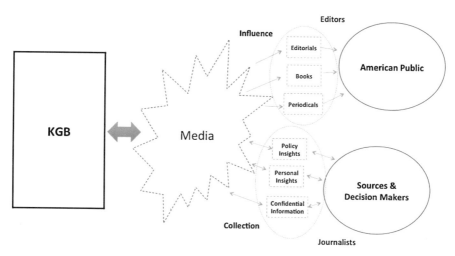

Fig. 4.11 The KGB, *US* media, and the American public

attention of security officials than would a KGB officer under Soviet diplomatic cover. And even if security officials did notice such a meeting, it would be much easier to provide a benign explanation for contact with a pesky American journalist than with a Soviet diplomat (Kalugin, 2009).

One example of General Kalugin's operative relationship between the Soviet KGB, the media, and American citizens is shown in Fig. 4.11.

Figure 4.11, showing how the media is leveraged by a foreign intelligence agency, includes:

1. Influencing the American public opinion through funding individual media outlets and placement of op-eds in established channels.
2. Collecting information on sources and decision makers through contact by journalists.

For the KGB, this was similar to their information operations in the Soviet Union, where the KGB often used journalists to publish their framing of an issue in *Literaturnaya Gazeta* (Kalugin, 2009). In addition, Kalugin goes on to describe affiliations with other media outlets to help with KGB messaging (Table 4.7).

Examples of KGB, and later Russian, inspired and funded print media in the United States is shown in Table 4.8.

As shown in Table 4.8, misinformation and disinformation venues change over time. However, active measure campaigns, similar to the IO example provided in Fig. 4.10, continued from the Cold War to the present. For example, Pacepa (Ion Mihai Pacepa, 2013) describes a series of information operations (IO) that were conducted against the United States from the 1950s to the present (Table 4.9).

Table 4.8 Time period and media venue for Russian messaging in the United States

Time period	Source name	Outlet description
1950s–present	*Workers world*	Funded and edited by the Soviet Union's KGB, *workers world* has an online presence promoting communist party issues (http://www.workers.org/wwp), (https://www.facebook.com/WorkersWorldParty/), (@workersworld)
1960s–1970s	*Ramparts magazine*	Originally a Catholic-based publication, it became the voice of the counter-culture movement (GARNER, 2009; Richardson, 2010); provided Soviets a credible platform for attacking Pope Pius XII and other senior members of the Catholic Church (Ion Mihai Pacepa, 2013). Ramparts' eventually became an anti-Vietnam War publication, receding from popularity when the United States left Vietnam (Richardson, 2010)
2001–present	*ANSWER*	Act now to stop war and end racism (ANSWER) was founded just after the 9/11 attacks to protest US military response to terrorism and is responsible for organizing some of the largest anti-war protests since its founding. ANSWER communicates through its Facebook (https://www.facebook.com/AnswerCoalition/) and twitter (@answercoalition) accounts

Each of the example information operations in Table 4.9 might be considered strategic, as their effects, however consequential, remain a challenge to quantify in terms of timeliness and ultimate breadth.

4.4.2   Area Versus Point Targeting: IO Campaigns and Social Media

The operations shown in Table 4.9 might also be looked at as "area targeting," with whole populations being swayed to sentiments that challenge a current government's policies (e.g., moving the US population to an anti-war position [e.g., Vietnam]). With the advent of the Internet, these area targeting approaches are complemented by point targeting techniques (Fig. 4.12).

Figure 4.12 brings out the difference between, and relative targeting placement of, IO campaigns and social media. An IO campaign is viewed as being primarily used for area targeting, with point targeting aligning to social media operations (e.g., 2016 US presidential campaign), and there are challenges in connecting the two. For example, many of the cited IO/Area campaigns (Fig. 4.11) had no specific follow-on, in terms of a set of point targets. For example, Operation Dragon, the Soviet disinformation campaign designed to frame the CIA for the Kennedy assassination, had the simple goal of creating enough noise around the subject to keep people guessing for decades.

Using this area vs. point targeting insight provides the following cyber operational modes:

1. Area targeting (e.g., distributed denial of service [DDoS]).

   (a) Jamming of nefarious transmitters in electronic warfare (EW).
   (b) Providing general messaging in classical information operations.

Table 4.9 Example Information Operations (IO) from the Cold War to the present

Time period	IO operation name	Objectives	Estimated effects
1950s	Operation walnut I and II	Soviet information operation (IO) designed to show air force superiority to the west	Billions of dollars invested in technologies for the development of allied reconnaissance capabilities to better understand soviet order of battle (e.g., U2, SR-71, CORONA, ...)
1960s	Operation dragon	IO campaign to frame the CIA as responsible for US President Kennedy's assassination	Decreased morale by confusing the American public and world opinion about President Kennedy's death; inspired multiple books and movies
	Operation Ares	IO campaign to convince Western Europe that the American military was using "Ghengis khan" type measures against the Vietnamese population	Turned world opinion against the United States, poisoned internal debate, damaged foreign policy, and radicalized students against the government
1970s	Operation SIG	Dissemination of Arabic and Persian versions of "*the protocols of the elders of Zion*" throughout the Middle East	Islamic republic of Iran, Islamic state, ongoing terrorism throughout the region
	Operation typhoon	Increase Western European terrorism	Red brigades (Italy), Baider Meinhof (Germany)
1980s	Glasnost	Characterize leaders of both soviet satellites (e.g., Romania) and the Soviet Union itself (i.e., Gorbachev) as liberally minded world leaders	Softened US diplomatic posture resulting in more aggressive soviet actions (e.g., Afghanistan invasion, fall of Iran, fall of Nicaragua, weakening of pro-American Philippine ally (i.e., Marcos)
2000s	ANSWER	US anti-war demonstrations	Formed 3 days after the 9/11/2001 attacks on New York and Washington; ANSWER coordinates anti-war rallies throughout the United States

2. Point targeting (e.g., outing, coordinating a rally, ...).

   (a) Focusing on specific user groups via social media.

   (b) Inducing confidentiality, integrity, and availability (CIA) effects.

One example of point targeting is the use of social media to focus on divisive issues. For example, an internal investigation by Facebook traced over 3000 ads costing $100,000 to Russian Internet trolls, containing messages about divisive issues, including race (Mike Isaac, 2017). The breadth of area operations, or IO, however, is a key differentiator between cyber and kinetic operations; and exemplifies how cyber is challenged to fit with kinetic analogies (e.g., Cyber JMEM (Gallagher, 2008; Mark Gallagher, 2013)) for effects evaluation.

Fig. 4.12 IO campaigns,
social media, and area/
point targeting

While point targeting via social media provided a few dozen rallies during the run up to the 2016 US presidential election (Mueller, 2019), the majority of current cyber operations are area, or "morale" operations. In addition, many of these IO operations are projected through mainstream media (MSM) outlets to simply frame stories to fit national narratives. For example, recently, China began information operations designed to "tell their side of the story" via mass media supplements.

4.4.3   Example: Chinese Information Operations Via Conventional Media

While Soviet era disinformation operations, during the pre-Internet age, are now well documented, China's recent foray into mass media is now just starting to be reported. For example, Beijing is mimicking Russia's RT News approach (Rutenberg, 2017) and hiring local journalists to "tell China's story well" as part of campaign to persuade international audiences on key Chinese policy positions (Louisa Lim, 2018):

> Since 2003, when revisions were made to an official document outlining the political goals of the People's Liberation Army, so-called "media warfare" has been an explicit part of Beijing's military strategy. The aim is to influence public opinion overseas in order to nudge foreign governments into making policies favorable towards China's Communist party. "Their view of national security involves pre-emption in the world of ideas," says former CIA analyst Peter Mattis, who is now a fellow in the China program at the Jamestown Foundation, a security-focused Washington think-tank. "The whole point of pushing that kind of propaganda out is to preclude or preempt decisions that would go against the People's Republic of China (Louisa Lim, 2018).

As shown in Fig. 4.13, China has multiple documented investments for influence operations into mainstream media (MSM). While exact amounts are hard to come by, the *Daily Telegraph* was reportedly paid £750,000 annually to carry the *China Watch* insert once a month (Hazelwood, 2016). Even the *Daily Mail* has an agreement with the government's Chinese-language mouthpiece, the *People's Daily* (Greenslade, 2016). Such content-sharing deals are one factor behind *China Daily*'s astonishing expenditures in the United States.; it has spent $20.8 million on US

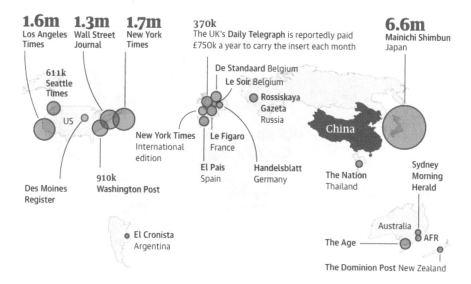

Fig. 4.13 The global reach of the China Watch newspaper supplement

influence since 2017, making it the highest registered spender that is not a foreign government.

Mainstream media (MSM) is a proven platform for messaging to a given population, providing specific messaging and advocating for a select platform. Social media compliments MSM in providing the ability to disseminate standard news. In addition, social media provides the ability to more clearly specify its audience, targeting interest groups, and even individuals, with messaging designed to raise their awareness, and potentially call them to action, concerning an issue of interest.

A key difference between current Chinese IO and legacy US approaches is that the Voice of America, Radio Free Europe, ..., were funded and staffed by the DoD. The Chinese are leveraging journalists from Western countries, via pay, to develop and disseminate their pro-Chinese messaging. Similar to traditional Soviet active measures, China is using information operations to adjust the climate in order to ease its interests forward (Daniel Kliman, 2020) resulting in four of their media outlets designated "state propaganda outlets" (Delaney, 2020).

4.5   Strategic to Tactical Cyber Influence Operations

Disinformation operations provide a look at both the conceptual model, translating falsehoods to "disinformation," and, through historical anecdotes, the operational techniques required to succeed. Understanding disinformation helps us to understand current translations of disinformation, via online trolls and interest groups, into the point targeting available with social media (Fig. 4.14).

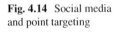

Fig. 4.14 Social media and point targeting

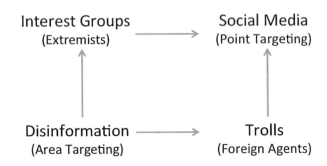

Table 4.10 Russian troll and social media productivity

Venue	Number of accounts	Activity	Measures of performance (MOPs)
Facebook	≥6	≥3 postings/day 2 news discussions/day	Expected to have ≥500 subscribers by the end of the first month
Twitter	≥10	≥50 tweets/day	≥2000 followers

As shown in Fig. 4.14, trolls, working together at a troll farm, which is an organization whose employees or members (trolls) attempt to create conflict and disruption in an online community by posting deliberately inflammatory or provocative comments. Trolls are key to the influence operations process, tailoring messages to susceptible audiences with the goal of having the effects and response of interest.

One challenge for disinformation operations is their strategic nature. It is a challenge to predict where/when/how trolls are active and the type of disinformation effects that they will generate.

4.5.1   Troll Farms: Chaos Creators

Russia is famous for its Troll Farm, where web workers at the Internet Research Agency (IRA), located in a neo-Stalinist building in St. Petersburg's Primorsky district, are paid approximately $1500 per month (twice as much for the "Facebook Desk"), to generate "fake news" and associated followers (Singer & E. T., 2018) (Table 4.10).

Trolls, shown operating through social media activity in Table 4.10, were said to be responsible for at least two clashes of race-based extremists (i.e., Charlottesville (Sales, 2020, Houston; Bertrand, 2017). In addition, as shown in Fig. 4.15, trolls use social media to stoke both sides of an issue; coordinating a day, time, and place for a clash to occur, all from the comfort of their St. Petersburg Troll Farm.

According to the Mueller Report (Mueller, 2019), dozens of Internet Research Agency (IRA)-inspired rallies occurred leading up to the 2016 US presidential election. In addition, Fig. 4.15 was effectively an "active measure," or textbook Cold

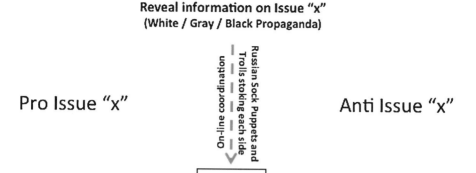

Fig. 4.15 Cyber-induced event

War tactic (Fig. 4.15). Online inducing of these events is usually performed on a social media site, where users provide trolls with their identifying attributes for targeting groups and individuals of interest.

4.5.2 Political Cyber Influence Operations: Election Tampering

Using social media to target specific interest groups and individuals, cyber was also used in the 2016 US presidential election for a combined media/political campaign to expose the private communications of the Democratic Party (BBC, 2016), where Ferguson (Ferguson, 2018) tells the story, that:

> There was, after all, a third network involved in the US elections of 2016, and that was Russia's intelligence network. At the time of writing (2017), it is clear that the Russian government did its utmost to maximize the damage to Hillary Clinton's reputation stemming from her and her campaign's sloppy email security, using WikiLeaks as the conduit through which stolen documents were passed to the American media. To visit the WikiLeaks website is to enter the trophy room of this operation. Here is the 'Hillary Clinton Email Archive', there are 'The Podesta Emails'. Not all the leaked documents are American, to be sure. But you will look in vain for leaks calculated to embarrass the Russian government. Julian Assange may still skulk in the Ecuadoran embassy in London, but the reality is that he lives, an honoured guest of President Vladimir Putin, in the strange land of Cyberia – the twilight zone inhabited by Russia's online operatives.

Russian participation in the 2016 US presidential election can be looked at as an extension of previous socio-political cyber operations, where effects included:

- Distorted messaging (i.e., in Georgia, modified government websites) via online personas masquerading as real people (e.g., Guccifer 2.0 (Sanger & J. R., 2018)).
- Denied campaigning capabilities of the Democratic Party (e.g., in Estonia, government/banks/telecommunications suffered from distributed denial of service (DDoS) attacks).

Each of the above provided a news-like, message altering, effect, with the addition of denying critical infrastructure operations, communicating to citizens that the government is no longer in control.

Due to the pervasive Russian use of active measures, especially in US and Western European political campaigns, the Alliance for Securing Democracy put together Hamilton 68 (Alliance for Securing Democracy, n.d.), which tracks 600 monitored Twitter accounts linked to Russian influence operations. These are accounts that include bots and humans and are run by troll factories in Russia and elsewhere, along with sites that clearly state they are affiliated with the Russian government. They also include accounts by the posers, a network of people who are not necessarily reporting to the Kremlin but jump on the pro-Russian/anti-American bandwagon and join forces. In addition, Hamilton 68 also tracks people who may be using a wide variety of tools, including information operations, cyber attacks, and financial influence, often through proxy networks. This helps identify folks who hide well and are a challenge to track.

4.6 Cyber Influence Operations Summary

Cyber's role in providing a platform for social media, including pictures and videos, makes it the premier twenty-first century influence platform. Social media provided the platform for both the RT News "making" the war in Eastern Ukraine and the Crimea for "outing" the irregular Russian troops taking part in both the annexation and support operations (e.g., using Russian social media, VKontakte). VKontakte was also reportedly used by the Russians to spy on the Ukrainians (Matlack, 2014).

In terms of phasing operations, Fig. 4.16 uses cyber for shaping (phase 0) through post-conflict peace (phase 5) operations. This was a rapid militarization of the seemingly innocent social media used to connect with friends and family only a decade or so before (e.g., the early 2000s).

A quick overview of the development of cyber influence operations includes answering the questions posed at the beginning of this chapter:

1. What are the mechanisms used in current cyber influence operations?

   As shown in Table 4.2, propaganda operations come in forms that can be generally categorized white (clear attribution), black (false attribution), and gray (unclear attribution).

   Influence operations, as described in this chapter, leverage mass media to get the message out to the greatest number of people (i.e., area targeting in Fig. 4.10). Using mass media was shown to be standard tradecraft for the Soviets (Sect.

4.1.4) and is currently being used by the Chinese to shape their message globally (Fig. 4.11). In addition, this broader messaging might be looked at as "area targeting," with more specific "point targeting" prosecuted via social media (Fig. 4.12). In addition, the Mueller report (Mueller, 2019) provides examples of cyber-induced events during the 2016 elections (Fig. 4.13).

2. How have influence operations evolved with changes in technology?

Using influence to motivate someone else to adopt a point of view dates back from time immemorial. The focus here has been mostly on the use of mass communications, and, more recently, social media, to limit the demographic to a select audience (e.g., via social media) to be persuaded on issues of interest to them.

3. How do influence operations work in commercial, government, and military applications?

Section 4.3.3 provides background on influence campaigns, using a marketing template, and comparing it to a cyber operation. This approach could be used to get a specific message out in commercial, government, or military channels. In addition, Fig. 4.3 provides a more specific military campaign structure (e.g., JP 5.0) to show how influence operations are used in the shaping through domination phases of a cyber campaign.

A novel aspect of cyber is the possible magnification of messaging (Fig. 4.5). This includes leveraging the one to many possibilities for developing online identities, and using each one to reinforce a message (Fig. 4.2) as implied online substantiation with the hope of validating a message of interest.

4. How did the Soviet Union use information operations during the Cold War? What are examples of more recent Russian cyber influence operations?

Section 4.1.4 provides examples of Soviet disinformation, including their development and dissemination. In addition, Sect. 4.4 uses documented Soviet TTPs for using journalists and financing media venues for the dissemination of disinformation (Fig. 4.9). Table 4.8 provides example Cold War information operations.

Section 4.5.1 provides recent examples of Russian Troll Farms, paid operatives that focus on divisive issues to create chaos on the Internet. In addition, Fig. 4.13 shows how trolls have the ability to induce a cyber event (Mueller, 2019). This includes election tampering (Sect. 4.5.2) or the inclusion of cyber influence into tactical military operations (Sect. 4.2.3).

5. How are cyber influence operations used in conjunction with kinetic operations?

Table 4.10 is a summary of Russian operations that include cyber. And Fig. 4.14 provides a process flow for cyber's inclusion into kinetic operations.

6. What are the dynamics, or moving parts, of influence operations? How are these dynamics changed by cyber technologies (e.g., social media)?

Figure 4.2 provides a basic description of washing falsehoods in the development of disinformation. Using influence is possible across the phases of a standard military campaign (Fig. 4.3).

Using Fig. 4.2, we see that increased message frequency via propaganda can be performed through magnification of messaging (Fig. 4.5). This, combined with a valid source, shortens the time to produce disinformation from a falsehood. Power law relationships (Fig. 4.6) are one way to portray the disproportionate number of followers that thought leaders have on the Internet.

Figure 4.8 provides the life cycle of an influence operation, comparing a standard marketing approach to a cyber operation via the cyber threat framework (CTF). In addition, Fig. 4.8 provides additional context for the execution of a cyber-induced event (Fig. 4.13), or the inclusion of cyber effects into conventional operations (Fig. 4.14).

7. How does targeting occur in cyber influence operations? For example, do ideas like area targeting vs. point targeting apply in cyber influence operations? How?

Targeting follows standard steps in identifying a target, its access approaches, and TTPs for holding it at risk or prosecuting it. In addition, cyber targets have the additional benefit of potentially perishable effects. For example, a cyber target may be prosecuted by simply making it unavailable for a desired time period (e.g., denial of service), with no need to destroy it. In addition, denial of service might be performed over a broad range of cyber targets with a single network access point, making for an area target. Point targeting would be the denial, or disabling, of just one of these targets.

Expanding on area and point targeting, an "area target" might be the broad dissemination of a message (Sect. 4.1.4), similar to what the former Soviet Union did with active measures during the Cold War. Refining a message to target a specific group is an example of point targeting (Sect. 4.4.1).

Using cyber for influence operations, however, was clear from the first online newspaper OpEd put on the web. Section 4.1.4 shows how Cold War active measures heavily leveraged authoritative people and publications to legitimize falsehoods and create uncertainty via disinformation (Fig. 4.3). Each of these operations (Table 4.9) required a handful (Fig. 4.8) of malevolent journalists and editors to write, place, and publish the falsehoods of interest. Cyber has the ability to amplify this messaging through multiple "personas" (Figs. 4.6 and 4.7) that are likable, exert some authority, or are known to us (Table 4.5). The use of cyber has therefore added a personal dimension to active measures of old, narrowing mass media area targeting to the point targeting of an interest group, or even an individual.

Bibliography

Alinsky, S. (1989). *Rules for radicals: A practical primer for realistic radicals.* Vintage.

All World Report. (2018, 7 20). *Ex-Playboy model: 'I've had Twitter sex with 12 Russian hackers'.* Retrieved November 26, 2020, from All World Report: https://allworldreport.com/world-news/ex-playboy-model-ive-twitter-sex-12-russian-hackers/

Allbright, C. (2017, 11 1). *A Russian Facebook page organized a protest in Texas. A different Russian page launched the counterprotest.* Retrieved February 1, 2019, from the Texas tribune: https://www.texastribune.org/2017/11/01/russian-facebook-page-organized-protest-texas-different-russian-page-l/

Alliance for Securing Democracy. (n.d.). *Hamilton 68*. Retrieved June 3, 2019, from Hamilton 68: https://securingdemocracy.gmfus.org/hamilton-68/

Attkisson, S. (2017). *The smear - how shady political operatives and fake news control what you see, what you think, and how you vote*. Harper Collins.

Barber, G. (2018, 7 13). *Someone found a use for bitcoin. Russian hackers!* Retrieved April 5, 2020, from WIRED: https://www.wired.com/story/russian-hackers-bitcoin/.

BBC. (2016, October 27). *18 revelations from Wikileaks' hacked Clinton emails*. Retrieved August 21, 2018, from BBC: https://www.bbc.com/news/world-us-canada-37639370

Ben Collins, K. P. (2017, 9 11). Exclusive: Russia used Facebook events to organize anti-immigrant rallies on U.S. soil. *Daily beast, 2019*(6), p. 3.

Bernays, E. (1923). *Crystallizing public Opionion*. Boni and Liveright.

Bernays, E. L. (1918). *Propaganda*. Desert.

Bertrand, N. (2017, 11 17). *Russia organized 2 sides of a Texas protest and encouraged 'both sides to battle in the streets'*. Retrieved November 23, 2020, from Business insider: https://www.businessinsider.com/russia-trolls-senate-intelligence-committee-hearing-2017-11

Boghardt, T. (2009). Operation INFEKTION: Soviet bloc intelligence and its AIDS disinformation campaign. *Studies in Intelligence, 53*(4), 1–24.

Bringuier, D. C. (2013). *Crime without punishment*. AuthorHouse.

Casey, M. (2017, 9 26). How Russia Keeps Exploiting Anti-Black Racism in the US. .

Christopher Andrew, V. M. (2015). *The Mitrokhin archive*. Penguin.

Cialdini, R. B. (2007). *Influence - the psychology of persuasion*. Collins Business.

Coldewey, D. (2018, 7 13). *Russian hackers used bitcoin to fund election interference, so prepare for FUD*. Retrieved April 5, 2020, from TechCrunch: https://techcrunch.com/2018/07/13/russian-hackers-used-bitcoin-to-fund-election-interference-so-prepare-for-fud/2018/07/13/russian-hackers-used-bitcoin-to-fund-election-interference-so-prepare-for-fud/2018/07/13/russian-hackers-used-bitcoin-to-fund-election-interference-so-prepare-for-fud/

Daniel Kliman, A. K.-T. (2020, 5). *Dangerous synergies - countering Chinese and Russian digital influence operations*. Retrieved July 6, 2020, from Center for new American Security (CNAS): https://s3.amazonaws.com/files.cnas.org/documents/CNAS-Report-Dangerous-Synergies-May-2020-DoS-Proof.pdf?mtime=20200506164642

Delaney, R. (2020, 6 23). *US designates 4 more Chinese media organisations as 'state propaganda outlets'*. Retrieved July 23, 2020, from South China morning post: https://www.scmp.com/news/world/united-states-canada/article/3090161/us-designates-four-major-chinese-media-outlets?utm_source=dailybrief&utm_medium=email&utm_campaign=DailyBrief2020Jun23&utm_term=DailyNewsBrief

Department of Defense. (2012). *Joint Publication 1–13.4 Military Deception*. Department of Defense.

Director of National Intelligence. (n.d.). *Cyber Threat Framework*. Retrieved February 15, 2019, from Cyber Threat Framework: https://www.dni.gov/index.php/cyber-threat-framework

DoD Joint Staff. (2009). *Strategic communication joint integrating concept. U.S. Department of Defense (DoD), joint staff*. DoD.

Edward Bernays, M. C. (2004). *Propaganda*. Ig Publishing.

Ferguson, N. (2018). *The square and the tower - networks and power, from the freemasons to Facebook*. Penguin.

FireEye. (2014). *APT28: A window into Russia's cyber espionage operations?* Retrieved September 9, 2018, from FireEye: https://www.fireeye.com/content/dam/fireeye-www/global/en/current-threats/pdfs/rpt-apt28.pdf

Friends Committee on National Legislation. (n.d.). *Understanding Drones*. Retrieved January 17, 2019, from https://www.fcnl.org/updates/understanding-drones-43

Gallagher, M. (2008). *Cyber analysis workshop*. MORS. Reston: MORS.

Garner, D. (2009, 10 6). *Back when ramparts did the storming*. Retrieved January 24, 2019, from New York times: https://www.nytimes.com/2009/10/07/books/07garner.html

Greenslade, R. (2016, 8 12). *What is mail online doing in partnership with the People's daily of China?* Retrieved February 19, 2019, from the Guardian: https://www.theguardian.com/media/greenslade/2016/aug/12/mail-online-goes-into-partnership-with-the-peoples-daily-of-china

Hadnagy, C. (2018). *Social engineering: The science of human hacking.* NY.

Hamilton, B. A. (2020). *Bearing witness: uncovering the logic behind russian military cyber operations.* Retrieved April 1, 2020, from Booz Allen Hamilton.

Harvey Klehr, J. E. (2009). I.F. Stone, soviet agent - case closed. *Commentary.*

Hazelwood, J. (2016, 4 3). *China spends big on propaganda in Britain... but returns are low.* Retrieved February 16, 2019, from Hong Kong free press: https://www.hongkongfp.com/2016/04/03/china-spends-big-on-propaganda-in-britain-but-returns-are-low/

Ho, A. K., Jim Garrison, J. M., & Stone, O. (1991). *JFK [motion picture].* Warner Bros.

IDS International. (n.d.). *SMEIR.* Retrieved January 3, 2019, from https://www.smeir.net/

Ioffe, J. (2017, 10 21). The history of Russian involvement in America's race wars. *Atlantic,* 2019(6), p. 3.

Ion Mihai Pacepa, R. J. (2013). *Disinformation - former spy chief reveals secret strategies for undermining freedom, attacking religion and promoting terrorism.* WND.

Jarvis, J. (2011, 3 17). *Revealed: US spy operation that manipulates social media - Military's 'sock puppet' software creates fake online identities to spread pro-American propaganda.* Retrieved January 3, 2019, from Guardian: https://www.theguardian.com/technology/2011/mar/17/us-spy-operation-social-networks

Jen Weedon, W. N. (2017, 4 27). *Information operations and Facebook.* Retrieved January 3, 2019, from Facebook: https://fbnewsroomus.files.wordpress.com/2017/04/facebook-and-information-operations-v1.pdf

Joint Chiefs of Staff. (2009, 10 7). *Strategic Communication Joint Integrating Concept.* Retrieved March 31, 2020, from https://www.jcs.mil/Portals/36/Documents/Doctrine/concepts/jic_strategiccommunications.pdf?ver=2017-12-28-162005-353

Joint Chiefs of Staff. (2012). *Information operations.* Joint Chiefs of Staff.

Joint Chiefs of Staff. (2013). *Joint Publication 3–12 Cyberspace Operations.*

Joint Chiefs of Staff. (2017, 6 17). *Joint planning.* Retrieved February 15, 2019, from Joint Publication 5-0: https://www.jcs.mil/Portals/36/Documents/Doctrine/pubs/jp5_0_20171606.pdf

Joint Chiefs of Staff. (2018, 7 25). *Joint Concept for Operating in the Information Environment (JCOIE).* Retrieved April 1, 2020, from JCS: https://www.jcs.mil/Portals/36/Documents/Doctrine/concepts/joint_concepts_jcoie.pdf?ver=2018-08-01-142119-830

Joint Staff. (2018, 6 8). *Joint publication 3-12 cyberspace operations.* Retrieved September 16, 2019, from joint publications: https://www.jcs.mil/Portals/36/Documents/Doctrine/pubs/jp3_12.pdf

Jones, S. G. (2018, 10 1). *Going on the offensive: A U.S. strategy to combat Russian information warfare.* Retrieved March 28, 2020, from https://www.csis.org/analysis/going-offensive-us-strategy-combat-russian-information-warfare

Joseph Farah, G. E., Ion Pacepa, D. K., & Moore, S. (2017). *Disinformation - The Secret Strategy to Destroy the West* [Motion Picture]. WND Films. Retrieved February 7, 2019, from YouTube: https://www.youtube.com/watch?v=LF9-xdj_8oQ&t=692s

Kalugin, O. (2009). *Spymaster : My thirty-two years in intelligence and espionage against the west.* Basic Books.

Kaplan, R. D. (2019, 1 7). *A new cold war has begun.* Retrieved January 23, 2019, from foreign policy: https://foreignpolicy.com/2019/01/07/a-new-cold-war-has-begun/

Kenney, M. (2008). From Pablo to Osama: Trafficking and terrorist networks, government bureaucracies, and competitive adaptation. Penn State University Press. Retrieved from https://www.amazon.com/Pablo-Osama-Trafficking-Bureaucracies-Competitive/dp/B010WIIJRK/ref=sr_1_fkmr0_1?ie=UTF8&qid=1547027596&sr=8-1-fkmr0&keywords=robb+2008+pablo+to+osama

Kotler, P. (2017). *Marketing Management*. Pearson.

Krebs, B. (2013, 6 13). *Iranian elections bring lull in Bank attacks*. Retrieved September 14, 2019, from KrebsonSecurity: https://krebsonsecurity.com/2013/06/iranian-elections-bring-lull-in-bank-attacks/

Lasswell, H. D. (1927, 8), The theory of political propaganda. *The American Political Science Review,* 21(3), 627–631.

Louisa Lim, J. B. (2018, 12 7). *Inside China's audacious global propaganda campaign*. Retrieved February 1, 2019, from the Guardian: https://www.theguardian.com/news/2018/dec/07/china-plan-for-global-media-dominance-propaganda-xi-jinping

Mark Gallagher, M. H. (2013). Cyber joint munitions effectiveness manual (JMEM). *Modeling and Simulation Journal*.

Matlack, C. (2014, 4 17). *The kremlin tried to use VKontakte—Russia's Facebook—To spy on Ukrainians*. Retrieved April 1, 2020, from Bloomberg Businessweek: https://www.bloomberg.com/news/articles/2014-04-17/the-kremlin-tried-to-use-vkontakte-russia-s-facebook-to-spy-on-ukrainians

McCarthy, T. (2017, 10 14). *How Russia used social media to divide Americans*. Retrieved February 1, 2019, from the Guardian: https://www.theguardian.com/us-news/2017/oct/14/russia-us-politics-social-media-facebook

Mike Isaac, S. S. (2017, 10 1). Facebook to deliver 3,000 Russia-linked ads to congress on Monday. *New York times, 2019*(6), p. 3.

Mueller, R. (2019). *Report on the investigation into Russian interference in the 2016 presidential election*. U.S. Department of Justice. Washington: U.S. Department of Justice.

Nakashima, E. (2018, 10 23). *Pentagon launches first cyber operation to deter Russian interference in midterm elections*. Retrieved March 12, 2019, from Washington post: https://www.washingtonpost.com/world/national-security/pentagon-launches-first-cyber-operation-to-deter-russian-interference-in-midterm-elections/2018/10/23/12ec6e7e-d6df-11e8-83a2-d1c3da28d6b6_story.html?utm_term=.8c47d573557b

Nakashima, E. (2019, 2 27). *US disrupted internet access of Russian troll factory on day of 2018 midterms*. Retrieved March 12, 2019, from Washington post.

Nasaw, D. (2001). *The chief*. Mariner Books.

NPR. (2017, 10 30). *Russians Targeted U.S. Racial Divisions Long Before 2016 And Black Lives Matter*. Retrieved February 1, 2019, from NPR: https://www.npr.org/2017/10/30/560042987/russians-targeted-u-s-racial-divisions-long-before-2016-and-black-lives-matter

Office of Strategic Services. (1943). *Morale operations field manual no. 2*. U.S. Army, Office of Strategic Services.

Parham, J. (2017, 10 18). *Russians posing as black activists on facebook is more than fake news*. Retrieved August 22, 2018, from wired: https://www.wired.com/story/russian-black-activist-facebook-accounts/

Paula Hamilton, L. S. (2008). *Oral history and public memories*. Temple University Press.

Pegues, J. (2018). *Kompromat - how Russia undermined American democracy*. Prometheus.

Peiss, K. (2020, 1 3). *Why the U.S. sent librarians undercover to gather intelligence during world war II*. Retrieved June 4, 2020, from time: https://time.com/5752115/world-war-ii-librarians/

Popken, B. (2018, 11 5). *Factory of lies: Russia's disinformation playbook exposed*. Retrieved March 22, 2020, from NBC news: https://www.nbcnews.com/business/consumer/factory-lies-russia-s-disinformation-playbook-exposed-n910316

Prosser, M. B. (2006). *MEMETICS—A GROWTH INDUSTRY IN US MILITARY OPERATIONS (MS thesis)*. Retrieved January 3, 2019, from DTIC.: https://apps.dtic.mil/dtic/tr/fulltext/u2/a507172.pdf

Richardson, P. (2010). *A bomb in every issue: How the short, unruly life of ramparts magazine changed*. The New Press.

River, C. (Ed.). (2018). *Horace Greeley: The life and legacy of 19th century America's Most influential editor*. Charles River Publishers.

Rivera, J. (2014, 3 2). *Has Russia Begun Offensive Cyberspace Operations in Crimea?* Retrieved April 5, 2020, from Georgetown Security Studies Review: https://georgetownsecuritystudiesreview.org/2014/03/02/has-russia-begun-offensive-cyberspace-operations-in-crimea/

Robb, J. (2007, June). The Coming Urban Terror. City Journal.

Robb, J. (2008). *Brave new war: The next stage of terrorism and the end of globalization.* Wiley.

Rutenberg, J. (2017, 9 17). *RT, sputnik and Russia's new theory of war - how the kremlin built one of the most powerful information weapons of the 21st century — And why it may be impossible to stop.* Retrieved July 2, 2020, from New York times: https://www.nytimes.com/2017/09/13/magazine/rt-sputnik-and-russias-new-theory-of-war.html

Sales, B. (2020, 12 16). *Online antisemitism peaks during moments of national tension. And it's being partly driven by Russian trolls.* Retrieved December 20, 2020, from the Jewish news of Northern California: https://www.jweekly.com/2020/12/16/online-antisemitism-peaks-during-moments-of-national-tension-and-its-being-partly-driven-by-russian-trolls/

Saletan, W. (2017, 03). *Hate makes us weak.* Retrieved June 3, 2019, from slate: http://www.slate.com/articles/news_and_politics/politica/2017/03/how_russia_capitalizes_on_american_racism_and_xenophobia.html

David E. Sanger, J. R. (2018, July 15). *Tracing Guccifer 2.0's many tentacles in the 2016 election.* Retrieved September 9, 2018, from New York times: https://www.nytimes.com/2018/07/15/us/politics/guccifer-russia-mueller.html

Sherman, E. (2011, 2 19). *So, why does the air force want hundreds of fake online identities on social media?* Retrieved January 9, 2019, from CBS news: https://www.cbsnews.com/news/so-why-does-the-air-force-want-hundreds-of-fake-online-identities-on-social-media-update/

Singer, P. W., & E. T. (2018). *LikeWar - the Weaponization of social media.* Houghton Mifflin.

Stoll, C. (2005). *The Cuckoo's egg: Tracking a spy through the maze of computer espionage.* Pocket Books.

Strategy and Tactics of Guerilla Warfare. (n.d.). Retrieved September 9, 2018, from Wikipedia: https://en.wikipedia.org/wiki/Strategy_and_tactics_of_guerrilla_warfare

Stu Woo, A. B. (2018, 12 24). *Huawei had a Deal to give Washington redskins fans free Wi-fi, until the government stepped in.* Retrieved February 1, 2019, from The Wall Street Journal: https://www.wsj.com/articles/huawei-had-a-deal-to-give-washington-redskins-fans-free-wi-fi-until-the-government-stepped-in-11545647401

Sydell, L. (2016a, 11 23). *We tracked down A fake-news creator in the suburbs. Here's what we learned.* Retrieved February 15, 2019, from NPR: https://www.npr.org/sections/alltechconsidered/2016/11/23/503146770/npr-finds-the-head-of-a-covert-fake-news-operation-in-the-suburbs

Sydell, L. (2016b, 11 23). *We tracked down A fake-news creator in the suburbs. Here's what we learned.* Retrieved November 20, 2020, from NPR: https://www.npr.org/sections/alltechconsidered/2016/11/23/503146770/npr-finds-the-head-of-a-covert-fake-news-operation-in-the-suburbs

Tauberg, M. (2018a, 6 28). *Power law in popular media.* Retrieved January 25, 2019, from medium: https://medium.com/@michaeltauberg/power-law-in-popular-media-7d7efef3fb7c

Tauberg, M. (2018b, 6 19). *Power Law in Popular Media .* Retrieved April 5, 2020, from Medium: https://medium.com/@michaeltauberg/power-law-in-popular-media-7d7efef3fb7c

The famous people. (n.d.). *Social media stars.* Retrieved 4 5, 2020, from social media stars.: https://www.thefamouspeople.com/list-of-social-media-stars.php

Theohary, C. (2020, 1 14). *Defense primer: Information operations.* Retrieved 3 28, 2020, from CRS: https://crsreports.congress.gov/product/pdf/IF/IF10771

Twitter. (2020, 6 12). *Disclosing networks of state-linked information operations we've removed.* Retrieved June 12, 2020, from Twitter: https://blog.twitter.com/en_us/topics/company/2020/information-operations-june-2020.html

U.S. Department of Defense. (2012, 11 27). *Information Operations (JP 3–13).* Retrieved 3 30, 2020, from Joint Chiefs of Staff: https://www.jcs.mil/Portals/36/Documents/Doctrine/pubs/jp3_13.pdf

U.S. Department of Defense (DoD). (2003, 10 30). *2003 Information Operations Roadmap.* Retrieved March 30, 2020, from GWU NSA Archive: https://nsarchive2.gwu.edu/NSAEBB/NSAEBB177/info_ops_roadmap.pdf

Unger, R. (2012, 8 20). *The dirtiest presidential campaign ever? Not even close!* Retrieved January 18, 2019, from Forbes: https://www.forbes.com/sites/rickungar/2012/08/20/the-dirtiest-presidential-campaign-ever-not-even-close/#23d819a53d84

US Information Agency. (1992). How soviet Active measures themes were spread. In S. Active (Ed.), *Measures in the 'Post-cold War' era* (pp. 1988–1991). US House of Representatives Committee on Appropriations.

Vaynerchuk, G. (n.d.). *Gary Vaynerchuk.* Retrieved February 12, 2019, from Gary Vaynerchuk: https://www.garyvaynerchuk.com/

Vera Zakem, M. K. (2018, 4 1). *Exploring the utility of memes for U.S. government influence campaigns.* Retrieved January 3, 2019, from Center for Naval Analyses (CNA): https://www.cna.org/cna_files/pdf/DRM-2018-U-017433-Final.pdf

Vietnam War With Walter Cronkite (2008). [Motion Picture].

Welch, C. (2018, 12 26). *LinkedIn co-founder says he unknowingly backed disinformation effort in Alabama senate race.* Retrieved January 23, 2019, from the verge: https://www.theverge.com/2018/12/26/18156702/reid-hoffman-backed-group-alabama-election-misinformation-roy-moore

Chapter 5
Cyber ISR and Analysis

The purpose of this chapter is to present a general background on cyber intelligence, surveillance, and reconnaissance (ISR) for analysis and targeting. This includes looking at the people, processes, and technologies used to develop a cyber-based intelligence product. In addition, we will look at specific cyber surveillance and reconnaissance tools (e.g., Stuxnet components) and discuss the practical implications of producing cyber intelligence products from current terabyte downloads, along with associated cyber analysis and targeting implications.

5.1 Background

In Chap. 4, we saw how analysis, and subsequent intelligence, is necessary for understanding a target system's message vulnerabilities. While Chap. 4's focus was on messaging, in Chap. 3 we reviewed a cyber threat intelligence framework, including approaches for exploiting known system vulnerabilities. This chapter is more straightforward in that we view cyber intelligence as simply information collection for analysis and understanding. Put this way, we can learn from historians of intelligence, and their analysis of spycraft, to develop a common sense understanding of cyber intelligence, surveillance, and reconnaissance (ISR) for analysis and targeting. For example, Dr. Warner points out that cyber technologies and techniques, in some respects, originated in the intelligence profession (Warner, 2017). Intelligence activities and cyberspace operations can look quite similar; what we call cyber is intelligence in many important ways.

The US Department of Defense's (DoD) intelligence process consists of six phases—direction, collection, processing, analysis, dissemination, and feedback (JP 2-0, 2013)—each of which may be concurrent, or may be skipped entirely,

---

"Intelligence activities and cyberspace operations can look quite similar; what we call cyber is intelligence in an important sense." —Dr. Michael Warner, US CyberCom Command Historian

© Springer Nature Switzerland AG 2022
J. M. Couretas, *An Introduction to Cyber Analysis and Targeting*,
https://doi.org/10.1007/978-3-030-88559-5_5

Fig. 5.1 Relationship of data, information, and intelligence

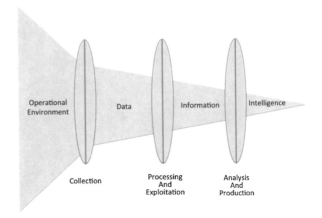

depending on the situation. Brian P. Kime's "Intelligence Preparation of the Cyber Operational Environment" (Kime, 2016) relates the DoD intelligence cycle to information security by providing a process that transforms threat data to information and intelligence (Fig. 5.1).

As shown in Fig. 5.1, the relationship of data, information, and intelligence provides a general picture of the cyber intelligence process. For example, cyber resources are used over the collection, processing/exploitation, and analysis/production phases to create an intelligence product. The process provided by Fig. 5.1 is used for any cyber ISR operation across the people, process, and technology elements of a target of interest.

5.2   Introduction

Cyber is novel in that it provides both a path to target and a means to collect on the target itself. Traditional ISR (e.g., imagery and signals) usually relies on a platform (e.g., aircraft and satellite) to carry the collecting sensor. In differentiating cyber, we will therefore ask a few introductory questions before diving into this chapter:

1. What is the overlap between cyber and human intelligence?
2. How different is cyber from open-source intelligence (OSINT)?
3. What does the cyber intelligence cycle look like?
4. What are the key technologies in cyber intelligence?
5. How did cyber analysis and targeting play a role in countering Islamic State of Iraq and Syria (ISIS) operations (e.g., counter terror)?
6. How does cyber help in countering nation state actors (e.g., countering the Iranian nuclear program)?
7. How does cyber compare as a means of technical intelligence?

In the abstract, each intelligence requirement starts out with the people, their operational processes, and any tool/technology signatures that can be used to identify them to understand their associated intent. An overview of intelligence collection is provided by the narrowing down of the initial collectibles (Fig. 5.2).

As shown in Fig. 5.2, the heavy lifting in cyber ISR starts with understanding the people. After determining the "who," and what they are doing (i.e., processes), we consider how they use information technology to perform their work. Of course, this process may be reversed. With only a technology signature from the attack, we may need to backtrack who was in possession of, or had the skills to use, the detected technical capability.

As background, an example network is shown in Fig. 5.3. This provides the standard cyber "target," including both business processes and industrial control—used for the controlling of electro-mechanical devices (e.g., power plants, hospitals, automated production, …).

In performing intelligence collection on systems such as Fig. 5.3, the analyst should adhere to standard cyber terminology (e.g., JP 3-12, discussed in Chap. 2) along with following both well-known collection practices and ensuring that all of this meets the requirements of the standard targeting cycle (e.g., JP 3-60, discussed in Chap. 2).

5.3   Cyber and Human Intelligence

Cyber amplifies both collection capabilities, through automation, and the signature of persons/networks participating online. This results in a net increase in intelligence (e.g., social network analysis and communication signatures) garnered from surveillance and reconnaissance activities on the web.

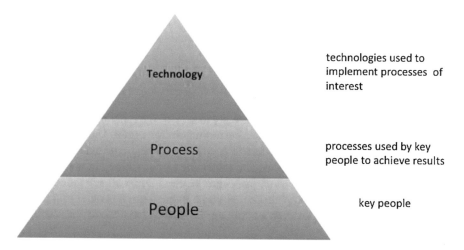

Fig. 5.2 Cyber collection pyramid

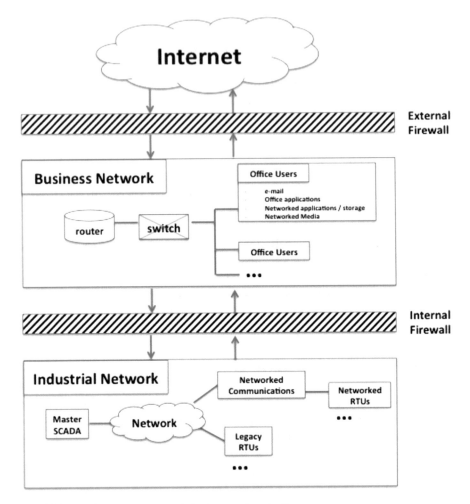

Fig. 5.3 Example computer network: business and production

5.3.1 Human Analogs: Automating Spies

Cyber capabilities have been applied directly to service the functions of spying and counter spying that human agents have performed for millennia. The scale that can be exploited, however, is vastly different. A key difference, in the cyber era, is that previously "the devices that secured and transmitted information did not also store it" (Warner, 2017). Today, however, past, current, and future data are vulnerable to spies and eavesdroppers in unprecedented ways. In addition, a parallel to a human spy is an implant, which can sit in a computer for weeks, months, or years, collecting secrets.

The change in scale with cyber is dramatic (Max Roser). For example, Cold War information transfer, often from a single spy, was likely on the order of kilobytes (e.g., information collected from a foe, condensed to a single micro dot, and then transferred to a friendly country). Cyber, however, produces terabyte downloads (e.g., Moonlight Maze (Clarke, 2012), the Equifax disclosure (Newman, 2017), counter-ISIS operations (Malcolm Nance, 2017) …) and are common in the current cyber era without a spy even needing to set foot on adversary terrain.

Therefore, a single cyber operation, possibly a single implant, is measurably more powerful than all the spy services in the world only a generation ago (i.e., 1 TB/1 KB ~ 1 billion multiplier in data collection). In addition, the ability to cross-correlate individual records has some people wondering whether the entire US population's private information is currently in a well-organized Chinese database due to the Office of Personnel Management (OPM), Anthem Insurance, Equifax Credit Agency, and multiple other data exfiltrations (Sanger, 2018). This is the kind of "big data" intelligence that cyber currently provides. In combination with graphical technologies, the intelligence can also show the type/level of connection between the respective individuals.

5.3.2 ISIS and Human Intelligence

Similar to the Jalil Ibn Ameer Aziz example discussed in Chap. 4 (Sect. 4.3), one of the early successes in using the web for counter-terror operations included the arrest of "Irhabi007," (Economist, 2007) an Al Qaeda in Iraq (AQI; i.e., ISIS predecessor) web operator moving money, propaganda videos, and facilitating operations from his mother's basement in E. London (Rita Katz, 2006). Irhabi007 effectively provided a remote back office for al Qaeda, managing money and information, along with performing command and control (C2), as needed.

In addition to performing command and control (C2), ISIS leveraged online sympathizers to promote its "brand." Carlin (Carlin, 2018) describes three distinct types of ISIL—Islamic State of Iraq and the Levant—sympathizers online:

1. "Nodes": the primary drivers of online engagement who were prominent voices and provided new material to ISIL's online followers.
2. "Amplifiers": sometimes real, sometimes bots, who primarily retweeted and spread material online.
3. "Shout-Outs": these were believed to be "a unique innovation and vital to the survival of the ISIL online scene. They primarily introduce new, pro-ISIS accounts to the community and promote newly created accounts of previously suspended users, allowing them to quickly regain their pre-suspension status." The "shout-outs" provided "little substantive content," but often had the largest online followings because they represented something like an online directory for other accounts, playing "a pivotal role in the resilience of ISIS' Twitter community" (Lorenzo Vidino, 2015).

Using cyber for human intelligence initially proved its worth during the 2008 surge of US military operations in Iraq, when the US Army was battling multiple insurgent/terror factions in Eastern Iraq. General David Petraeus, leader of the surge, credits cyber with removing 4000 enemy combatants from the battlefield due to both individual and group targeting (Harris, 2014).

5.3.2.1 Financial Intelligence

While social media, and subsequent social network analysis (SNA), have the ability to see a picture of the human terrain, financial intelligence clarifies the relationships in terms of how resources are being proportioned for an organization to effectively achieve its objectives; in threat finance, financial structure is organizational policy. Financial intelligence, therefore, has the potential to provide the detailed relationships through which resources flow in a target organization (Cassara, 2015), complementing legacy census, and medical and open-source data, with real-time feeds to provide a comprehensive cyber intelligence picture (Fig. 5.4).

As shown in Fig. 5.4, cyber intelligence combines both historical and real-time "big data" into reporting that, if assembled correctly, may provide the best information product possible to operations and targeting professionals. Acquiring the information for these products sometimes warrants dangerous, special forces operations to assemble the information picture of interest.

2015 Special Forces Raid on ISIS Finance Minister

On the morning of May 16, 2015, American Special Forces mounted a raid deep into ISIS territory (Fig. 5.5), with the goal of apprehending ISIS' financial mastermind, the equivalent of capturing Al Capone's accountant.

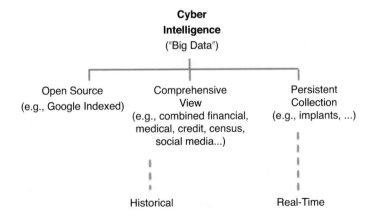

Fig. 5.4 Cyber intelligence ("big data")

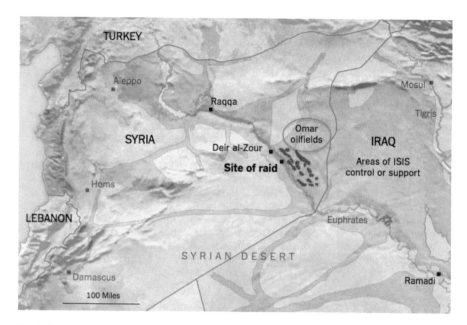

Fig. 5.5 Site of Abu Sayyaf raid (May 16, 2015)

As shown in Fig. 5.5, the Euphrates Oil Company at Al-Omar was the largest oilfield in Syria. With Al-Omar being key to ISIS' oil revenue, this is where Abu Sayyaf, ISIS' treasurer, kept his offices. Abu Sayyaf collected and distributed hundreds of millions of dollars in profits throughout the caliphate, money obtained from selling illicit oil, antiquities, slaves, and levying taxes on Christians (Malcolm Nance, 2017). While Abu Sayyaf and his crew were not taken alive, the ensuing sensitive site exploitation provided four to seven terabytes of computer data that gave US intelligence a treasure trove of information about the financial workings of ISIS (Helene Cooper, 2015).

5.3.2.2 ISIS and Census Information

As early as 2007, Al Qaeda in Iraq (AQI), ISIS' predecessor, kept detailed records of its fighters and population, documented in the Sinjar Report (Combatting Terrorism Center (West Point), 2007). This capability was expanded during the development of ISIS' internal security database, which included:

> The cyber equivalent to "intelligence 'keys to the caliphate,' a softcopy database, not linked to the Internet, containing the personal data of every man, woman, child, and slave in and under the control of ISIS, as well as the communications and financial links to its affiliates worldwide." (Malcolm Nance, 2017)

This database was constructed at the direction of Samir Abd Muhammad Al-Khlifawi, a.k.a., Haji Bakr, shadow commander of ISIS' military wing and its

Chief of Spies. An experienced spy under Saddam Hussein's Baathist Mukhabarat, Haji Bakr's vision of a police state impenetrable by foreign intelligence was managed by a large, mostly paper, filing system, with detailed records on ISIS members and the local population. This policy was implemented by Emirs, supported by lower-level lieutenants, to cross-reference data for all levels of ISIS society to maintain the organizational trust and loyalty of its subjects.

> "A second component of the ISIS personnel database was already well known to US intelligence since Americans had created it at a cost of billions during its occupation of Iraq. The Iraqi government implemented computerized databasing and biometrics of all citizens in 2005. It recorded personal information including digital photos, fingerprints, and even some retina scans of anyone who registered to vote, served in government or military, collected pensions, and received a passport or the new digital Iraqi national identity card. Additionally, anyone with a terrorist or criminal background or held in a detention center was entered into a national criminal database. During the battle for Mosul, ISIS had either rapidly seized or already had in their possession these databases. Additionally, the army and police biometric databases for each soldier were located in three major headquarters lost to ISIS in Mosul, Tikrit, and Ramadi. To this end, Haji Bakr employed everything he had learned as a Baathist to create a new intelligence and security network for the religious terror nation." (Malcolm Nance, 2017)

In addition to this massive paper database on the ISIS membership, a financial system for the collecting and disbursing of billions of dollars included a modern computer database used to enter each dossier, written by a member of the internal security arm, that was recorded to ensure that people collecting and spending money were closely monitored.

The ISIS financial database provided mappings of its organization, how it allocated resources (i.e., effectively policy prescription), along with a census data that included population details, including biometrics and family connections (Fig. 5.6).

As shown in Fig. 5.6, ISIS used US military efforts to provide the Iraqis with a census, along with ISIS personnel documentation, to develop the internal security database that was retrieved during the May 16, 2015, special forces mission.

5.4   Cyber Collection Processes

Capturing terabytes of enemy data, as in the ISIS case, is an ideal outcome. The work of cyber ISR, however, is often painstaking, laborious, and requires countless hours of collection, analysis, and processing to develop useful insights. While the collection cycle is relatively static (Table 5.1), the target sophistication and periodicity will affect specific techniques used (e.g., countering denial and deception), and collection cycles employed.

Targeting specific networks, their nodes, or known actors/activities, as shown in Fig. 5.3, is one example of cyber analysis for targeting. Social media, focused on human operations, is another key area for performing cyber ISR and analysis. Network analysis is a common application for finding out who is talking to whom. Social Network Analysis (SNA) is an implementation of network analysis for social media.

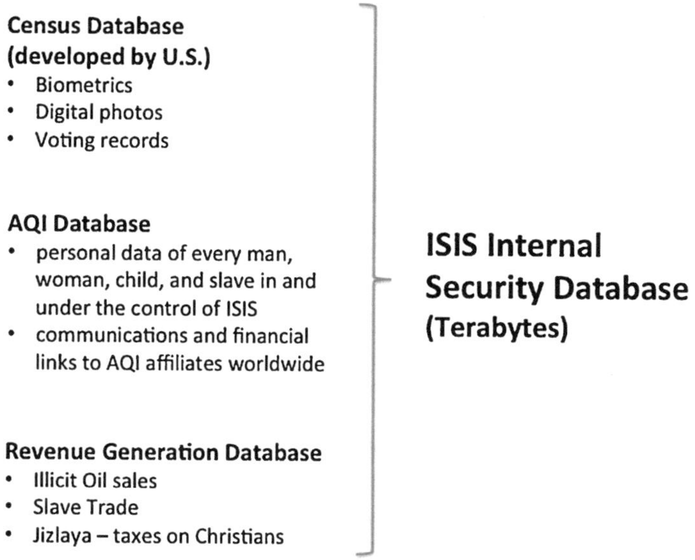

**Census Database
(developed by U.S.)**
• Biometrics
• Digital photos
• Voting records

AQI Database
• personal data of every man,
  woman, child, and slave in and
  under the control of ISIS
• communications and financial
  links to AQI affiliates worldwide

Revenue Generation Database
• Illicit Oil sales
• Slave Trade
• Jizlaya – taxes on Christians

**ISIS Internal
Security Database
(Terabytes)**

Fig. 5.6 ISIS big data platform (terabytes)

Table 5.1 Cyber collection cycle

Tasking	Collection		Processing, exploitation, analysis (PEA)
Understand organization	Online	Download key data (e.g., Equifax, Anthem, OPM …)	Terabytes for each downloaded database
Identify key players	Off-line	SOF Raid (Sect. 5.3.2.2)	4–7 TB
People	**Process/technology**		

5.4.1   Cyber and Social Network Analysis (SNA)

The US Army developed and implemented social network analysis (SNA) to fight improvised explosive device (IED) networks and target associated terrorists. This tool may offer solutions for performing cyber ISR and analysis (Timur Chabuk, 2018).

As a social science method, SNA provides mapping and quantification of human relationships. SNA really came to the forefront with the advent of social media platforms. Facebook, Snapchat, Twitter, and many other social media platforms provide rich data sets profiling human interactions with key words, hash-tags, and memes. In addition, SNA is useful for detecting influencers and influencees, by issue. This is especially important if the influencer is determined to be fostering social/ideological divides, as described in the Mueller report (Mueller, 2019).

The sheer volume of data in current social media platforms drives a requirement for automated techniques to intelligently collect and process the data to provide meaningful analytical reporting. Example uses of SNA for social media processing are described:

"… SNA can help detect subgroups within a broader network that may represent social/ideological divides and discern if certain members of the network are disproportionately targeted or elevated by bot activity. SNA techniques also can help identify which human voices are being promoted or trolled by bots the most (i.e., echo chambers). Through monitoring upticks and patterns in online behavior, analysts can better understand how actors such as Russia tie their online activity to tactical activities." (Timur Chabuk, 2018)

As discussed, the initial use of SNA for building people-centric connection diagrams really took off with the application to social media, which is key to the cyber collection cycle.

5.4.2 Cyber Collection Cycle

In performing social network analysis (SNA), the actual content of social media posts contains a wealth of often-overlooked data that is incorporated into more advanced cyber collections. For example, integrated multimedia content includes video, audio, text overlays, and even special effects that are available via relatively simple open-source search. However, the volume of social media data is often too large to examine manually, and requires automated algorithms to provide support.

The cyber collection cycle is key to combining the open source and directed collections to achieve the understanding (i.e., intelligence) for the objective of interest. One place to start is to evaluate the people, processes and technologies over the tasking, collection and processing /exploitation/analysis ISR cycle in terms of a technical collection conceptual model (e.g., T(C)PEAD; Table 5.1).

As shown in Table 5.1, cyber ISR initially focuses on the people of interest, similar to the cyber collection pyramid in Fig. 5.2. The T(C)PEAD steps that compose the collection cycle are parts of well-known conceptual models used for both information processing and tactical engagement.

5.4.3 Open-Source Intelligence (OSINT)

With the development of the Internet, and especially social media, open-source intelligence (OSINT) is increasingly valuable. For example, in "LikeWar" (Singer, 2018), Singer shows how Bellingcat used open-source intelligence to narrow down the suspects who shot down the Dutch airliner MH-17 on July 17, 2014, within 2 years. Bellingcat, a loosely organized research team, leveraged Russian social media (i.e., VKontakte [VK]) to combine family chatter, soldier postings, and equipment pictures to perform their analysis (Fig. 5.7).

As shown in Fig. 5.7, the MH-17 example, open-source intelligence provides the ability to use imagery of soldiers and equipment, online personal information tying soldiers to anti-aircraft units, and social media chatter to develop intelligence reports that would have required a nation state intelligence agency to develop only a decade ago. In addition, Fig. 5.8 provides an OSINT flow example, inspired by the MH-17 incident, to show how an example Internet-based OSINT report is generated.

Figure 5.8 shows an example of how a focused manual search is performed, using the Internet, to produce an OSINT report that led to a recent conviction of Russian military personnel (Kramer, 2019). This is an example of an OSINT report, on the downing of flight MH-17, that would have required a nation state intelligence agency to develop only a few years ago.

The online analysis of the downing of flight MH-17 included identifying the personnel, from technical skills involved in perpetrating an event, which is a common sense method of mapping the OSINT associated with a cyber event. This is a forensic approach for identifying key individuals based on specific skills required to make an event happen.

Cyber can also be used to answer broader questions about a population. For example, how many people work in defense and are members of the Ashley Madison dating site for married people (Zetter, 2015b)? How many of these defense workers using Ashley Madison to coordinate their extra-marital affairs have a US government security clearance (KOERNER, 2016)? And how many of these cleared players have financial issues that the Equifax breach (Newman, 2017) would show them to be susceptible to being approached by a foreign intelligence service? These are the kind of questions that can be answered by "big data" systems that result from combining the data heists over the last few years (e.g., Fig. 5.4 and Table 5.1).

5.4.3.1   Cyber Espionage: Big Data and Recent Downloads

Open-source collections, combined with directed collection, or espionage, have the ability to effectively provide a census of US security professionals:

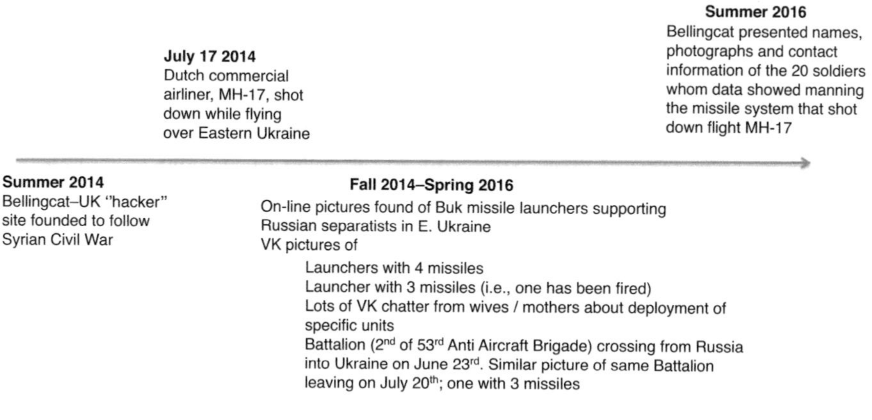

July 17 2014
Dutch commercial airliner, MH-17, shot down while flying over Eastern Ukraine

Summer 2016
Bellingcat presented names, photographs and contact information of the 20 soldiers whom data showed manning the missile system that shot down flight MH-17

Summer 2014
Bellingcat–UK "hacker" site founded to follow Syrian Civil War

Fall 2014–Spring 2016
On-line pictures found of Buk missile launchers supporting Russian separatists in E. Ukraine
VK pictures of
Launchers with 4 missiles
Launcher with 3 missiles (i.e., one has been fired)
Lots of VK chatter from wives / mothers about deployment of specific units
Battalion (2nd of 53rd Anti Aircraft Brigade) crossing from Russia into Ukraine on June 23rd. Similar picture of same Battalion leaving on July 20th; one with 3 missiles

Fig. 5.7 Open-source intelligence example: Bellingcat and MH-17 attribution to Russia's 53rd anti-aircraft brigade

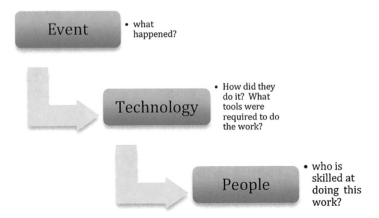

Fig. 5.8 OSINT flow (event - > technology - > people)

> "… Privately, American intelligence officials concluded that the Chinese were assembling a giant database of who worked with whom, and on what, in the American national security sphere, and were applying "big data" techniques to analyze the information. The C.I.A. could not move some officers to China, for fear their cover had been blown. Publicly, Obama administration officials offered millions of Americans credit protection for a few years in the wake of the data breach — as if Mr. Xi's agents were looking for credit card numbers." (Sanger, 2018)

OSINT, as a form of cyber intelligence, has therefore come into its own, threatening even a first-rate intelligence service with the ability to identify clandestine officers. This is an example of combining OSINT with espionage, directing collection to target specific information items that lead to uncovering a target demographic.

5.4.4 Directed Collection

Another example of people-focused collection includes operation Aurora (Carlin, 2018), where Chinese agents leveraged existing US law enforcement taps on Internet accounts (e.g., Google, Microsoft, …) in order to see who US authorities were interested in and the type of information that they were collecting. This approach, in line with the T(C)PEAD approach shown in Table 5.1, has the foreign intelligence service (i.e., China) using US law enforcement to determine the targets and tasking and then using their own resources for the collection, processing, and exploitation.

5.4.4.1 Post-Event Forensics

In addition to the tracking of people from a physical event as shown in Fig. 5.8, this same process might be used from a cyber event, as well. This includes understanding a network intruder's objectives (i.e., from collections and areas of interest), discerning identity, and following the intruder back to a home location or machine. Similar to criminal forensics, this is the approach used by the more famous cyber security firms (Table 5.2).

As shown in Table 5.2, each of the cyber security firms uses forensics to discern the attack type, tools used, and the organization/players involved. Some would argue that Table 5.2's firms have made the largest contributions to cyber understanding over the last decade. For example, it was only a few years ago that experts proclaimed that attribution in cyber was not possible (Newman, 2016).

One of the common ways for security firms to categorize threat actors is by determining who they target, the type of information that they acquire, and the tools that they use to access this information. One way to categorize these threat actors is to designate them as Advanced Persistent Threats (APTs), with a number that corresponds to when the APT was discovered. In Chap. 2, we briefly discussed how MITRE CARET (MITRE) is used to characterize APTs over the cyber threat framework attack life cycle. MITRE's CARET is an additional tool for characterizing APT groups (i.e., characteristic attack steps and tools used at each step).

5.4.5 Manual vs. Automated Search

The current generation of Internet users expects automated search. "Googling" any topic of interest brings back a myriad of results, bringing an everyday user to a level that would compete with a trained researcher's capability in her father's time. An unsaid caveat to this use of information as an appliance is that an organization has taken the time to index the information so that it instantaneously appears. This painstaking research, indexing the available Internet knowledge, is not done for the sensitive security information associated with cyber ISR and analysis.

Table 5.2 Cyber security firms

Cyber security firm	Primary role	Key capability	Famous for
FireEye (FireEye)	Forensics	Analysis	Ongoing: Identifying significant Advanced Persistent Threat (APT) groups
Crowdstrike (CrowdStrike)	Forensics	Analysis	2016 Democratic Party network compromise by Russian actors
Dragos (Dragos)	Industrial control systems (ICS)	ICS network understanding	2014 Ukrainian ICS attack investigation
Symantec (Symantec)	Forensics	Analysis	2012 Stuxnet investigation

5.4.5.1   Defensive Cyber Operations (DCO)

While we cannot completely rely on automated search for cyber analysis and target-ing applications, automated searches are useful to provide cues to inform research-ers of "where to look" for recently posted information of interest. For example, automated search capabilities are provided by each of the public search engines and social media platforms, usually with additional scripting capability to refine search fields, in order to complement the hard work that analysts do during the forensics and analysis phases of cyber ISR (Fig. 5.9).

As shown in Fig. 5.9, similar to Fig. 5.8, post-event forensics provide the analyst with a starting point to use both manual and automated search/evaluation in order to understand the event and the people/organizations behind the event. In addition, as shown in Fig. 5.9, the post-event forensics and person/organization identification require human processing in order to analyze the data and determine the type and responsible party for a network anomaly. This is even more challenging when pro-actively analyzing the data in the case of "offensive" cyber operations.

5.4.5.2   Active Cyber Operations

For cyber ISR, active operations include developing an understanding of target net-works and extracting usable information, often called computer network exploita-tion (CNE) (Zetter, 2016). Similar to the labor-intensive example of performing forensics (Fig. 5.9), it is a challenge to process the terabyte downloads (e.g., Abu Sayyaf raid in Sect. 5.3.2). With a single person able to read one page per minute, for 6 hours per day, it would take over 10,000 years to read 1 TB of text (Table 5.3).

As shown in Table 5.3, the current "standard" terabyte-level downloads will challenge even the largest of collection organizations. Table 5.3 also brings out the need for using technology, as a labor-saving approach, to complement both passive and active cyber ISR (Warner, 2017).

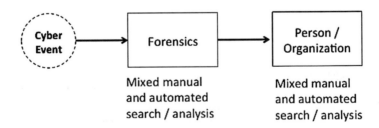

Fig. 5.9 Cyber ISR's combination of manual and automated search

Table 5.3 Processing-time example for cyber collection

Pages	Characters	Bytes	Time to read (human estimate)
1 page	1000 char	1 KB	1 min
10	10,000	10	10 min
100	100,000	100	100 min (1.7 hr)
1000	1,000,000	1000 (1 MB)	17 hours (2.8 days)
1,000,000	1,000,000,000	1000 MB (1 GB)	11.6 years
1,000,000,000	1,000,000,000,000	1,000,000 GB (1 TB)	11,574 years

5.5 Technology: Passive and Active Cyber ISR

Passive and active cyber ISRs, as the terms are used here, include passive ISR being used to collect aggregate data sources, similar to the downloading of sensitive records (e.g., Fig. 5.5), with active cyber ISR requiring more skills to develop sensitive accesses, exploits, and collections on a target of interest. As introduced in Sect. 1.3.1, passive ISR, for example, takes the form of web storage sites like Wikileaks, Open Secrets, …, where hackers managed to access, and disclose, sensitive information that was already aggregated based on the target organization's uses of the data. Follow-on cyber effects in using the collected information might be to smear an organization's leadership. For example, the release of the Podesta emails during the 2016 US presidential election (Mueller, 2019) aired issues internal to the DNC strategy for winning the election, which the DNC then needed to spend time explaining, likely diverting energy from the campaigning required to win the election.

Forcing the DNC to take actions outside the scope of their purpose of winning the election is effectively controlling their actions. In military terms, managing an adversary's scheme of maneuver is strategic success. In addition, traditional tools for controlling an adversary's scheme of maneuver usually include capital ships, earthworks (e.g., trenches), artillery/fires, massing of forces, or the persistent use of aerial bombardment—all very expensive options. Changing an organization's focus and maneuver priority via outing their e-mails is a very inexpensive method of achieving the same objective.

5.5.1 Passive Reconnaissance: Voluntary Reporting Sites (E.g., Wikileaks)

As discussed throughout Chap. 4, covertly extracted information, especially when reframed, is a potent technique for putting an opponent off balance. Recently, this has been performed via WikiLeaks, Open Secrets, and a variety of other "news" sources that exist based on voluntary information alone, for strategic effect. As shown in the 2016 US presidential election, voluntary submissions of Democratic

National Committee (DNC) emails, just before the Democratic convention, was timed so as to change the discussion from core issues to internal organization debates. For example, the revelations prompted the resignation of DNC chair Debbie Wasserman Schultz before the Democratic National Convention (Martin, 2016). The DNC issued a formal apology to Bernie Sanders and his supporters "for the inexcusable remarks made over email" that did not reflect the DNC's "steadfast commitment to neutrality during the nominating process" (Reuters, 2016).

Outing attacks on the DNC during the 2016 US presidential election were only the most recent uses of cyber to put an opponent off balance. As early as 1991, the pre-Internet was used to coordinate counter rallies to the Glasnost Putch (Andrei Soldatov, 2016), which had a goal of halting the opening up of Russia and returning it to the closed society of the former Soviet Union. With conventional phone systems unavailable, the early Internet was one of the few communications media available to coordinate mobilization of Glasnost counter-Putch movement.

While the Glasnost Putch failed due to pre-Internet coordination of counter rallies in 1991, only 20 years later the Internet became the center of gravity for the Arab Spring. Festering grievances, combined with Wikileaks' publishing of US diplomatic cables that showed tepid US government regime support (e.g., Tunisia), exploded into demonstrations across North Africa, bringing down governments in Tunisia, Libya, and Egypt. Similar protests emerged in Syria, causing the government to lose footing in key locations that soon became synonymous with ISIS.

The Arab Spring included the outing of US State Department cables, secret diplomatic communications that included US policy positions favoring the military over the regime (Coles, 2011). Wikileaks became famous for hosting the site with these once secret communications. In addition, Wikileaks exerted implicit control over the ops tempo of protestor advances with the posting of releases; an example of information velocity paralleling the physical advance of protestors in key capitols across the Maghreb. As discussed in Chap. 4, mainstream media organizations (e.g., *Guardian*) picked the Wikileaks stories and further disseminated the developing stories.

The Arab Spring is a good example of the transition from social commentary, through cyber, to the development of strategic effects—regime change for governments across the Middle East. Similarly, the Panama Papers, which also included Wikileaks and mainstream media organizations, was more about targeting the rich and powerful. Overall, 11.5 million documents, stored at Panamanian law firm Mossack Fonseca, included the names of highly placed Russian and Chinese political leaders involved in off-shore corporations designed to hide money and evade taxes (Bernstein, 2017; Bernstein & Soderbergh, 2019). While this outing of world and business leaders arguably had less of a strategic effect than the Arab Spring, the Panama Papers outed information that people had spent a lot of time and effort to keep private.

While the Panama Papers showed the private finances of world leaders and celebrities, the Arab Spring and counter to the Glasnost Putch provide examples of the mobilizing effects that Internet-based communications (i.e., cyber) can provide (Table 5.4).

Table 5.4 Internet dissemination and strategic political examples (1991 Glasnost Putch, 2011 Arab Spring, and 2016 Panama Papers)

Event	Open-source news effects	Internet use
1991: Failure of the post Glasnost Putch	Objective was to remove Gorbachev from power and revert to the Soviet Union (Russia) (Andrei Soldatov, 2015)	In 1991, the programmers used Russia's nascent Internet (i.e., USEnet and emails) to share information about troops movements, and coordinate counter-Putch rallies (Andrei Soldatov, 2016).
2011: The Arab Spring (Middle East)	Beginning in Tunisia, mass demonstrations against unelected governments across North Africa and throughout the Middle East	WikiLeaks releases accelerated the information velocity of the social media postings and anti-government coordination in key Arab capitals (Council on Foreign Relations)
2015: Panama Papers	Release of off-shore account information results in the downfall of Pakistani President Nawaz Sharif, Icelandic President Gunnlaugsson, and the embarrassment of multiple International government officials (Bernstein, 2017)	WikiLeaks releases broadened the reach of information concerning senior leadership involvement in off-shore accounts; China's sensors took "Panama" off of the official China Internet for a few days (Singer, 2018)

As shown in Table 5.4, cyber targeting, via the Internet and through its horizontal dissemination capabilities, has provided an ability to achieve strategic information effects from the outset. The early Internet's use to counter the 1991 Glasnost Putch, for example, was only the first example of the strategic effect that the Internet can have on maintaining, or changing, a ruling regime. The 2011 Arab Spring and 2015 Panama Papers provide further evidence of the kinds of regime-changing effects that the Internet can facilitate.

While passive ISR, the extracting and outing of sensitive information, can have powerful effects on an adversary's ability to maneuver, active ISR operations differ, in that they can border on "mission impossible." Cifford Stoll's outing of the KGB—*Komitet Gosudarstvennoy Bezopasnosti* (Committee for State Security)—trying to extract US military "Star Wars" missile defense program information in 1980s from pre-Internet government computers was described in *"The Cuckoo's Egg"* (Stoll, 2005). Similarly, the more recent Stuxnet operation to slow down the Iranian nuclear weapons program provided an example of how technology has matured for remote reconnaissance (Zetter, 2014). These are examples of active cyber ISR operations.

5.5.2 Active ISR: Bots and Searching the Net

Active cyber ISR includes the focused collections that lead up to controlling, or affecting, a specific process when doing a cyber integrity attack. For example, the collections and planning used to target the Iranian centrifuge computer network and component technologies, more commonly called the Stuxnet operation, included a detailed reconnaissance of the system used to refine low-concentration uranium into bomb-grade material. Active cyber ISR therefore provided insight concerning the computer network, its configuration, and the supervisory control and data acquisition system (SCADA) coordinating the nuclear refinement process. This active cyber ISR collection was implemented via bot searches to collect engineering drawings, building schematics, and operational technology specifications.

While the Internet can get very complex, in terms of overall mapping (Cheswick), sites of interest will usually have associated people, policies, processes, and technologies that provide reliable signatures of their operations. This kind of active analysis was a key part of the Stuxnet campaign (GATES) (Fig. 5.10).

As shown in Fig. 5.10, the many steps to successfully execute Stuxnet included phases that spanned from open-source assessment to the development of specific technologies to perform the mission (Table 5.5).

As shown in Fig. 5.10, and detailed in Table 5.5, the majority of the tools, and effort, were applied in the Cyber ISR portion of mapping the network and operations of the Iranian nuclear refinement facility. While SPE, Gauss, and possibly

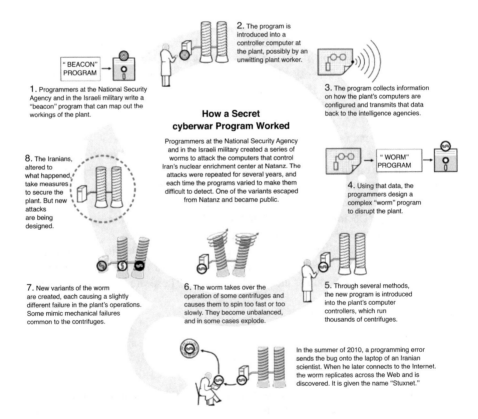

2. The program is introduced into a controller computer at the plant, possibly by an unwitting plant worker.

" BEACON" PROGRAM

1. Programmers at the National Security Agency and in the Israeli military write a "beacon" program that can map out the workings of the plant.

3. The program collects information on how the plant's computers are configured and transmits that data back to the intelligence agencies.

How a Secret cyberwar Program Worked

Programmers at the National Security Agency and in the Israeli military created a series of worms to attack the computers that control Iran's nuclear enrichment center at Natanz. The attacks were repeated for several years, and each time the programs varied to make them difficult to detect. One of the variants escaped from Natanz and became public.

8. The Iranians, altered to what happened, take measures to secure the plant. But new attacks are being designed.

" WORM" PROGRAM

4. Using that data, the programmers design a complex "worm" program to disrupt the plant.

7. New variants of the worm are created, each causing a slightly different failure in the plant's operations. Some mimic mechanical failures common to the contrifuges.

6. The worm takes over the operation of some centrifuges and causes them to spin too fast or too slowly. They become unbalanced, and in some cases explode.

5. Through several methods, the new program is introduced into the plant's computer controllers, which run thousands of centrifuges.

In the summer of 2010, a programming error sends the bug onto the laptop of an Iranian scientist. When he later connects to the Internet, the worm replicates across the Web and is discovered. It is given the name "Stuxnet."

Fig. 5.10 How a secret cyber program worked (Stuxnet) (GATES)

Table 5.5 Stuxnet components and descriptions (Zetter, 2014)

Bot name	Type	Exploit	Size	C2	Life span	Misc
Stuxnet (Fig. 5.10, Steps 5 ~ 7)	Siemens targeted malicious code	.LNK (thumb drive)	500 KB 100,000 machines			
Flame (Fig. 5.10, Steps 1 ~ 3)	Surveillance and reconnaissance code		≥ 10,000 installations 20 MB (> 650,000 lines of code [loc])	≥ 80 domains		
SPE (mini-flame) (Fig. 5.10, Steps 1 ~ 3)	Info stealer Back door					
Gauss (Fig. 5.10, Steps 1 ~ 3)		- USB thumb drive delivery				Stored in Palida Narrow font
Duqu (Tilde-d Platform) (~DQ) (Fig. 5.10, Steps 1 ~ 3)	Remote Access Trojan (RAT) Back door Dropper Recce	.LNK exploit (USB flash drive) TrueType font (MS Word) Zero Day	≤ 36 installations	AES JPEG comms	≤ 36 days on infected network	TrueType font-parsing engine

Table 5.6 Stuxnet T(C)PEAD

Tasking	Collection		Processing, exploitation, analysis (PEA)	Dissemination
Siemens controller Network access	Off-line	Open-source research	Tool specific Open source	Information products
	Online	Flame/SPE GAUSS Duqu		

other tools were used in performing this cyber ISR, Duqu and Flame were the primary reconnaissance tools used to gather the information to develop the Stuxnet worm that performed the integrity attack (Table 5.6).

As shown in Table 5.6, even the most advanced bot collection techniques can be described with known conceptual models for technical intelligence collection (e.g., from Table 5.1)

5.5.2.1 Duqu and Flame

According to Zetter (Zetter, 2014), Kaspersky (Kaspersky, 2014) believes that Flame and its development platform were developed by the United States, while Israel created Duqu and the Tilde-d platform. The two respective platforms were both used to build their portions of Stuxnet (Fig. 5.11).

As shown in Fig. 5.11, there are big differences between Duqu and Flame in terms of file size, life span, and distribution. While Duqu and Stuxnet are alleged to be based on the same source code (Matrosov, 2012; Boldizsar Bencsath, 2012), FLAME is an order of magnitude larger, at 20 MB.

Duqu (~DQ)

Duqu got its name from the Laboratory of Cryptography and System Security (CrySyS) at Budapest University of Technology and Economics due to its using "~DQ" as a prefix for the files it creates (Symantec, 2011). Duqu 1.0 was essentially a remote-access Trojan, or RAT, providing a back door to give attackers a persistent foothold on targeted machines. RATs are usually downloaded invisibly with a user-requested program such as a game, or downloaded as an email attachment (Rouse). After compromising a host system, the attacker can use it to distribute RATs to other vulnerable computers and establish a botnet (Fig. 5.12).

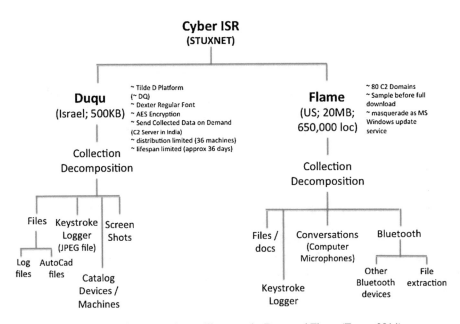

Fig. 5.11 Cyber reconnaissance and surveillance tools: Duqu and Flame (Zetter, 2014)

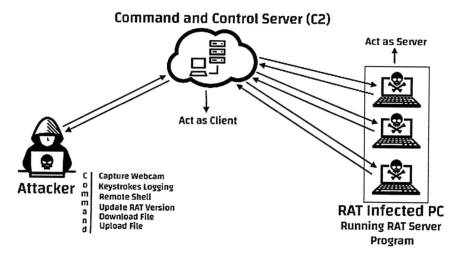

Fig. 5.12 Attackers, remote-access Trojan (RAT) infected PCs, and the command and control (C2) system

As shown in Fig. 5.12, once the back door was installed, Duqu contacted a command-and-control server (U.S. Department of Homeland Security (DHS), 2017), from which the attackers would download additional modules to give their attack code more functionality, such as key stroke logging (Zetter, 2014).

Duqu is believed to have provided reconnaissance to selectively probe networks, in search of data (e.g., log files), that would lead to a better understanding of the target environment (Boldizsar Bencsath, 2012a, b) (Zetter, 2014).[1] For example, the log files of interest were searched specifically for the AutoCAD files, or blueprints, related to industrial control systems, computer networks, and plant floor machinery for Stuxnet's ultimate target. In addition, Duqu was found to (Zetter, 2014):

- Record passwords via a combination keystroke logger/infostealer
- Steal documents (e.g., AutoCAD files)
- Take screenshots
- Catalog devices/machines in the overall network architecture
- Wait until it was summoned by command and control systems in India to send its collected data

Duqu infected other machines only if the attackers sent instructions from their command server to do so (INFOSEC, 2019). In addition, Duqu was also much stealthier in communicating with its command servers than Stuxnet.[2] Duqu

---

[1] Both the encryption routines and the six kernel hooks (i.e., machine functions that the malware used to successfully attack) used by Duqu were identical to those of Stuxnet, providing evidence for their common provenance.

[2] Team Duqu also stored some of the scripts for controlling the operation at other locations, rather than on the command servers, so that anyone who seized control of these front-end servers could

communication was encrypted and concealed. Using a strong encryption algorithm, AES, prevented anyone from reading Duqu transmissions. Further, Duqu communications were masked in a .JPEG image file for concealment (U.S. Department of Homeland Security (DHS), 2017). This focus on security led to a near successful attempt to infiltrate Kaspersky (INFOSEC Institute, 2018). Duqu's tight control was also shown by its infecting only 3 dozen machines, much less than the approximate 100,000 machines that Stuxnet was found in (Zetter, 2014). Additional security was due to Duqu's limited life span, often deleting all traces of itself after 36 days on an infected machine (U.S. Department of Homeland Security (DHS), 2017).[3]

While it is well known that Stuxnet used the .LNK exploit (Goodin, 2017; NVD, 2019), embedded on the USB flash drive, to drop its malicious cargo, Duqu was revealed to use a Microsoft Word attachment as a dropper (U.S. Department of Homeland Security (DHS), 2017) when infecting a Hungarian software certification authority (Zetter, 2014). Symantec found this dropper to be exploiting a zero-day buffer overflow vulnerability in the TrueType font-parsing engine for Windows (Symantec, 2011).

At 20 MB, and an estimated 650,000 lines of code (loc), Flame is much bigger than the 500 KB Duqu (Constantin, 2013). Flame originally was noticed for its masquerading as a Windows update service (Sotirov). Upon inspection, however, Flame was believed to be a key member of the Stuxnet family, principally as a reconnaissance tool.

Flame

While Stuxnet was a large 500 kilobytes when compressed, Flame was at least 20 megabytes with all its components combined (Zetter, 2014), consisting of more than 650,000 lines of code that consisted of multiple modules that:

- Siphoned documents from infected machines
- Recorded keystrokes and captured screenshots every 15–60 seconds
- Eavesdropped on conversations by turning on the computer's internal microphone
- Used the computer's Bluetooth to extract files and data from discoverable smartphones and other Bluetooth-enabled devices in the area

In addition, the infrastructure supporting Flame included over 80 domains operating at command servers in Germany, the Netherlands, Switzerland, and other countries by which the attackers controlled infected machines and downloaded documents from them (Lee, 2012). Fake identities were used to register these domains, purchased with pre-paid credit cards so that they could not be traced. Kaspersky, upon discovering these domains, redirected traffic for about 30 of them

---

not seize and examine the scripts to determine what Duqu was doing.

[3] Researchers eventually uncovered multiple versions of Duqu, with varying removal times. In some cases, Duqu removed itself after 30 days; in other versions it was 36 days. One case was reported to last 120 days before deletion (Zetter, 2014).

to their own "sinkhole," collecting stolen files being sent back from Iran (Zetter, 2014).

Flame was controlled, in its machine placement, not spread randomly. In addition, Flame did sample downloads at its installed locations of about 1 KB, used to determine whether there were files of interest for further downloading.

As an example of performance, the Kaspersky team found that one command server (out of approximately 80) communicated with 5000 infected machines in just 10 days. Realizing that the malware had been in the wild since 2007 or 2008, the total number of installations had to be in the tens of thousands, building up innumerous stores of cached collections. One Malaysian server was found to contain 5.7 GB believed to be collected over a 10-day period (Zetter, 2014). If this was a consistent download rate, it was estimated that Flame produced 500 MB per day. In addition, Flame was found configured to communicate with four pieces of malware, identified as SP, SPE, FL, and IP (GReAT, 2012). Each of these components was designed to communicate with the command server using a custom protocol that the attackers had written (Zetter, 2014).

Flame's capabilities, at 500 MB per day, while amazing from a surveillance standpoint compared to collection rates only a generation ago, are a challenge to process (Sect. 5.4.3). Thinking about the ability to collect this much information provides context on the "big data" discussion of Sect. 5.3. Understanding both the scale of data, the terabytes available with cyber, and the detail (e.g., turning on cameras and microphones to get real-time operator feeds (Zetter, 2014)) adds both a breadth and a depth not previously imagined in an intelligence collection world designed to process static images and episodic human communications.

5.6 Summary

While cyber provides, through active collection, the "big data" (Figs. 5.4 and 5.6) feeds, via remote access (e.g., Duqu and Flame: Fig. 5.11), that exponentially expand on what a human spy could provide only a generation ago, processing this data remains a challenge. As shown in Sect. 5.3, today's common terabyte downloads require automated tools to analyze and process the data. The funnel in Fig. 5.1 is therefore a great depiction of how all these data are winnowed down into finished intelligence.

Collecting information stores are the cyber ISR examples that we looked at in Sect. 5.5. The acquisition of ISIS' population database (i.e., with biometrics, financial supporters, and operators) is a great example of "cyber ISR," providing one of the most comprehensive enemy order of battle packages ever developed (Fig. 5.6).

Leveraging the people, processes, and technologies that make up a cyber target includes using Social Network Analysis (SNA) to provide a first step, in the way of human network description, and then using process and technology signatures to validate the identity of specific cyber actors when doing analysis and targeting— Fig. 5.2's people->processes->technology. In addition, this can be modified slightly and reversed, going from event->processes->people (Fig. 5.8), and combined with social media, to discern, down to the individual operators, how a foreign actor shot down a Dutch airliner (Fig. 5.8). Bots, "big data," and social media add quantity and speed dimensions that uniquely characterize Cyber ISR.

Finally, going back to the original questions posed at the beginning of the chapter:

1. What is the overlap between cyber and human intelligence?

    While cyber looks similar to human intelligence in some key ways (e.g., social network analysis described in Sect. 5.3), the scale of cyber sets it apart from any of the other intelligence disciplines. For example, in Sect. 5.3 we introduced the idea that cyber was used to produce 4–7 terabytes of data during the 2015 Abu Sayyaf raid. This volume of data would have been a challenge for an entire spy organization to collect only a generation or so ago, and can now be done remotely, possibly through a bot (Sect. 5.5).

2. How different is cyber from open-source intelligence (OSINT)?

    In Fig. 5.8, we looked at a use of cyber to perform OSINT. In this example, we saw how the scope of cyber research, from news reporting to social media, helped an open-source firm to pinpoint Russian military operatives involved in the 2014 downing of the Dutch airliner MH-17. This is a use of cyber to perform open-source analysis.

3. What does the cyber intelligence cycle look like?

    Table 5.1 provides the generic cyber intelligence cycle, with Fig. 5.8 providing a generalization of how cyber-based systems are used for open-source analysis (i.e., Fig. 5.7).

4. What are the key technologies in cyber intelligence?

    In Chap. 4, we looked at how cyber operators obfuscate and multiply their online behavioral operations (Figs. 4.6 and 4.7). In this chapter, we looked at the use of bots for performing reconnaissance (Figs. 5.11, 5.12, and 5.13; Table 5.6) to take control of an adversary's system. Cyber technologies therefore include available data, operational personas, and automated entities for the collection and control of adversary systems.

5. How did cyber analysis and targeting play a role in counter-ISIS operations (e.g., counter terror)?

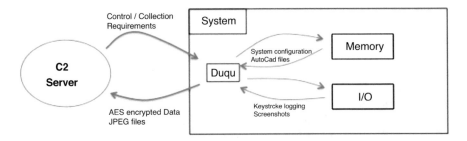

Fig. 5.13 Duqu and system collection

Cyber was used over the life cycle of counter-ISIS operations. For example, the 2015 Abu Sayyaf raid (Fig. 5.5) yielded 4–7 TB of organization data (Fig. 5.6), making cyber a key element in early counter-ISIS operations.

6. How does cyber help in countering nation state actors (e.g., countering the Iranian nuclear program)?

As shown in Fig. 5.10, Stuxnet was used to slow down the Iranian nation state nuclear program for approximately 2–3 years (Hayden, 2016). Evidence shows that a few different tools (e.g., Fig. 5.11) were used for the remote surveillance and control of the Iranian nuclear centrifuges of interest.

7. How does cyber compare as a means of technical intelligence?

The potential for collecting over 500 MB of data from an automated "bot" provides for an unprecedented scale of intelligence collection. Performing these large-scale collections (e.g., Fig. 5.6) introduces new technical issues in processing the data (i.e., Sect. 5.4). A saving grace is that the overall collection cycle leverages tried and true conceptual models (e.g., T(C)PEAD; Tables 5.1 and 5.6) to pinpoint where technical development can help winnow massive data collections to usable knowledge, in order to support intelligence-based decision-making.

Bibliography

Akamai. (n.d.). *Global traffic management*. Retrieved 11 14, 2018, from Akamai: https://www.akamai.com/us/en/multimedia/documents/product-brief/global-traffic-management-product-brief.pdf

Anderson, N. (2010, 11 7). *How China swallowed 15% of 'Net traffic for 18 minutes*. Retrieved 11 13, 2018, from Ars Technica: https://arstechnica.com/information-technology/2010/11/how-china-swallowed-15-of-net-traffic-for-18-minutes/

Andrei Soldatov, I. B. (2015). *The red web - the struggle between Russia's digital dictators and the new online revolutionaries*. Public Affairs.

Andrei Soldatov, I. B. (2016, 8 19). *An act of courage on the Soviet Internet*. Retrieved 11 29, 2018, from SLATE: https://slate.com/technology/2016/08/the-1991-soviet-internet-helped-stop-a-coup-and-spread-a-message-of-freedom.html

Bernstein, J. (2017). *Secrecy world - inside the Panama papers investigation of illicit money networks and the global elite*. New York, NY, USA: Henry Holt and Company.

Boldizsar Bencsath, G. P. (2012, 11 6). The Cousins of Stuxnet: Duqu, Flame, and Gauss. Retrieved 5 2, 2020, from Research Gate.

Boldizsar Bencsath, G. P. (2012a, 11 6). *Future internet*. Retrieved 11 30, 2018, from future internet. www.mdpi.com/1999-5903/4/4/971/pdf

Boldizsar Bencsath, G. P. (2012b, 11 6). *The cousins of Stuxnet: Duqu, flame, and gauss*. Retrieved 5 2, 2020, from Research Gate: https://www.researchgate.net/publication/263606038_The_Cousins_of_Stuxnet_Duqu_Flame_and_Gauss

Carlin, J. P. (2018). *Dawn of the code war – America's battle against Russia, China and the rising global cyber threat*. PublicAffairs.

Cassara, J. A. (2015). *Trade-based money laundering: The next frontier in international money laundering enforcement*. John Wiley and Sons.

Cheswick, W. (n.d.). *The internet mapping project*. Retrieved 11 28, 2018, from the internet mapping project: http://cheswick.com/ches/map/

Clarke, R. A., Robert, K. (2012). *Cyber war: The next threat to National Security and what to do about it*. Ecco.

Coles, I. (2011, 3 15). *Assange: Wikileaks' cables spurred Arab uprisings*. Retrieved 5 2, 2020, from Reuters: https://www.reuters.com/article/us-britain-assange/assange-wikileaks-cables-spurred-arab-uprisings-idUSTRE72E9LO20110315

Combatting Terrorism Center (West Point). (2007). *Al Qaeda's foreign fighters in Iraq - a first look at the Sinjar records*. Retrieved 11 15, 2018, from DTIC: http://www.dtic.mil/dtic/tr/fulltext/u2/a474986.pdf

Constantin, L. (2013, 5 28). *Researchers identify Stuxnet-like malware called 'Flame'*. Retrieved 5 4, 2020, from Computerworld: https://www.computerworld.com/article/2505035/researchers-identify-stuxnet-like-malware-called%2D%2Dflame-.html

Correlates of War Project. (n.d.). Retrieved 4 29, 2020, from https://correlatesofwar.org/

Council on Foreign Relations. (n.d.). *The Arab uprisings five years on*. Retrieved 11 28, 2018, from https://www.cfr.org/interactives/arab-uprisings-five-years?gclid=EAIaIQobChMIiobysPal2wIVCyaGCh1-Kw7VEAAYASAAEgL6-PD_BwE#!/?cid=ppc-Google-grant-Arab_Spring_Timeline-121715

CrowdStrike. (n.d.). Retrieved 11 29, 2018, from https://www.crowdstrike.com/

Dragos. (n.d.). *Dragos*. Retrieved 29 2018, 11, from https://dragos.com/

Economist. (2007, 7 12). *A world wide web of terror*. Retrieved 5 1, 2020, from Economist: https://www.economist.com/briefing/2007/07/12/a-world-wide-web-of-terror

FireEye. (n.d.). *FireEye*. Retrieved 29 2018, 11, from https://www.fireeye.com/

Gates, G. (n.d.). *How a secret cyberwar program worked*. Retrieved 12 5, 2018, from New York Times: https://archive.nytimes.com/www.nytimes.com/interactive/2012/06/01/world/middleeast/how-a-secret-cyberwar-program-worked.html

Goodin, D. (2017, 4 20). *Windows bug used to spread Stuxnet remains world's most exploited*. Retrieved from Ars Technica: https://arstechnica.com/information-technology/2017/04/windows-bug-used-to-spread-stuxnet-remains-worlds-most-exploited/

GReAT. (2012, 9 17). *Full Analysis of Flame's Command & Control servers*. Retrieved 5 5, 2020, from SECURELIST: https://securelist.com/full-analysis-of-flames-command-control-servers/34216/

Harris, S. (2014). *@War - the rise of the military-internet complex*. Eamon Dolan/Mariner Books.

Hayden, M. V. (2016). Playing to the edge: American intelligence in the age of terror. Peguin.

Heckman, K. E. & Frank J. S. (2015a). Cyber Counterdeception: How to Detect Denial & Deception (D&D). In P. S. Sushil Jajodia. *Cyber Wrfare: Building the scientific foundation*. Springer.

Heckman, K. E., Frank J. S. (2015b, April). Denial and deception in cyber defense. *Computer, 48*(4), 36–44.

Helene Cooper, E. S. (2015, 5 17). *ISIS official killed in U.S. Raid in Syria, Pentagon says*. (N. Y. Times, Producer, & New York Times) Retrieved 11 10, 2018, from New York Times: https://www.nytimes.com/2015/05/17/world/middleeast/abu-sayyaf-isis-commander-killed-by-us-forces-pentagon-says.html

INFOSEC. (2019, 8 25). *Duqu 2.0: The most sophisticated malware ever seen [Updated 2019]*. Retrieved 5 4, 2020, from INFOSEC: https://resources.infosecinstitute.com/duqu-2-0-the-most-sophisticated-malware-ever-seen/#gref

INFOSEC Institute. (2018, 9 1). *Duqu 2.0: The most sophisticated malware ever seen [Updated 2018]*. Retrieved 11 30, 2018, from https://resources.infosecinstitute.com/duqu-2-0-the-most-sophisticated-malware-ever-seen/#gref

Jake Bernstein, S. Z. (Writer), & Soderbergh, S. (Director). (2019). *The Laundromat* [Motion Picture].

Kaspersky. (2014, 11 11). *Stuxnet patient zero: First victims of the infamous worm revealed*. Retrieved 5 2, 2020, from Kaspersky: https://www.kaspersky.com/about/press-releases/2014_stuxnet-patient-zero-first-victims-of-the-infamous-worm-revealed

Keith Collins, S. F. (2018, September 4). *Can you spot the deceptive Facebook post?* Retrieved November 15, 2018, from New York Times: https://www.nytimes.com/interactive/2018/09/04/technology/facebook-influence-campaigns-quiz.html

Kime, B. P. (2016). *Threat intelligence-planning and direction.* Retrieved 37, 2019, from SANS: https://www.sans.org/reading-room/whitepapers/threats/threat-intelligence-planningdirection-36857

Koerner, B. I. (2016, October 23). *Inside the cyberattack that shocked the US government.* Retrieved September 7, 2018, from Wired: https://www.wired.com/2016/10/inside-cyberattack-shocked-us-government/

Kramer, A. (2019, 6 19). *Four to face murder charges in downing of Malaysia airlines flight 17.* New York Times.

Launius, S. (2019). *Evaluation of comprehensive taxonomies for information technology threats.* Retrieved 3 7, 2019, from SANS: https://www.sans.org/reading-room/whitepapers/threatintelligence/evaluation-comprehensive-taxonomies-information-technology-threats-38360

Lee, D. (2012, 6 12). *Flame: Attackers 'sought confidential Iran data'.* Retrieved 5 4, 2020, from BBC: https://www.bbc.com/news/technology-18324234

Long, S. B. (n.d.). *Loss exchange ratio database project (LERD).* Retrieved 04 29, 2020, from https://facultystaff.richmond.edu/~slong/LERD.html

Lorenzo Vidino, S. H. (2015, 12). *ISIS in America: From retweets to Raqqa.* (G. W. University, Producer) Retrieved 6 6, 2019, from Program on Extremism: cchs.gwu.edu/sites/cchs.gwu.edu/files/downloads/ISIS%20America%20-%20Full%20Report.pdf

Malcolm Nance, C. S. (2017). *Hacking ISIS - how to destroy the cyber Jihad.* Skyhorse Publishing.

Martin, J. (2016, 7 24). Debbie Wasserman Schultz to Resign D.N.C. Post. *The New York Times.*

Matrosov, A. (2012, 7 20). *Flame, Duqu and Stuxnet: In depth code analysis of mssecmgr.ocx.* Retrieved 5 2, 2020, from ESET: https://www.welivesecurity.com/2012/07/20/flame-in-depth-code-analysis-of-mssecmgr-ocx/

Max Roser, H. R. (n.d.). *Technological progress.* Retrieved 12 05, 2018, from https://ourworldindata.org/technological-progress

MITRE. (n.d.). *CARET.* Retrieved 9 30, 2018, from CARET: https://car.mitre.org/caret/#/

Mueller, R. (2019). *Report on the investigation into Russian interference in the 2016 presidential election.* U.S. Department of Justice. Washington: U.S. Department of Justice.

Nakashima, E. (2018, 10 23). *Pentagon launches first cyber operation to deter Russian interference in midterm elections.* Retrieved 3 12, 2019, from Washington Post: https://www.washingtonpost.com/world/national-security/pentagon-launches-first-cyber-operation-to-deter-russian-interference-in-midterm-elections/2018/10/23/12ec6e7e-d6df-11e8-83a2-d1c3da28d6b6_story.html?utm_term=.8c47d573557b

Nakasone, P. M. (2019). A cyber force for persistent operations. *Joint Forces Quarterly, 92*(1), 10–15.

Newman, L. (2016, 12 24). *Hacker lexicon: What is the attribution problem?* Retrieved 7 9, 2019, from Wired: https://www.wired.com/2016/12/hacker-lexicon-attribution-problem/

Newman, L. H. (2017, September 8). *The Equifax breach exposes America's identity crisis.* Retrieved September 7, 2018, from Wired: https://www.wired.com/story/the-equifax-breach-exposes-americas-identity-crisis/

NVD. (2019, 2 26). *CVE-2010-2568.* Retrieved from https://nvd.nist.gov/vuln/detail/CVE-2010-2568

Orr, A. (2018, 11 7). *China re-routed US internet traffic for 2.5 years.* Retrieved 11 13, 2018, from https://www.macobserver.com/link/china-reroute-internet-traffic/

Osinga, F. P. (2007). *Science, strategy and war: The strategic theory of John Boyd.* Routledge.

Reuters. (2016, 7 25). Democratic National Committee apologizes to Sanders over emails.

Rita Katz, M. K. (2006, 3 26). Terrorist 007, Exposed. Washington Post.

Rouse, M. (n.d.). *RAT (remote access Trojan).* Retrieved 5 4, 2020, from SearchSecurity: https://searchsecurity.techtarget.com/definition/RAT-remote-access-Trojan

Sanger, D. E. (2018, 11 29). *New York Times.* Retrieved 12 1, 2018, from New York Times: https://www.nytimes.com/2018/11/29/us/politics/china-trump-cyberespionage.html

Singer, P.W. (2018). *LikeWar - the Weaponization of social media.* Houghton Mifflin.

Sotirov, A. (n.d.). *Analyzing the MD5 collision in flame.* Retrieved 5 2, 2020, from https://speaker-deck.com/asotirov/analyzing-the-md5-collision-in-flame

Stoll, C. (2005). *The cuckoo's egg: Tracking a spy through the maze of computer espionage.* Pocket Books.

Symantec. (2011, 11 23). *W32.Duqu - The precursor to the next Stuxnet (Version 1.4 (November 23, 2011)).* Retrieved 11 30, 2018, from Symantec.com/content/en/us/enterprise/media/security_response/whitepapers/w32_duqu_the_precursor_to_the_next_stuxnet.pdf

Symantec. (n.d.). *Symantec.* Retrieved 29 2018, 11, from https://www.symantec.com/

Syverson, P. (n.d.). *Paul Syverson web page.* Retrieved 11 29, 2018, from http://www.syverson.org/

Temple-Raston, D. (2008, 7 5). *FBI surveillance team reveals tricks of the trade.* Retrieved 4 29, 2020, from NPR: https://www.npr.org/templates/story/story.php?storyId=92207687

Temple-Raston, D. (2016, 9 12). *Cyber bombs reshape U.S. battle against terrorism.* Retrieved 9 3, 2019, from NPR: https://www.npr.org/2016/09/12/493654985/cyber-bombs-reshape-u-s-battle-against-terrorism

Temple-Raston, D. (2019, 1 11). *Hacks are getting so common that companies are turning to 'Cyber Insurance'.* Retrieved 9 3, 2019, from NPR: https://www.npr.org/2019/01/11/684610280/hacks-are-getting-so-common-that-companies-are-turning-to-cyber-insurance

Timur Chabuk, A. J. (2018, 9 1). *Understanding Russian information operations.* Retrieved 9 9, 2018, from Signal (AFCEA): https://www.afcea.org/content/understanding-russian-information-operations

U.S. Department of Homeland Security (DHS). (2017, 4 18). *More ICS-CERT JSAR documents ICS Joint Security Awareness Report (JSAR-11-312–01)(W32.Duqu-Malware).* Retrieved 5 4, 2020, from ICS-CERT Landing: https://www.us-cert.gov/ics/jsar/JSAR-11-312-01

Warner, M. (2017). Intelligence in cyber - and cyber in intelligence. In A. E. G. Perkovich (Ed.), *Understanding cyber conflict - 14 analogies.* Georgetown University Press.

Zetter, K. (2014). *Countdown to zero day - Stuxnet and the launch of the world's first digital weapon.* Crown.

Zetter, K. (2015, 8 18). Hackers Finally Post Stolen Ashley Madison Data. Retrieved 5 1, 2020, from Wired: https://www.wired.com/2015/08/happened-hackers-posted-stolen-ashley-madison-data/.

Zetter, K. (2015a, 10 15). *Darpa is developing a search engine for the dark web.* Retrieved 11 15, 2018, from WIRED: https://www.wired.com/2015/02/darpa-memex-dark-web/

Zetter, K. (2015b, 8 18). *Hackers finally post stolen Ashley Madison data.* Retrieved 5 1, 2020, from Wired: https://www.wired.com/2015/08/happened-hackers-posted-stolen-ashley-madison-data/

Zetter, K. (2016, 7 16). *Hacker lexicon: What are CNE AND CNA?* Retrieved 12 1, 2018, from WIRED: https://www.wired.com/2016/07/hacker-lexicon-cne-cna/

Zurasky, M. W. (2017). *Methodology to perform cyber lethality assessment.* Old Dominion University.

Chapter 6
Cyber Security and Defense for Analysis and Targeting

The purpose of this chapter is to present a general background on cyber security and defense for analysis and targeting. We will begin with a description of current thinking on cyber security systems, including layered defense. This will be followed by a review of end points, connections, and key network nodes in terms of their vulnerabilities and possible technical tools to secure them. Included in this survey will be a review of the cyber attack cycle, a look at organized malware (e.g., botnets) construction and its detection, and a reflection on the importance of understanding defended network terrain in order to develop a successful network defense strategy. In addition, we will look at the broader context of cyber defense, and how we might incorporate conceptual security architecture approaches (e.g., denial and deception) in order to improve the likelihood of success in securing and defending a network.

6.1 Background

As initially discussed in Chap. 2, cyber defense is the cornerstone of current cyber policy and is designed to secure national and international use of the Internet. The concept of cyber security entered American pop culture with the release of the film *WarGames* (Lasker & Badham, 1983). The film caught the attention of the Reagan administration by showing how a teenage hacker could penetrate national defense systems and nearly cause a nuclear war with the Soviet Union. After viewing the film, President Reagan signed National Security Decision Directive 145 (NSDD-145), "National Policy on Telecommunications and Automated Information Systems Security" (Reagan, 1984). NSDD-145 was the first of the "cyber" presidential directives, with Executive Orders continuing to the present day (e.g., Table 2.3).

NSDD-145, similar to the Executive Orders and policy prescriptions that followed (Chap. 2), spelled out the goal of deterring adversaries, and being resilient when attacked, as key pillars of information system defense. A layered cyberspace

J. M. Couretas, *An Introduction to Cyber Analysis and Targeting*, https://doi.org/10.1007/978-3-030-88559-5_6

defense, as shown in Fig. 6.1, is the goal for protecting cyberspace equities in the United States.

As shown in Fig. 6.1, building partnerships and defending forward is the first line of defense. This includes leveraging non-military instruments of power (e.g., via diplomacy) and fortifying partner defenses. Layer 2, denying benefits, includes securing systems so as to raise the cost of attacking US systems. This may include technology (e.g., encryption) or legal (e.g., bad actor identification and conviction) means. Layer 3, imposing costs, includes preserving the military option to respond in-kind for cyber attacks that are effectively "above the line" in the spectrum of conflict (Fig. 2.2) and register as an armed attack by the Schmitt criteria (Sect. 2.3.3).

While Layer 1 in Fig. 6.1 is primarily a function of policy, fortifying Layer 2 is helped by good cyber hygiene procedures. And the US Department of Defense (DoD) is still being called out for lack of reporting on cyber hygiene, resulting in seven recent US Government Accountability Office (GAO) recommendations (GAO, 2020). It is not just the DoD, however, that is having issues with maintaining secure information technology (IT) systems. The well-known 2016 Democratic National Committee hacks (Poulsen, 2018) and Republican National Committee cloud data leak (Reilly, 2017) have even led to an HBO documentary, "Kill Chain" (Morales, 2020), on inadequate election defense. Cyber hygiene issues include a

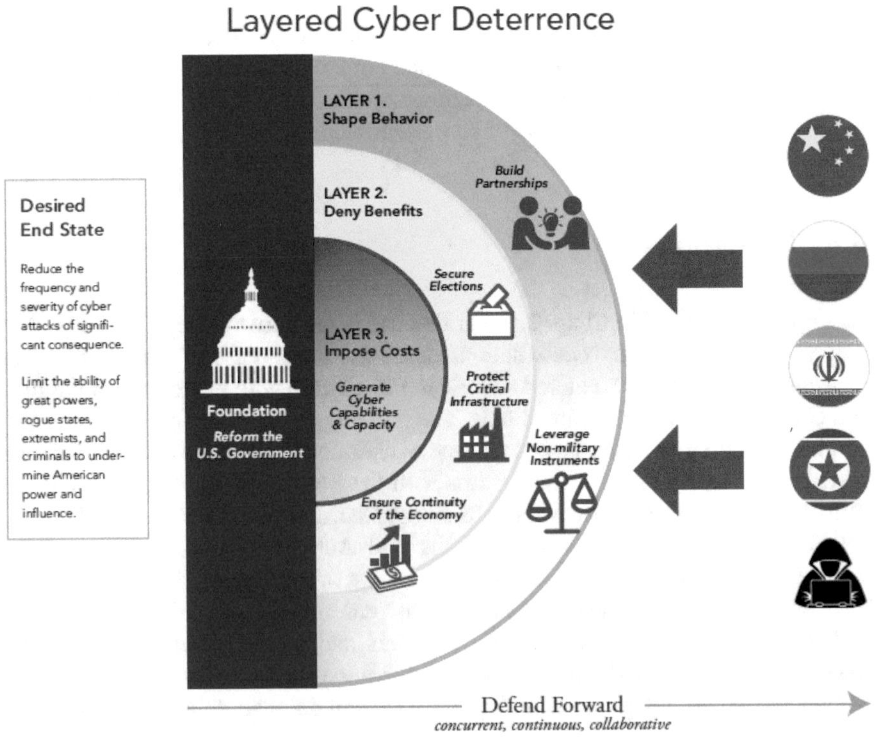

Fig. 6.1 Layered cyber defense (Angus King, 2020)

general lack of cyber security tactics, techniques, and procedures (TTPs) used in election security, including voting machines still connected to the Internet, with examples of white hat hackers penetrating voting systems in 3 days or less.

Shoring up our computer IT security, initially discussed in Sect. 2.4.2, includes cyber security controls (CSCs), which distill best practices into a prioritized list for defending a network of interest. Cyber security and defense are often thought of in terms of technical solution. The more basic cyber defenses include firewalls, authentication approaches, and keeping passwords updated. However, the multiple challenges to successful cyber defense require more than simply buying a tool. Adversary understanding, including their attack cycle, is key to a successful defense. Key questions for cyber security and defense for cyber analysis and targeting therefore include the following:

1. Who is primarily responsible for understanding cyber security and defense when performing cyber analysis and targeting?
2. What are the key issues that provide an intersection between cyber defense and cyber targeting?
3. What are the elements of an IT system to be secured?
4. What is the role of countermeasures for the targeting analyst in cyber security and defense?
5. Why is cyber security and defense important to analysis and targeting?
6. How do cyber attack cycles help the cyber analyst to understand target behavior?
7. How do current security tools work on cyber security threats?
8. How does infrastructure help to provide system-level security?
9. How do tools factor into cyber analysis and targeting?

6.2 Security and Defense Process

(Cyber) Defenders think in terms of lists, Attackers think in graphs.
  John Lambert (Lambert, 2018).

Chapter 2 (Sect. 2.3) provides background on cyber defense policy with examples from governments around the world. In addition, Chap. 2 provides doctrine and example tactics, techniques, and procedures (TTPs) as background for the cyber defender to better understand the threat. Chapter 3 continues the discussion by covering a taxonomy of cyber threats. This includes a review of the NIST (US National Institute of Standards and Technology) cyber taxonomy, cyber security data standards and the DREAD (damage/reproducibility/exploitability/affected users/discoverability), STRIDE (spoofing/tampering/repudiation/information disclosure/denial of service/elevation of privilege), and CVSS (Common Vulnerability Scoring System) frameworks for identifying and scoring system vulnerabilities.

As a defensive process, we will now expand on the use of attack frameworks (e.g., Lockheed Martin attack cycle) to show how they are equally important, and

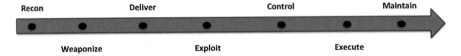

Fig. 6.2 Lockheed Martin attack cycle as a threat model

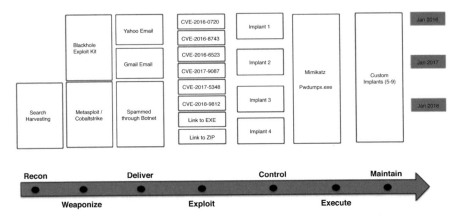

Fig. 6.3 Notional 2-year collection campaign (Heckman & F. J., 2015a, b, c)

used for, defensive resiliency evaluation. For example, Fig. 6.2 (Heckman & F. J., 2015a, b, c) shows the respective phases that an attacker will go through.

The attack cycle shown in Fig. 6.2 has wide acceptance as the basis for MITRE ATT@CK, and CARET, making it a good candidate to discuss the phases that a defender experiences on the other side of an attack.

6.2.1 Attacker and Cyber Kill Chain

Heckman (Heckman & F. J., 2015a, b, c) used the attack cycle shown in Fig. 6.2 to describe the actions that an attacker takes at each of the steps. These are the actions to be defended against (Fig. 6.3).

As shown in Fig. 6.3, a persistent actor may work in time scales of years to realize his goal of network access. In addition, Fig. 6.3 provides each of the tool types used across an attack cycle. Reconnaissance includes search harvesting, or doing open-source analysis, to learn about the targeted organization, and people who work there. Having a clear target understanding helps the attacker refine, or weaponize, an exploit. For example, the Blackhole Exploit Kit (Abandon, 2013) is known to be used in conjunction with spam campaigns (Jon Oliver, 2012). In addition, Metasploit, briefly discussed in Chap. 3, is a freely available penetration testing tool that attackers can buy on the open market.[1] As a penetration testing tool, Metasploit has predefined exploits for many targets of interest. At the "exploit phase" of Fig. 6.3, the attacker will either buy or build an exploit that often corresponds to a known vulnerability documented in the common vulnerability enumeration (CVE) (NIST). In addition, attacks are commonly delivered through email (i.e., delivery in Fig. 6.3), which provides defenders one of their first lines of defense, using firewalls to filter out malicious email messages.

Implants, shown at the control step in Fig. 6.3, are the actual beacons planted on the target machine for the collection of data. In Chap. 5, we looked at Duqu and Flame as examples of beacons planted to collect reconnaissance information on the Iranian centrifuge system (Zetter, 2014). Similarly, beacons might be used to collect files, log keystrokes, or change control code in an operating system or industrial process. Implants can also provide a backdoor for additional operations. For example, as shown in the "execute" phase in Fig. 6.3, Mimikatz (Porup, 2019) is a well-known tool for dumping passwords. When Mimikatz is implanted, the attacker then extracts the passwords and uses them to access additional accounts on the system (i.e., privilege elevation). Combined with a method of maintaining system access (i.e., a backdoor), Mimikatz is an "implant" that can be used to maintain the flow of targeted information. This access/extraction "implant," combined with a command and control (C&C) system, completes the cyber attack cycle shown in Fig. 6.3, providing the attacker with a system that he can enter and continue to exploit at will.

Implicit in the tools used across the notional attack cycle provided in Fig. 6.3 are the different skills required. For example, specialized research, programming, and development activities are likely required to harvest requisite information, develop implants, and maintain operations on a target of interest (Table 6.1).

As shown in Table 6.1, there are multiple skills involved in a successful cyber attack. Narrowing down the target is a key skill that analysts hone to ensure that the correct tools are constructed for the target of interest. Tool development, a specialized software development capability, relies on the accuracy of the analyst to ensure that the target is relevant. In addition, expert operators provide the maneuver capability to both get on the target and maintain presence.

Table 6.1 Threat actor attributes

Threat actor capability	Example
Software development ability	Custom implants
Expert operators	Ability to convert CVEs to access via implants
Research capability	Search harvesting

---

[1] An example attack construction and execution is shown in https://blog.cobaltstrike. com/2015/04/23/user-defined-storage-based-covert-communication/ and https://youtu.be/ yFXVVuK0mYY

This section illustrates the people, processes, and technologies that threat actors use to analyze a cyber target. This includes the time required, persistence, and specialty skills that might be required for a threat actor to breach a target of interest. While training (e.g., identifying spear-phishing e-mails) and individual technologies (e.g., firewalls) will help cyber defenders understand the different phases of the attack cycle, an organization should be ready at multiple levels to maintain a secure network.

6.3 Cyber Defense: End Points, Connections, and Data

Securing a network includes understanding, monitoring, and detecting a threat by analysts and technologies for near real-time alerting of anomalous behavior. Threat understanding is an awareness of how a threat might make an end point (e.g., personal computer), connection (e.g., virtual private network [VPN]), or data center (e.g., cloud) an addressable target.

Attackers are commonly known as Advanced Persistent Threats (APTs). APTs target the people supporting target networks (Zachary Cohen, 2021), known defense processes (e.g., software updates), or characteristic tools (e.g., network monitoring (Oladimeji, 2021)). Threat actor processes and tools are described in open-source reporting. For example the network security company FireEye (FireEye., 2017) reports on identified APT targeting approaches (e.g., spear phishing) and tools for planting an exploit.

6.3.1 End-Point Security

In many configurations, end points can be one of the most vulnerable parts of a computer system. While there are several reasons for this, a few that we will cover include the following: (1) there is a human involved and (2) the attack surface provided by the multiple applications on an end point can be very large, with an attack surface defined as follows:

> Attack Surface: "The set of points on the boundary of a system, a system element, or an environment where an attacker can try to enter, cause an effect on, or extract data from, that system, system element, or environment." (NIST)

In addition, users sometimes fail to keep their application and/or system licensing up to date. This results in the software being unsupported by commercial patches. Unsupported end points are vulnerable to exploits that would be obsolete if they updated their system with the most recent commercially available patches.

Having a human involved automatically makes an end point subject to social engineering. It is often human actions that can compromise security. For example, the operator may click on malware while visiting a "watering hole." A watering hole[2] is an infected website where a targeted group is known to frequent. Similarly, it is the operator who is targeted by spear-phishing attacks, where the computer user is lured into clicking on an e-mail attachment. While training is likely the best approach for educating users on how to maintain security when navigating the Internet, antivirus (AV) systems are a key technical approach for managing the underlying computer system. AV systems are used to patch known exploits and are looked at as a first line of defense for computer security.

6.3.1.1   Antivirus Systems

While classic antivirus systems use pre-calculated signatures to diagnose the binary footprint of recorded malware specimens, modern antivirus solutions leverage a set of techniques that detect previously unknown specimens as well. Using empirical rules, antivirus products attempt to understand the purpose of sample code. These data are then used to discern the behavior of the code samples, based on known program snippets, with the idea of determining what the code is attempting to accomplish.

However, it remains a challenge for AV systems to catch recently developed exploits with less defined signatures. For example, even if the user maintains the system with the most up-to-date patches, a determined attacker can use a "zero day," or exploit for which no known patch currently exists.

6.3.1.2   Zero Days

A zero-day exploit has no known vulnerability reporting and therefore the AV companies have neither a signature to detect, nor a patch to remediate the issue. Google, Microsoft, and other companies attempt to remediate zero days through white market bounty programs where they pay hackers to find security holes in their software (Zetter, 2013).

Table 6.2   Zero-day exploit examples with cost (Zetter, 2013)

Exploit	Estimated cost
Mac OS	$50,000
Windows	$100,000
Apple iOS	$100,000
Adobe reader	$5000 to $30,000
Firefox	$3000

---

[2] https://csrc.nist.gov/glossary/term/watering_hole_attack

Zero days are famous due to their use in cyber attacks. For example, one of the most famous uses of zero-day exploits was Stuxnet, with five known zero days in the most advanced version of the attack vector (Zetter, 2014).

There is a market for zero days with both vendors and cyber actors looking to purchase them (Lepore, 2021; Perlroth, 2021). In addition, as Zetter (Zetter, 2014) points out, the sale of exploits is legal and largely unregulated, with widely varying prices depending on the rarity of the vulnerability (Table 6.2).

Zero-day prices, as shown in Table 6.2, vary due to the ubiquity of the software and the rarity of the vulnerability. In addition, zero-day exploits are perishable, in that their shelf life is limited as to how long it takes for another hacker to release the exploit and have the antivirus vendors patch it. In fact, a recent study surmises that zero day use more likely shows that threat actors have deeper pockets than expert knowledge (Kathleen Metrick, 2020).

For a zero day to be effective, however, the attacker needs access to the end point of interest. Honey files, first described by Stoll (Stoll, 2005), or the more generalized honeypots and honey nets, are one way to keep an attacker guessing when they are trying to navigate a target network.

6.3.1.3   Honeypots

Honeypots can take the form of both client and server systems. For example, client honeypots mimic the behavior of potentially vulnerable applications, like web portals, in order to analyze suspect files. In addition, client honeypots may therefore browse the web or read spam emails like a regular user and open the received content. When suspect files are analyzed, client honeypots use rule-based approaches to determine if the content is safe. If the attack focused on downloading and installing malware, client honeypots can capture the malware.

Honeypots with a high degree of interaction feature a complete network and associated operating system, including computer systems with advanced logging and analysis capabilities. One of the major benefits of high-interaction honeypots is that entire attack tactics, along with malware samples, are collectible. All of the structure of a high-interaction honeypot comes at the cost of a relatively static configuration, however, resulting in the generating of attack information on only one version, or patch-level, of the associated honeypot system software and vulnerabilities.

In addition, honey files have been used since the beginning of "cyber" to track hacker activity and to better understand the types of information that hackers are interested in. For example, in "*The Cuckoo's Egg*," Clifford Stoll used honey files to monitor what the KGB—*Komitet Gosudarstvennoy Bezopasnosti* (Committee for State Security)—controlled hackers were trying to discern from probing his research network (Stoll, 2005).

An advantage of honey files is the possibility of embedding components into documents so as to present the attacker with a cyber version of "marked bills" to catch bank robbers of old. Additional measures for monitoring attacker patterns include log file analysis, where each system traversed maintains a record, or log, of the attackers' virtual footsteps.

Honeypots are virtual targets that are designed to monitor an attacker's tools, tactics, and techniques. In addition, from a cyber defense standpoint, the time that an attacker spends maneuvering in a honeypot is time not spent in other, more valuable, parts of a defended network. A next step from a honeypot is to make the target environment dynamic, or a moving target defense (MTD).

6.3.1.4 Moving Target Defense (MTD)

Moving target defense (MTD) has the potential to controllably change network terrain, across multiple system dimensions, in order to obfuscate attack paths and increase attacker uncertainty. In addition, MTD can reduce the window of opportunity, increasing the attacker's intelligence collection costs and obfuscating target identification.

One example implementation of MTD is to reduce zero-day susceptibility in memory-based attacks. For example, many zero-day attacks exploit memory vulnerabilities. As reported by Polyverse (Polyverse), of the 7217 CVEs in 2017 of high or medium severity, an estimated 80% involve memory exploitation—a vulnerability whose security can be increased through the use of MTD. In addition, MTD has potential applications across the attack life cycle (Hamed Okhravi, 2016; Table 6.3).

As shown in Table 6.3, moving target defenses might be used across the different steps of the cyber attack life cycle. In terms of moving target domains, dynamic networks change network features such as Internet Protocol (IP) addresses, routes, or service locations. These defenses challenge normal system administrators and multi-file applications, in that files may not be in expected locations when performing maintenance, or during runtime, respectively (Brian Van Leeuwen, 2016). Dynamic platforms, on the other hand, include mechanisms that result in modifying the operating system or central processing unit (CPU) version.

Dynamic runtime environments include instruction set randomization (ISR) (Georgios Portokalidis, 2011). Instruction set randomization is used to defeat code injection attacks. These are often attacks that exploit a buffer overflow vulnerability, injecting malicious code into a running program. Similarly, dynamic software

Table 6.3 Moving target defense and attack phases (Hamed Okhravi, 2018)

Moving target (MT) domains	Attack phases				
	Reconnaissance	Access	Development	Launch	Persistence
Dynamic networks	✔			✔	
Dynamic platforms		✔	✔		✔
Dynamic runtime environments			✔	✔	
Dynamic software			✔	✔	
Dynamic data			✔	✔	

techniques change application code and dynamic data change the format or representation of data.

The promise of MTD is accompanied by criticism that the overall technique requires a central key, acquisition of which will defeat the approach (Ana Nora Sovarel, 2005). Therefore, while MTD helps prevent most attackers from accessing a system, log file analysis is a method for tracking the movements of determined attackers when they do.

6.3.1.5   Log File Analysis

When a system is infected by malware, it usually changes specific system files to ensure that it gets loaded during the next boot cycle. These activities should be captured in log files, thereby providing a trace for defenders to find unintended system file changes. However, log files can be a challenge to meaningfully interpret. Therefore, additional tools, like the host-based intrusion detection system (HIDS), are used to monitor log file changes to detect malware. The disadvantage in HIDS is that a flag is only raised after a successful infection, additionally providing malware the possibility for applying active countermeasures for preventing detection. One approach might be for the malware to shut down host-based security mechanisms or modify recently created log files.

Log files also provide clear indications of specific program activity. For example, firewall log files may show a bot trying to talk to its command and control (C&C) server through a port blocked by a firewall. Multiple log file entries showing blocked firewall port traversal attempts may be an indication of a bot infection (Heli Tiirmaa-Klaar, 2014). In addition, a key challenge in analyzing log files is that there is a large volume of information to sift through in order to detect and classify an incident. The large amount of data in log file analysis both obfuscates threat identification and provides a large number of false alarms.

6.3.2   Connection Security

While system log files can indicate malware activity on a system, we can also use them to provide a view of communications between agents in a network and the outside world. For example, using firewall log files to see how data are flowing out of a network provides a clue concerning possible control or file data that should not be flowing either within a network, or to nodes outside of a protected network. Bots, or botnets, are one example of organized malware that might be detected with connection security applications.

We looked at a practical application of botnets in the active intelligence section of Chap. 5 (Sect. 5.5.2), showing how Duqu, Flame, Gauss, and SPE were used to develop intelligence for the Stuxnet operation. Each of these ISR capabilities might be looked at as a "bot," autonomously searching, and propagating itself, across the web, looking for a specific Siemens controller for future control system modification (Zetter, 2014). In the context of this chapter, we will now look at bots, and botnets, to better understand how and when botnets are detectible for cyber security applications.

6.3.2.1   Background: Bots and Botnets

A key and growing threat that requires securing end points and connections is the use of bots and their aggregate botnets to monitor and disrupt commercial and government operations. Each botnet has four entities (Fig. 6.4).

As shown in Fig. 6.4, each botnet consists of four basic elements that include the botmaster, the command and control (C&C) server, the C&C channel, and the bots, distributed throughout the Internet (HONG, 2012; Table 6.4).

As shown in Table 6.4, the botmaster controls the botnet's operation, which includes executing commands, or sending instructions to the bots to achieve an operation's goals. While each bot is an independent piece of malware, they are often designed to continuously increase by trying to migrate to additional host computers. For example, a bot may include computers on the same network. This might be done by obtaining each account's email address list, and sending itself, as malware attachments, in emails from the user's trusted account to colleagues and friends.

The population size of a botnet is one measure of its danger. Upon instruction by the botmaster, or periodically, each bot downloads and updates itself to a new version that is more resilient, and might even be updated with new exploits and attack methods. This is the method through which botnets evolve. Botnet communication with the botmaster is performed through the C&C server, where the updates are stored.

Fig. 6.4 Botnet components: botmaster, C&C server, C&C channel, and bots

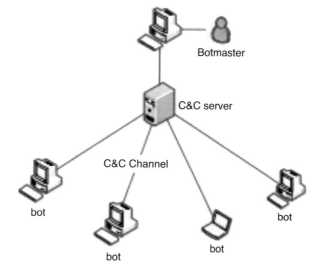

Table 6.4 Botnet component descriptions

Botnet component	Description
Botmaster	The botmaster is the entity controlling the botnet's operation. The botmaster executes commands to achieve organizational goals (e.g., denial of service, data exfiltration) by sending instructions to the bots. In addition, the botmaster may rent the botnet to others for their use.
Bots	Each bot is a relatively small program that is implanted into a computer to create a controlled node, or "zombie," that is then connected to a botnet. In addition, bot is the term that refers to both the malware/program and the computer infected by that malware; bots are the nodes through which the botmaster directly conducts malicious activities.
Command and control (C&C) servers	C&C servers both host the malicious software to be downloaded by the bots and are intermediaries that control the botnet, coordinating the bots when performing an attack.
C&C channel	C&C channels are the communications through which information and control signals are sent from the C&C servers to the bots. The quality of C&C channels, including their ability to hide their communication activities, is key to both the survivability of the botnet and the botnet's ability to carry out attacks.

C&C server placement and configuration depends on the botnet's C&C model. For example, botnets are known to have both centralized and decentralized architectures:

– **Centralized model:** C&C servers are hosted on protected sites, free from external policies or content−/service-level controls (e.g., from a hosting service).
– **Decentralized model:** more dynamic, in that a range of communication protocols might be employed with possibly varying architectural structures (e.g., a variation on the peer-to-peer [P2P] model). In addition, the C&C servers might employ different deployments, depending on available communications servers:

  • C&C server installed as an Internet Relay Chat (IRC) server: botnet communicates via Internet Relay Chat (IRC) protocol.
  • C&C server installed on a web server: botnet communicates via HyperText Transfer Protocol (HTTP).

Botnet C&C servers are the botnet components that communicate with the botmaster. Using denial and deception to mask C&C server identity is therefore the focus of many botnet evasion techniques for preventing detection. This is performed through the botnet's communication channel.

Either push or pull channels are used for botnet communication:

– **Push channel:** C&C servers send commands and code updates to bots, who are waiting for a message.
– **Pull channel:** bots communicate with C&C servers at regular intervals to obtain instructions and updates.

Many of the early botnets used the encrypted IRC protocol, which is still a popular botnet communication channel. However, as IRC is less commonly used, and with defenders increasing their botnet detection ability, botnets are transitioning to communication methods that can be disguised as ambient network traffic, which includes using HTTP or Peer-to-Peer (P2P) protocols. More advanced, some botnets communicate via social network site postings. In addition, C&C servers increasingly use public key cryptography to encrypt their code updates and instructions, challenging network defenders to authenticate C&C servers and take control of a given botnet.

C&C servers are the means by which the botmaster controls his botnet. A botnet can therefore be dismantled by taking down the C&C servers. To prevent C&C server detection, botnet designers add proxies, one or more layer(s), between the bots and C&C servers, to help mask C&C server identities. These proxies are optional in providing an extra layer of obfuscation, making it a challenge for network defenders to track bots back to their C&C servers.

6.3.2.2 Botnets as a Security and Cyber Defense Threat

The ability of bots, and overall botnets, to infect and multiply in networks makes them a formidable challenge for current cyber security defenders. For example, botnets are used to perform denial of service, exfiltrate information, and/or perform financial attacks. Over 1,000,000 bots are currently estimated to be active (Spamhaus, 2020). These bots are used by operators that span from criminals to nation states. For example, North Korea's Lazarus Group, widely identified with the "Hidden Cobra" botnet (Gallagher, 2019), is known for perpetrating theft, extortion, and cryptojacking; or usurping another user's machine to mine for crypto currency. The United States recently accused North Korea of a number of financially based attacks using the Hidden Cobra botnet (Sanger & N. P., 2020):

- "Cyber-enabled Financial Theft and Money Laundering": Democratic People's Republic of Korea (DPRK) performs a lot of operations that involve stealing altcoin, a cryptocurrency.
- "Extortion Campaigns": cyber protection rackets. The DPRK uses "long-term paid consulting arrangements to ensure that no such further malicious cyber activity takes place."
- "Cryptojacking": taking over user machines and using their resources to mine crypto currency.

Detecting bots and botnets is performed by detecting possible malicious behavior on a network of interest. As discussed in Sect. 6.3.3.1, botnet C&C communications provide the network defender with a clear target for detecting network botnet activity. Finding these botnets is done by analyzing network flows, either directly, with network-based intrusion detection system (NIDS) that analyzes the entire packet, or with abstracted versions that simplify the problem to network connectivity analyses.

6.3.2.3   Network Analysis

Network traffic analysis is used to find botnet activity. This approach also helps with detecting the spread of botnets, for example, by monitoring their command and control (C&C) communications, and subsequent failed domain name system (DNS) queries, that result from botnets' obfuscating their communications (HONG, 2012). Network-based intrusion detection system (NIDS) is used to inspect the data sent over networks, also known as deep packet inspection, in order to decompose communications to discern malicious activity.

NIDS emulates the protocol stack to organize received packets in the same way that a standard communication end point would receive the message. For example, if a targeted system also disables potentially harmful connections (e.g., using a white list) to avoid infections, this defense approach is also referred to as an intrusion prevention system (IPS).

One advantage of NIDS is that botnets do not use active countermeasures to avoid their detection (Heli Tiirmaa-Klaar, 2014). A drawback of NIDS is their limited scalability in observing networks with high traffic loads, inability to reconstruct encrypted traffic, and a weakness in detecting seemingly random communications.

One approach to counter pattern-matching deficiencies is to look for communications anomalies to find malware, especially botnet, communications in regular traffic. Anomaly-based NIDS (A-NIDS) observes network characteristics, like protocols used, point-to-point system communications, and the network load of individual systems, in order to develop patterns and look for anomalies. A training phase is usually a first step in deploying an A-NIDS system, developing a baseline of benign network behavior.

6.3.2.4   Netflow

While both NIDS and A-NIDS are challenged by the scale that accompanies current Internet connected networks, with multiple Internet connected nodes, netflow uses less detailed connection information. Looking directly at routing devices to understand network traffic related information, netflow analyzes node connections, via IP-addresses, port numbers, and protocol data. Since the payload of individual packets is not individually parsed, netflow can process larger amounts of traffic.

One example of using netflow to find botnet behavior is to detect port scanning (Heli Tiirmaa-Klaar, 2014). In some cases infected systems use port scanning to find available services on target hosts via communication calls to multiple ports. If the service is unavailable, the probed system rejects the incoming connection, which can produce a lot of rejected connections, which are an indicator of malicious software behavior. Another example, the distribution of spam emails, is the creation of a lot of mail server connections to a mail server, typically on Port 25, which is also an indicator of malicious activity.

Combining malicious activity detections into a common operational picture (COP) is used for executive-level monitoring from military to business operations.

In the case of security, a Security Operations Center (SOC) is a place to house the detecting technologies, defending teams, and other policy and process resources for network defense.

6.3.3 Data Security

One way to integrate cybersecurity professionals with the required tools is a Security Operations Center (SOC). SOCs are organizations, usually with their own building, that combine activities that use tools like Netflow, NIDS, and A-NIDS to monitor and analyze network activity to understand anomalous behavior on end points, applications, databases, websites, servers, and other systems to detect potential security issues early. In addition, the SOC may be monitoring the computing resources in a central location, often called a "cloud."

> Cloud computing is defined as: "The practice of using a network of remote servers hosted on the Internet to store, manage, and process data, rather than a local server or a personal computer." (lexico).

Cloud computing provides the advantages of a centralized location for performing computation and storing data. However, a centralized, "always on," node also provides a fixed target for cyber attackers. Managed Service Security Providers (MSSPs) are an example of a data center with a fixed attack surface.

6.3.3.1 Security Operation Center (SOC)

Due to the rapid pace of technology, new organizations, along with developing leadership roles, have accompanied the development of SOCs. For example, the Chief Information Security Officer (CISO) is a relatively new title for the person in charge of ensuring that an organization's data are secure. In addition, the CISO provides guidance on the goals and objectives for an organization's Security Operations Center (SOC).

Constructing and maintaining an SOC requires a set of decisions in terms of the type of staff to keep in-house versus what can be outsourced. It is a challenge to balance the cost and security objectives that meet an enterprise's risk objectives (Joseph Muniz, 2015). Some example SOC approaches are shown in Table 6.5.

As shown in Table 6.5, in a dedicated SOC the CISO relies on an internal security team that manages all aspects of computer security. This includes developing strategy, designing the architecture, determining how to perform risk management, and prosecuting day-to-day operations. Keeping this all in-house requires a heavy investment in staff expertise, as well as building a special facility for the SOC. The SOC then becomes the principal location for all enterprise IT security issues. A dedicated SOC is likely the highest-cost option, due to the people, equipment, and facility cost. This option also provides the most organizational control, and best security.

Table 6.5 Security Operations Center (SOC) models

SOC type	Description
Dedicated SOC	The business maintains control over each element of the security life cycle. This is the most expensive, due to the high labor, equipment, and rent costs. A dedicated SOC is also likely the most secure option.
Virtual SOC	The lowest-cost approach and likely the least secure. SOC responsibilities are added to another team (e.g., network operations). The virtual SOC operating model seems like a good approach for small businesses, but may cost more when an event occurs due to the lack of key skills needed to handle a security incident.
Outsourced SOC	Less expensive than an SOC that is staffed in-house; an outsourced SOC is potentially less secure. The challenge with this option is to ensure that the managed service security provider (MSSP) is operating at an acceptable service level, based on enterprise security requirements.
Hybrid SOC	This approach is used by organizations that maintain their own security, while outsourcing low-risk elements of their operations to an MSSP.

On the opposite end of the security spectrum is the virtual SOC, where the organization attempts to add SOC tasking as secondary responsibilities for existing staff members. For example, this might be an additional duty for the network operations team. In this case, the term SOC is simply the set of processes used by the team when handling a security event. The lack of experience and focus usually results in a high cost for the organization when a breach occurs.

While the virtual SOC adds security processes as additional tasking for an organization's existing staff, an outsourced SOC includes using defined processes to contract with an external security provider. This might include using internally defined security requirements to develop a contract with a Managed Security Service Provider (MSSP) to completely outsource the security requirement. Service-level agreements (SLAs) will be negotiated, which may include measures of performance (MOPs), in the contract to ensure that the outsourced security capability meets the needs of the overall enterprise. For example, in the outsourced SOC, the MSSP provides the SOC, the technology, and the people with the right skills to ensure the response needed to keep the business safe. A potential downside to this approach is a bad actor penetrating the MSSP, or computing and data center that the enterprise relies on.

An additional method of outsourcing, as needed, is the Hybrid SOC. In this model, the CISO keeps a core team of capable security professionals, while outsourcing computing and data protection capabilities, as required. The choice of a hybrid SOC may be due to challenges imposed by time zone shifts due to global operations, specialized skill requirements (e.g., security architects, penetration testers, digital forensics experts, …) at the MSSP, or simply a tiered management structure with the in-house personnel focused on ensuring that the business needs that the SOC is protecting have the required level of protection.

6.3.3.2   Cloud Computing

Cloud computing was one of the first steps in outsourcing enterprise IT. Developing cloud-based solutions includes provisioning for the three layers of cloud software (e.g., Information as a Service [IaaS], Platform as a Service [PaaS], and Software as a Service [SaaS]) that the enterprise requires (Thomas Erl, 2013). The quality of cloud providers is judged based on their level of investment in each of the three levels (Evans, 2017). To make shopping for cloud services easier, each of the large cloud providers that sell cloud services might be part of an MSSP enterprise computing solution whose outsourcing considerations were discussed in Sect. 6.3.3.1.

Cloud computing, a key component of the MSSPs that businesses rely on when outsourcing their information infrastructure, has a disadvantage in that certain threat actors are actively targeting them. In addition, the "always on" availability makes certain cloud configurations a static, persistent, target. A recent example of cloud compromise is reported by *The Wall Street Journal* story of the Cloud Hopper campaign, which was estimated to have compromised over a dozen cloud providers (Rob Barry, 2019):

> The Journal found that Hewlett Packard Enterprise Co. was so overrun that the cloud company didn't see the hackers re-enter their clients' networks, even as the company gave customers the all-clear.
>
> Inside the clouds, the hackers, known as APT10 to Western officials and researchers, had access to a vast constellation of clients. The Journal's investigation identified hundreds of firms that had relationships with breached cloud providers, including Rio Tinto, Philips, American Airlines Group Inc., Deutsche Bank AG, Allianz SE, and GlaxoSmithKline PLC.
> …
> They came in through cloud service providers, where companies thought their data was safely stored. Once they got in, they could freely and anonymously hop from client to client, and defied investigators' attempts to kick them out for years.

Seeing the vulnerabilities in using third-party providers challenges the idea of secure outsourcing of data and compute capability (Ricker, 2019). For example, one financially motivated threat actor, dubbed "Fin9," is widely known for attacking MSSPs to spread through these systems due to trusted access (BlackBerry® Cylance®, 2020). In addition, it is widely known that some of the MSSPs use the Linux operating system, trusting it to provide "always on" availability. The Winnti group, as an example, has been developing remote access Trojans (RATs) to specifically target the Linux operating system, which runs many of the MSSP clouds, for up to 10 years (BlackBerry., 2020). A current example is the Amazon GovCloud, which uses Linux/UNIX in its elastic compute cloud (EC2) web service for managing Windows Server instances in Amazon's data centers (Amazon Web Services (AWS)).

One of the key reasons for outsourcing computing is to save on staff costs. Moving data to an insecure location should raise eyebrows on the part of CISOs, and corporate leadership, who depend on the data to run their enterprise. Therefore, it would be natural to look for solutions designed to secure the data regardless of its location. Blockchain advertises the promise of secure storage.

6.3.3.3 Blockchain: Example Data Security Technology

A developing area for IT security is to use blockchain's ledger-based security to protect data, with cloud being one example. While this is an overview section, details concerning how blockchain works are included in courses and background tutorials (JOHN KOLB, 2020).

A key design feature for blockchain is that it converts its protected data into an immutable, tamper-free state. This is performed through a combination of public and private keys. The public key is for the address of where the data are stored in the public ledger. For example, a blockchain transaction uses distributed peering to improve security. In securing data, blockchain is currently used in applications that include election security to prevent election rigging and fraud. For example, the well-known 198 million voter records outed in a 2016 Republican National Committee cloud compromise (Reilly, 2017) might have been avoided with the security of a good blockchain implementation. Sierra Leone, for example, is reported to have successfully deployed a blockchain-based capability to secure its 2018 elections (Biggs, 2018).

6.4 System-Level Security and Defense Approaches

While blockchain is a promising way of passively protecting our data, one form of active protection is deception. Reginald Jones, the British scientific military intelligence scholar, described the relationship between security and deception as a "negative activity, in that you are trying to stop the flow of clues to an opponent" in order to maintain a competitive advantage in a conflict (Jones, 1989; Sushil Jajodia, 2016). Reginald Jones also referred to deception as the "positive counterpart to security" that provides false clues to opponents (Sushil Jajodia, 2016).

6.4.1 Defensive Countermeasures

Three general categories of deceptive obfuscation, or signaling, can be used to know whether a deception had an effect on an attacker:

- The attacker acts in the wrong time and/or place.
- The attacker acts in a way that is wasteful of its resources.
- The attacker delays acting or stops acting at all.

While it is a challenge to see how an attacker is investing his resources, a few examples of successful deception include the following:

- Attacker use of distributed denial of service (DDoS) capabilities (e.g., botnet investment) on a defensive honeypot.

- Attacker use of "burnable" assets (e.g., zero-day exploits, downloaders, …) on non-critical attack surface elements.

Additional work factor indicators might include the time an attacker spends in a surveillance and reconnaissance phase; or the attacker's reverting to a reconnaissance phase after having moved to an attack phase.

6.4.1.1 Denial and Deception

A key element in the development of cyber security and defense includes a focus on deception. One of the first documented cyber infiltration operations, described in *"The Cuckoo's Egg"* (Stoll, 2005), shows the author, Clifford Stoll, using "honey files," as a means to track down the West German hackers working for the KGB infiltrating his network. Honey files, the equivalent of marked bills in bank robberies of old, are an example of how deception is used in cyber operations. In addition, deception operations are one way to frame the sometimes cat-and-mouse engagement between cyber attackers and defenders (Eric Gartzke, 2015).

6.4.1.2 Use of Denial and Deception (D&D) across the Lockheed Martin Cyber Kill Chain® Methodology

Adapting denial and deception (D&D) for cyber, Heckman (Heckman & F. J., 2015a, b, c) laid out possible denial and deception techniques across the Lockheed Martin cyber kill chain in Table 6.6.

Each of the measures provided in Table 6.6 has the potential to change a target's vulnerability estimate during that phase of the cycle. This might be used to determine how to apportion resources, via D&D measures, over the respective phases of the targeting and attack cycle.

6.4.1.3 Cyber Kill Chain and Deception Elements

As described by Bridkin (Greg Bridkin, 2016), deception is most effective in the early stages of cyber kill chains, leveraging the attacker's lack of familiarity with the target system. This provides the defender with multiple deception options, deflecting their attention from important subjects (Fig. 6.5).

As shown in Fig. 6.5, each stage of an attack provides alternative deception approaches for misleading an attacker (Table 6.7).

As shown in Table 6.7, applying deception requires a thorough understanding of the defended network technologies and configuration, network setup and maintenance processes, and, by implication, administrators associated with the technologies and processes. This level of understanding is one of the challenges that make securing cyber systems difficult.

Table 6.6 Cyber denial and deception (D&D) techniques (Rowe, 2007; Rowe & H. R., 2004)

Kill Chain Phase and Description	Denial and Deception Techniques
Reconnaissance: The adversary develops a target	Setup a decoy site that parallels the real site
	Lie about reasons when asking for an additional password
	Plant disinformation; redefine executables; give false system data
	Deliberately delay processing commands
	Pretend to be an inept defender or to have easy to subvert software
	Craft false personae with a public web footprint
	Operate low interaction honeypots
Weaponize: the adversary puts the attack in a form to be executed on the defender's computer / network	Camouflage key targets or make them look unimportant; disguise software as different software
	Redefine executables; give false file-type information
	Associate false times with files
	Do something in an unexpected way
	Falsify file creation times
	Falsify file modification times
	Reveal facts regarding the enterprise and/or its network
	Modify network to reveal / conceal activity
	Hide software using stealth methods
	Deploy low interaction honeypots
	Deploy sandboxes
Deliver: the adversary puts in place the means by which the vulnerability is weaponized	Transfer attack to a safer machine such as a honeynet
	Transfer Trojan horses back to an attacker
	Ask questions, with some intended to locate the attacker
	Try to frighten attacker with false messages from authorities
	Plan an operation for any false persona targeted by spear phishing
	Expose a past deception
	Confirm malicious email delivery expectations
	Visit drive-by download site with a honeyclient

(continued)

Table 6.6 (continued)

Kill Chain Phase and Description	Denial and Deception Techniques
Exploit: the adversary executes the initial attack on the defender	Pretend to be a naïve consumer to entrap identity thieves
	Lie that a suspicious command succeeded
	Lie about what a command did
	Lie that you can't do something, or do something not asked for
	Swamp attacker with requests
	Send overly large data or requests too hard to handle to the attacker
	Send software with a Trojan horse to attacker
	Induce attacker to download a Trojan horse
	Conceal simulated honeypot information
	Misrepresent the intent of software
	Allow malicious email to run in sandbox
Control the adversary: employ mechanisms to manage the initial victim's feedback	Deploy a high interaction honeypot with infrastructure
	Systematically misunderstand attacker commands
	Operate a high interaction honey pot
	Falsify internal data
Execute: leverage numerous techniques, the adversary executes the attack plan	Reveal technical deception capabilities
	Tamper with the adversary's commands
	Timeout the adversary's connections
	Emplace honeytokens in defender controlled environment
Maintain: the adversary establishes a long term presence on the defender's networks	Maintain false personae and their pocket litter
	Add / retire false personae
	Allow partial enumeration of fictional files
	Expose fictional systems that simulate additional services and targets

6.5 Summary

While deception may provide a "cheap" defensive countermeasure for cyber, it is also the most challenging to implement, due to the necessity of clearly understanding the attacker, his objectives, and how to (mis)guide his traversal of a blue network (Alperovich, 2018). However, the use of a Security Operations Center (SOC), as discussed in Sect. 6.3.3.1, provides the common operational picture (COP) that helps with the implementation of policy concepts, like deception, to help the cyber security defender control the flow of information about a defended system to potential attackers.

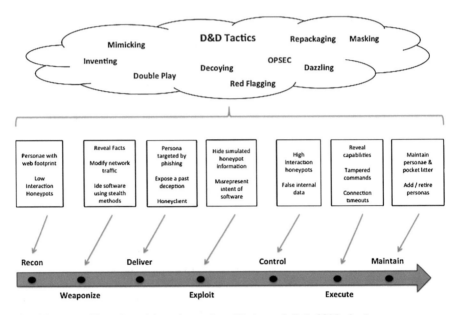

Fig. 6.5 Attack life cycle and deception options (Heckman & F. J., 2015a, b, c)

Table 6.7 Deception scenarios and scenario exploitation stages

Stage	Description
Pre-exploit (Recon, Weaponize, Deliver)	The attacker conducting reconnaissance to perform host detection, service enumeration, network topology mapping, access control lists (ACL) and packet filter detection and operating system (OS) fingerprinting, deception scenarios are centered on network topology and network setup deception (e.g., false topology, hidden and fictitious nodes / subnets, false network parameters). Therefore, providing false data on host discovery and network services, OS misidentification gateway / firewall shielding and ACL falsification are a few approaches to deceive attackers
Exploitation	Deception scenarios for this stage are focused on exposing false vulnerabilities and simulating exploitation processes within a decoyed sandboxing environment, misreporting system information
Post exploitation (Control, Execute, Maintain)	Deception scenarios for this stage are mostly devoted to post exploit deception protection against internal threats emanating from compromised nodes located inside network segments. Malicious software planted by an attacker can attempt an automated network discovery of nodes, devices, hosts, services, users, groups and network shares, as well as data harvesting from various network hosts. Other potential threats and hence opportunities for deception include horizontal malware propagation, attempts to access, modify or delete data and software, data collection and data exfiltration

Deception policies are implemented by processes that use technology solutions. For example, as discussed in Sect. 6.4, Stoll used honey files to track Russian KGB cyber attack operatives in a 1980s US defense network hacking attempt (Stoll, 2005). Other technical security measures include protecting files with blockchain encryption or obfuscating the environment (e.g., honeypots, moving target defense) to protect against a motivated attacker.

Technical solutions, however, do not ensure a successful cyber defense. Zero-day exploits and private key compromise are examples of vulnerabilities that defeat strong technical solutions. Cyber analysis is therefore key to providing network defenders with a clear structuring of how a system is protected at each phase of an attack cycle (Figs. 6.2, 6.3, and 6.5; Table 6.6), which help us to answer the questions posed at the beginning of the chapter.

1. Who is primarily responsible for understanding cyber security and defense when performing cyber analysis and targeting?

   While the roles are still developing, the Chief Information Security Officer (CISO) is primarily responsible for the overall network and its security. CISO staff would use the topics discussed in this chapter to improve network security (e.g., attack cycle understanding).

   In addition to attack cycle understanding, the defender is well served to clearly understand both what is being protected and how the adversary may mount an attack. Crown Jewels Analysis (MITRE) and Collaborative Research into Threats (CRITS) (MITRE, 2014) are a few example approaches to provide defenders with a clear picture of what is to be prioritized, in terms of a cyber protection and defense strategy. The Lockheed Martin attack cycle (Sect. 6.2.1) provides a conceptual framework for organizing the threats, technologies, and possible strategies for defending a cyber environment.

2. Why is cyber security and defense important to analysis and targeting?

   Cyber security and defense are important to analysis and targeting due to their conceptual intermingling. For example, the cyber defense terrain is the maneuver space for cyber targeting. Therefore, each of the target system analysis (TSA) steps will need to account for current or proposed cyber defenses for maneuver and exploit planning.

   In Chap. 2, we noted that the overriding policy interest and documentation is defensive, with any cyber evaluation including a discussion of cyber security and defense. This might be a description of the general threat, as discussed in Chap. 3, or the specifics of a tool (Chap. 5). In addition, cyber security and defense is an issue that includes the failure to protect data, as shown in Chap. 4's information operations outing attacks. The cyber analyst will therefore use cyber security and defense as a backdrop to almost every task that he performs. Section 6.3 decomposes the elements of a cyber system (e.g., end points, connections, and data) as a structure for placing the respective security tools in a layered defense.

3. What are the key issues that provide an intersection between cyber defense and cyber targeting?

Because both analysis and targeting span the traditional confidentiality, integrity, and availability (CIA) cyber security elements, along with covering information operations, security may involve everything from ensuring that the analytical work is not obvious to the target, to keeping a technical low profile during an operation. For example, it is common for penetration testers to get a waiver from a targeted company to ensure that there will be no legal repercussions for unlawful entering, surveilling, or gathering data from a paying target. This case assumes that defenses will not be able to detect and thwart an attacker.

4. What are the elements of an IT system to be secured?

In Sect. 6.3, we simplified the security tools and TTPs to end points (Sect. 6.3.1), connections (Sect. 6.3.2), and key nodes (Sect. 6.3.3) in order to look at how contemporary tools are used to detect both machine specific and distributed malware (i.e., botnet).

5. What is the role of countermeasures for the targeting analyst in cyber security and defense?

We reviewed the importance of understanding defended network terrain in developing a successful defense strategy. In addition, we looked at the broader context of cyber defense, and how we might incorporate honeypots (Sect. 6.3.1.3) and denial and deception (Sect. 6.4.1.2) to improve the likelihood of success in securing and defending a network.

6. How do cyber attack cycles help the cyber analyst to understand target behavior?

In the short term, cyber attack cycles (Sect. 6.2.1) provide perspective concerning the attacker's stage and likely tool types, and from there the defense can formulate a course of action with alternative mitigation strategies. For longer-term considerations, the attack cycle helps the defender focus investments on where the attacker is most vulnerable to ensure that the network security goals are best met.

7. How do current security tools work on cyber security threats?

Antivirus (AV) (Sect. 6.3.1.1) and log file analysis (Sect. 6.3.1.5) are well-known tools for understanding machine-specific threats. However, these tools are challenged by "zero days," (Sect. 6.3.1.2) or exploits for which a countermeasure does not currently exist.

AV tools and log file analysis can also be used to find anomalous activity that may be due to a more advanced threat (e.g., botnets: Sects. 6.3.2.1 and 6.3.2.2).

8. How does infrastructure help to provide system-level security?

Cloud computing, as a method of centralizing computing resources, also provides a means of providing very strong security to protect an organization's compute and data resources. Security Operation Centers (SOCs) (Sect. 6.3.3.1) provide executive management, especially the Chief Information Security Officer (CISO), with the ability to match outsourced infrastructure with their policy objectives. In addition, Amazon, Google, and Microsoft all provide cloud services (Sect. 6.3.3.2) that make it easy for an organization to move these key resources to the cloud.

Centralizing services via the cloud can also create vulnerabilities due to the "always on" nature of the service. Section 6.3.3.2 also describes how APT10 has focused on deploying RATs to Linux, specifically targeting MSSPs in order to capture client data. Section 6.3.3.3 talks about how developing technology (e.g., blockchain) might help with data security.

9. How do tools factor into cyber analysis and targeting?

End point and connection tools are used to detect malware, with Network Analysis (Sect. 6.3.2.3) and NetFlow (Sect. 6.3.2.4) being used to increase resolution concerning possible command and control of bots on a system of interest.

Looking across the attack cycle (Fig. 6.2), tools play a role in the successful completion of each step, for either a defender or an attacker. While there are multiple open-source tools for the reconnaissance, surveillance (i.e., Shodan [industrial control systems (ICS) evaluation]), and exploitation (e.g., Metasploit) of a target system, open-source forums also provide off-the-shelf applications and services to tailor exploits (Carr, 2012). In military terms, tools are the form of maneuver in cyberspace, much like what ships, tanks, and planes provide in the traditional domains of warfare.

Just as the attackers innovate their maneuver capabilities, defenders adjust the terrain to add complexity and risk to attacker behaviors. For example, honey files and honeypots (Sect. 6.3.1.3), techniques that make the attacker believe that seemingly important files and server locations are readily available for the attacker to purloin, have been used from the beginning of cyber defense (Stoll, 2005). Similarly, moving target defenses (MTD) (Sect. 6.3.1.4) have the ability to put an attacker in a virtual hall of mirrors.

One of the key takeaways from the evaluation of end points, connections, and clouds is how vulnerable some of the most sophisticated data security methods are (Sect. 6.3):

- Private keys: a vulnerability for blockchain, MTD, and cloud security solutions.

In addition, even the most advanced cyber attacks (e.g., bots) leave footprints (Sect. 6.3.3):

- Anomalous communications:
  - Port scans for propagating malware.
  - Connection requests for bots "phoning home".
  - DNS query failures to expose a bot obfuscating it callback to a C&C node.

These are just a few examples of how cyber security and defense analysts might use Sect. 6.3's techniques to detect anomalous behavior. In addition, the security technologies presented in this chapter help with both the analysis and targeting of cyber systems. For example, understanding deployed security technologies provides a network defender with a clear picture of strengths and weaknesses of his current system.

Each of these security technologies reviewed in this chapter can be positioned, at the correct step of a cyber attack cycle (Fig. 6.5), to clearly understand how the respective security measures can help defeat an attacking adversary. In addition, understanding the network terrain is key for red team attackers and penetration testers to capture the key information elements required at each step of a cyber target's development for attack planning.

Bibliography

Abandon, I. C. (2013, 9 26). *Blackhole exploit kit spam runs: A threat vortex?* Retrieved June 10, 2020, from TrendMicro: https://www.trendmicro.com/vinfo/us/threat-encyclopedia/web-attack/128/blackhole-exploit-kit-spam-runs-a-threat-vortex

Alperovich, D. (2018, November 13). *10 Cyber Myths.* 2018 CyCon keynote. Washington, DC, USA.

Amazon Web Services (AWS). (n.d.). *Amazon AWS GovCloud Glossary.* Retrieved June 8, 2020, from https://docs.aws.amazon.com/govcloud-us/latest/UserGuide/glossary.html

Amoroso, E. (2011). *Cyber attacks: Protecting National Infrastructure.* Elsevier.

Ana Nora Sovarel, D. E. (2005). Where's the FEEB? The effectiveness of instruction set randomization. In *SSYM'05: Proceedings of the 14th conference on USENIX Security* (p. 10). USENIX.

Angus King, M. G. (2020, 3). *United States of America cyberspace solarium commission.* Retrieved May 9, 2020, from The Cyberspace Solarium Commission: https://www.solarium.gov/

Australian Cyber Security Centre. (n.d.). *Essential eight.* Retrieved April 8, 2021, from Australian Government - Australian Signals Directorate: https://www.cyber.gov.au/acsc/view-all-content/essential-eight

Barth, A., & C. J. (2008). Robust defenses for cross-site request forgery. In *Proceedings of the 15th ACM Conference on Computer and Communications Security (CCS '08).*

Bennett, M., & E. W. (2007). *Counterdeception principles and applications for national security.* ArTech House.

Bernier, M. (2015). *Cyber effects categorization - the MACE taxonomy.* DRDC Center for Operational Research and Analysis. TTCP JSA TP3 Cyber Analysis.

Biggs, J. (2018, 3 8). *Sierra Leone just ran the first blockchain-based election.* Retrieved July 9, 2019, from tech crunch: https://techcrunch.com/2018/03/14/sierra-leone-just-ran-the-first-blockchain-based-election/

BlackBerry. (2020). *Decade of the RATS - cross-platform APT espionage attacks targeting Linux.* Windows and Android.

BlackBerry® Cylance®. (2020). BlackBerry® Cylance® 2020 threat report. *BlackBerry® Cylance®. BlackBerry® Cylance®.*

Blank, S. (2013). Russian information warfare as domestic counterinsurgency. *American Foreign Policy Interests, 35*(1), 31–44.

Blank, S. 2017Cyber war and information war a la Russe. In A. E. G. Perkovich (Ed.), *Understanding cyber conflict.* Georgetown.

Bodmer, S., & M. K. (2012). *Reverse Deception: Organized cyber threat counter-exploitation.* McGraw-Hill.

Bowden, M. (2011). *Worm - the first digital world war.* Atlantic Monthly Press.

Bradley Greaver, L. R. (2017). CARVER 2.0: Integrating the analytical hierarchy Process's multi-attribute decision-making weighting scheme for a center of gravity vulnerability analysis for US special operations forces. *Journal of Defense Modeling and Simulation, 15*(1), 111–120.

Brian Van Leeuwen, W. S. (2016). *MTD Assessment Framework with Cyber Attack Modeling.* Retrieved June 10, 2020, from https://www.osti.gov/servlets/purl/1408371

Carr, J. (2012). *Inside Cyber Warfare.* O'Reilly.

Chairman of the Joint Chiefs of Staff (CJCS). (2019, 3 8). *METHODOLOGY FOR COMBAT ASSESSMENT (CJCSI 3162.02).* Retrieved June 22, 2020, from Joint Chiefs of Staff: https://www.jcs.mil/Portals/36/Documents/Doctrine/training/jts/cjcsi_3162_02.pdf?ver=2019-03-13-092459-350

Claire Le Goues, A. N.-T. (2013). *Moving target defenses in the helix self-regenerative architecture*. Springer.

Clapper, J. R. (2015). *Statement for the record: Worldwide threat assessment of the US intelligence community. Senate armed services committee*. Senate Armed Services Committee.

Clare Maathuis, W. P. (2020, 4 11). *Decision support model for effects estimation and proportionality assessment for targeting in cyber operations*. Retrieved June 22, 2020, from science direct: https://www.sciencedirect.com/science/article/pii/S2214914719309250

Richard A. Clarke, R. K. (2012). Cyber war: The next threat to National Security and what to do about it. .

Cohen, J. (1988). *Statistical power analysis for the behavioral sciences*. Lawrence Earlbaum Associates.

Constantin, L. (2020, 12 15). *SolarWinds attack explained: And why it was so hard to detect*. Retrieved December 20, 2020, from CSO online: https://www.csoonline.com/article/3601508/solarwinds-supply-chain-attack-explained-why-organizations-were-not-prepared.html

Couretas, J. M. (2018). *Introduction to cyber modeling and simulation*. John Wiley and Sons.

Cox, M. (2018, 5 25). *Military.com*. Retrieved from US, Coalition Forces Used Cyberattacks to Hunt Down ISIS Command Posts: https://www.military.com/dodbuzz/2018/05/25/us-coalition-forces-used-cyberattacks-hunt-down-isis-command-posts.html?utm_source=Mike%27s+Daily+Blast&utm_campaign=8d4307f510-EMAIL_CAMPAIGN_2018_04_27_COPY_02&utm_medium=email&utm_term=0_beae3dbeb1-8d4307f510

Cybercrime to Cost Over $10 Trillion Annually by 2025. (2021, 2 21). *The Taiwan Times*.

David Evans, A. N.-T. (2011). Effectiveness of moving target defenses. In A. G. S. Jajodia (Ed.), *Moving target defense*. Springer.

David Leigh, L. H. (2011). *WikiLeaks - inside Julian Assange's war on secrecy*. Public Affairs.

Department of Defense. (2002). *Joint doctrine for targeting. Department of Defense, joint chiefs of Staff Washington, DC*. Department of Defense.

Department of Defense. (2012). *Joint Publication 1–13.4 Military Deception*. Department of Defense.

Director of National Intelligence. (n.d.). *Cyber Threat Framework*. Retrieved October 20, 2018, from https://www.dni.gov/index.php/cyber-threat-framework

Drolet, M. (2018, 1 26). *What does stolen data cost [per second]? 58 data records are stolen every second. Guess what the average cost is*. Retrieved September 13, 2018, from CSO online: https://www.csoonline.com/article/3251606/data-breach/what-does-stolen-data-cost-per-second.html

Dusan Repel, S. H. (2015). *The ingredients of cyber weapons. 10th international conference on cyber warfare and security* (pp. 1–10). ICCWS15.

Edward Waltz, M. B. (2007). *Counterdeception principles and applications for national security*. ArTech.

Emily, O., & Goldman, M. W. (2017). Why a digital Pearl Harbor makes sense. In A. E. G. Perkovich (Ed.), *Understanding cyber conflict*. Georgetown.

Eric Gartzke, J. L. (2015, 6 22). *Weaving tangled webs: Offense, defense, and Deception in cyberspace*. Retrieved March 31, 2020, from security studies: https://deterrence.ucsd.edu/_files/Weaving%20Tangled%20Webs_%20Offense%20Defense%20and%20Deception%20in%20Cyberspace.pdf

Evans, N. (2016, 3 3). *Multiple operating system rotation environment moving target defense* . Retrieved December 12, 2018, from US Patent and Trademark Office (USPTO): http://appft1.uspto.gov/netacgi/nph-Parser?Sect1=PTO1&Sect2=HITOFF&d=PG01&p=1&u=/netahtml/PTO/srchnum.html&r=1&f=G&l=50&s1=20160065612.PGNR

Evans, B. (2017, 11 7). *The top 5 cloud-computing vendors: #1 Microsoft, #2 Amazon, #3 IBM, #4 salesforce, #5 SAP*. Retrieved June 9, 2020, from Forbes: www.forbes.com/sites/bobevans1/2017/11/07/the-top-5-cloud-computing-vendors-1-microsoft-2-amazon-3-ibm-4-salesforce-5-sap/#409863346f2e

FireEye. (2017). *APT 29*. (FireEye, Producer) Retrieved September 9, 2018, from APT 29: https://www.fireeye.com/current-threats/apt-groups.html#apt29

FireEye. (2017). *M-Trends 2017 - A view from the front lines*. FireEye.

Fox, W. P., & J. H. (2018). Methodology for targeting analysis for minimal response. *Journal of Defense Modeling and Simulation, 15*(3).

Gallagher, M. (2008). *Cyber analysis workshop. MORS*. MORS.

Gallagher, S. (2019, 1 30). *FBI, air force investigators mapped north Korean botnet to aid shutdown*. Retrieved June 8, 2020, from Ars Technica: https://arstechnica.com/information-technology/2019/01/fbi-air-force-investigators-mapped-north-korean-botnet-to-aid-shutdown/

Ganesh, V. (2015, 12). *Parking lot monitoring system using an autonomous quadrotor UAV*. Retrieved September 19, 2018, from Clemson University: https://pdfs.semanticscholar.org/9651/9650a1a734cd1feac6f33b7858f476e14f80.pdf

GAO. (2020, 4). *DOD needs to take decisive actions to improve cyber hygiene*. Retrieved June 9, 2020, from GAO: https://www.gao.gov/assets/710/705886.pdf

Geers, K., & e. (2015). *Cyber war in perspective: Russian aggression against Ukraine*. NATO Cooperative Cyber Defence Centre of Excellence.

George Cybenko, G. S. (2016). Quantifying covertness in deceptive cyber operations. In V. S. S. Jajodia (Ed.), *Cyber Deception: Building the Scientific Foundation*. Springer.

Georgios Portokalidis, A. D. (2011). Instruction set randomization. In S. Jajodia (Ed.), *Moving Target Defense*. Springer.

Gibson, W. (1984). *Neuromancer*. Ace.

Gladwell, M. (2010, 1 10). *The sure thing*. Retrieved September 21, 2019, from the new Yorker: https://www.newyorker.com/magazine/2010/01/18/the-sure-thing

Godson, R., & J. W. (2002). *Strategic denial and Deception*. Transactio Publishers.

Greg Bridkin, D. F.-S. (2016). Design considerations for building cyber deception systems. In S. Jajodia (Ed.), *Cyber Deception* (pp. 69–95). Springer.

Gus, D., & D. D. (2011). *Cultural difference in dynamic decision making strategies in a non-lines, time delayed task* (Vol. 12). Cognitive Systems Research.

Hamed Okhravi, W. W. (2016). Moving target techniques: Leveraging uncertainty for cyber defense. *Lincoln Laboratory Journal, 22*(1), 10.

Hamed Okhravi, W. W. (2018, 5 22). *Moving target techniques: Leveraging uncertainty for cyber defense*. Retrieved June 10, 2020, from MIT Lincoln lab: https://www.ll.mit.edu/sites/default/files/page/doc/2018-05/22_1_8_Okhravi.pdf

Hastings, M. (2012, 1 18). *Julian Assange: The rolling stone interview*. Retrieved December 12, 2018, from rolling stone: https://www.rollingstone.com/politics/politics-news/julian-assange-the-rolling-stone-interview-234403/

Hayden, M. V. (2016). Playing to the edge: American intelligence in the age of terror. *Peguin*.

Heckman, K. (2013). *Active cyber network defense with denial and Deception. MITRE*. http://goo.gl/Typwi4

Heckman, K. E., & F. J. (2015a). Cyber Counterdeception: How to Detect Denial & Deception (D&D). In P. S. S. Jajodia (Ed.), *Cyber Warfare: Building the Scientific Foundation*. Springer.

Heckman, K. E., & F. J. (2015b). Cyber Counterdeception: How to Detect Denial & Deception (D&D). In P. S. S. Jajodia (Ed.), *Cyber Wrfare: Building the Scientific Foundation*. Springer.

Heckman, K. E., & F. J. (2015c April). Denial and deception in cyber defense. *Computer, 48*(4), 36–44.

Heickero, R. (2013). *Emerging cyber threats and Russian views on information warfare and information operations*. Swedish Defense Research Agency.

Heli Tiirmaa-Klaar, J. G.-P. (2014). *Botnets*. Springer.

Henderson, S. (2007). The dark visitor. .

Hofstede, G., & G. H. (2010). *Cultures and organizations* (3rd ed.). McGraw-Hill.

Hong, L. V. (2012, 4 16). *DNS traffic analysis for network-based malware detection*. Retrieved June 8, 2020, from KTH information and communication technology: https://people.kth.se/~maguire/DEGREE-PROJECT-REPORTS/120416-Linh_Vu_Hong-with-color-cover.pdf

Humphrey, W. (1989). *Managing the software process.* Addison Wesley.

Institute, S. A. N. S. (2006). *A guide to security metrics.* SANS.

Ivanov, S. (2007). *NTV.* Open Source Center.

Jaquith, A. (2007). *Security metrics: Replacing fear, uncertainty, and.* Pearson Education.

Jenna McLaughlin, Z. D. (2018, 12 8). *At the CIA, a fix to communications system that left trail of dead agents remains elusive.* Retrieved December 10, 2018, from yahoo news: https://news.yahoo.com/cia-fix-communications-system-left-trail-dead-agents-remains-elusive-100046908.html

Jervis, R. (1976). *Deception and misperception in international politics.* Princeton University Press.

John Kolb, M. A. (2020, 2). *Core concepts, challenges, and future directions in Blockchain: A centralized tutorial.* Retrieved June 8, 2020, from ACM: https://dl.acm.org/doi/fullHtml/10.1145/3366370

Joint Staff. (2016). CJCSI 3370.01B target development standards. Joint Staff, .

Joint Staff. (2017, 6 16). *Joint planning.* (J. Staff, producer) Retrieved February 11, 2019, from www.jcs.mil/Doctrine/Joint-Doctrine-Pubs/5-0-Planning-Series

Jon Oliver, S. C. (2012). *Blackhole exploit kit: A spam campaign, not a series of individual spam runs.* Retrieved May 16, 2020, from trend micro: https://www.trendmicro.de/cloud-content/us/pdfs/security-intelligence/white-papers/wp_blackhole-exploit-kit.pdf

Jones, R. (1989). *Reflections on intelligence.* William Heinemann Ltd.

Jonsson, E., & T. O. (1997). A quantitative model of the security intrusion process based on attacker behavior. *IEEE Transactions on Software Engineering.*

Joseph Muniz, G. M. (2015). *Security operations center: Building.* Cisco Press.

Kathleen Metrick, P. N. (2020, 4 6). *Zero-day exploitation increasingly demonstrates access to money, rather than skill — Intelligence for vulnerability management, part one.* Retrieved June 10, 2020, from FireEye: https://www.fireeye.com/blog/threat-research/2020/04/zero-day-exploitation-demonstrates-access-to-money-not-skill.html

Koerner, B. I. (2016, 10 23). *Inside the Cyberattack That Shocked the US Government.* Retrieved from Wired: https://www.wired.com/2016/10/inside-cyberattack-shocked-us-government/

Kohen, I. (2017, 8 15). *Cost of insider threats vs. investment in proactive education and technology.* Retrieved September 13, 2018, from CSO online: https://www.csoonline.com/article/3215888/data-protection/cost-of-insider-threats-vs-investment-in-proactive-education-and-technology.html

Lambert, J. (2018, 11 6). *@JohnLaTwC.* Retrieved December 14, 2018, from Twitter: https://twitter.com/johnlatwc/status/1059841882086232065?lang=en

Lamothe, D. (2017, December 16). *How the Pentagon's cyber offensive against ISIS could shape the future for elite U.S. forces.* Retrieved from Washington Post: https://www.washingtonpost.com/news/checkpoint/wp/2017/12/16/how-the-pentagons-cyber-offensive-against-isis-could-shape-the-future-for-elite-u-s-forces/?noredirect=on&utm_term=.88d87f70ceea

Lasker, L., & Badham, J. (1983). WarGames [motion picture]. .

Lepore, J. (2021, 2 1). *The next cyberattack is already under way.* Retrieved February 5, 2021, from New Yorker: https://www.newyorker.com/magazine/2021/02/08/the-next-cyberattack-is-already-under-way

Leversage, D. (2007). Comparing electronic battlefields: Using mean time-to-compromise as a comparative security metric. In I. K. V. Gordodetsky (Ed.), *Computer Network Security.* Springer.

lexico. (n.d.). *lexico.* Retrieved June 9, 2020, from https://www.lexico.com/definition/cloud_computing

Li, W., & R. V. (2006). Cluster security research involving the modeling of network exploitations using exploitation graphs. In *Sixth IEEE International Symposium on Cluster Computing and Grid Workshops.*

Lowrey, A. (2014, 12 16). *Sony's very, very expensive hack.* Retrieved September 25, 2018, from daily intelligencer: http://nymag.com/daily/intelligencer/2014/12/sonys-very-very-expensive-hack.html

Manadhata, P. K. (2008). *An attack surface metric*. CMU.

Max Boot, M. D. (2013). *Political warfare*. Council on Foreign Relations.

Merriam-Webster. (n.d.). *Target (noun, often attributive)*. Retrieved 7 8, 2019, from Merriam-Webster: https://www.merriam-webster.com/dictionary/target

MITRE. (2014, 7 3). *Cybersecurity*. Retrieved February 10, 2018, from Collaborative Research Into Threats (CRITs): https://www.mitre.org/capabilities/cybersecurity/overview/cybersecurity-blog/collaborative-research-into-threats-crits

MITRE. (n.d.-a). *CARET*. Retrieved 9 30, 2018, from CARET: https://car.mitre.org/caret/#/

MITRE. (n.d.-b). *Crown jewels analysis*. Retrieved December 14, 2018, from https://www.mitre.org/publications/systems-engineering-guide/enterprise-engineering/systems-engineering-for-mission-assurance/crown-jewels-analysis

Morales, A. (2020, 5 3). *What to make of HBO's 'kill chain: The cyber war on America's elections'*. Retrieved June 10, 2020, from The Fifth Domain: https://www.fifthdomain.com/opinion/2020/05/03/what-to-make-of-hbos-kill-chain-the-cyber-war-on-americas-elections/

Mullin, R. (2014, 11 24). *Cost to develop new pharmaceutical drug now exceeds $2.5B*. Retrieved September 13, 2018, from Scientific American: https://www.scientificamerican.com/article/cost-to-develop-new-pharmaceutical-drug-now-exceeds-2-5b/

Newman, L. H. (2017, September 8). *The equifax breach exposes America's identity crisis*. Retrieved September 7, 2018, from Wired: https://www.wired.com/story/the-equifax-breach-exposes-americas-identity-crisis/

Nilsson, N. (1987). *Principles of artificial intelligence*. Springer-Verlag.

NIST. (n.d.-a). *Computer security resource center*. Retrieved December 20, 2020, from Information Technology Laboratory: https://csrc.nist.gov/glossary/term/attack_surface

NIST. (n.d.-b). *National Vulnerability Database (NVD)*. Retrieved December 20, 2020, from NIST: https://nvd.nist.gov/

O.W.A.S.P. (OWASP). (2013). *OWASP Top 10*. Retrieved May 26, 2018, from http://owasptop10.googlecode.com/files/OWASPTop10-2013.pdf

Oladimeji, S. (2021, 2 9). *SolarWinds hack explained: Everything you need to know*. Retrieved February 10, 2021, from TechTarget: https://whatis.techtarget.com/feature/SolarWinds-hack-explained-Everything-you-need-to-know

Ortalo, R., & Y. D. (1999). Experimenting with quantitative. *IEEE Transactions on Software Engineering, 25*, 633–650.

Osinga, F. P. (2007). *Science, Strategy and War: The Strategic Theory of John Boyd*. Routledge.

Paul, K. (2021, 2 23). *SolarWinds hack was work of 'at least 1,000 engineers', tech executives tell senate*. Retrieved February 24, 2021, from Guardian: https://www.theguardian.com/technology/2021/feb/23/solarwinds-hack-senate-hearing-microsoft

Perlroth, N. (2021). *This is how they tell me the world ends: The Cyberweapons arms race*.

Polyverse. (n.d.). *Polyverse - How it Works*. Retrieved December 12, 2018, from https://polyverse.io/how-it-works/

Popular Mechanics. (2018, March 13). *How Long Does It Take Hackers To Pull Off a Massive Job Like Equifax?* Retrieved from Popular Mechanics: https://www.popularmechanics.com/technology/security/a18930168/equifax-hack-time/

Porup, J. (2019, 3 5). *What is Mimikatz? And how to defend against this password stealing tool*. Retrieved May 16, 2020, from CSO: https://www.csoonline.com/article/3353416/what-is-mimikatz-and-how-to-defend-against-this-password-stealing-tool.html

Poulsen, K. (2018, 10 25). *'Lone DNC Hacker' Guccifer 2.0 slipped up and revealed he was a Russian intelligence officer*. Retrieved July 8, 2019, from the daily beast: https://www.thedaily-beast.com/exclusive-lone-dnc-hacker-guccifer-20-slipped-up-and-revealed-he-was-a-russian-intelligence-officer

Reagan, R. (1984, 9 17). *National Security Decision Directive 145*. Retrieved May 9, 2020, from GWU National Security Archive: https://nsarchive2.gwu.edu/dc.html?doc=2778589-Document-01-Ronald-Reagan-National-Security

Reilly, K. (2017, 6 19). *Nearly 200 million U.S. voters' personal data accidentally leaked by data firm contracted by RNC*. Retrieved June 8, 2020, from Forbes: https://fortune.com/2017/06/19/deep-root-analytics-voter-data-exposed/

Rick Nunes-Vaz, S. L. (2011). A more rigorous framework for security-in-depth. *Journal of Applied Security Researh, 23.*

Rick Nunes-Vaz, S. L. (2014). From strategic security risks to National Capability Priorities. *Security Challenges, 10*(3), 23–49.

Ricker, T. (2019, 12 31). *Go read this 'cloud hopper' hacking investigation by the WSJ - much worse than originally reported*. Retrieved June 8, 2020, from TheVerge: https://www.theverge.com/2019/12/31/21044173/cloud-hopper-apt10-china-hackers

Rid, T. (2012, 2 27). Cyber War: Think Again.

Rob Barry, D. V. (2019, 12 30). *Ghosts in the clouds: Inside China's major corporate hack A journal investigation finds the cloud hopper attack was much bigger than previously known*. Retrieved June 8, 2020, from wall street journal: https://www.wsj.com/articles/ghosts-in-the-clouds-inside-chinas-major-corporate-hack-11577729061

Romanosky, S. (2016). Examining the costs and causes of cyber incidents. *Journal of Cybersecurity, 2*(2).

Rowe, N. (2007). Deception in defense of computer systems from cyber attack. In L. J. M. Andrew (Ed.), *Colarik, cyber war and cyber terrorism*. The idea Group.

Rowe, N., & H. R. (2004). Two taxonomies of deception for attacks on information systems. *Journal of Information Warfare, 3*(2), 27–39.

Youssef, N. A., S. H. (2016, 11 25). Why Did Team Obama Try to Take Down Its NSA Chief? Retrieved from Daily Beast: https://www.thedailybeast.com/why-did-team-obama-try-to-take-down-its-nsa-chief

Sailik Sengupta, A. C. (2019, 5 2). A Survey of Moving Target Defenses for Network Security. CoRR.

Sallhammar, K., & B. H. (2006). On stochastic modeling for integrated security and dependability evaluation. *Journal of Networks.*

Sanger, D. E. (2016, April 24). *U.S. cyberattacks target ISIS in a new line of Combat*. Retrieved from New York Times: https://www.nytimes.com/2016/04/25/us/politics/us-directs-cyberweapons-at-isis-for-first-time.html

Sanger, D. E. (2018). *The perfect weapon - war, sabotage and fear in the cyber age*. Crown.

Sanger, D. E., E. S. (2017, June 12). *U.S. Cyberweapons, used against Iran and North Korea, are a disappointment against ISIS*. Retrieved from New York times: https://www.nytimes.com/2017/06/12/world/middleast/isis-cyber.html

Sanger, D. E., N. P. (2020, 4 15). *U.S. accuses North Korea of cyberattacks, a sign that deterrence is failing*. Retrieved 6 3, 2020

Scott Stewart, F. B. (2009, 5 20). *A counterintelligence approach to controlling cartel corruption*. Retrieved June 22, 2020, from Stratfor: https://worldview.stratfor.com/article/counterintelligence-approach-controlling-cartel-corruption

Segura, V. (2009, June 25). *Modeling the economic incentives of DDoS Attacks*: Retrieved from Semantic Scholar: https://pdfs.semanticscholar.org/afdf/d974bc68dc05c48020e-f07a558a61ab94f8a.pdf.

Senate Armed Services Committee. (2014). *Inquiry in cyber Instrusions affecting US transportation command contractors* (113th Congress, 2d sess ed.). Government Printing Office.

Simon Pirani, J. S. (2009, 2 15). *The Russo-Ukrainian gas dispute of January 2009: a comprehensive assessment*. Retrieved December 11, 2018, from Oxford Institute for Energy Studies : https://www.oxfordenergy.org/wpcms/wp-content/uploads/2010/11/NG27-TheRussoUkrainianGasDisputeofJanuary2009AComprehensiveAssessment-JonathanSternSimonPiraniKatjaYafimava-2009.pdf

Simonsson, M., & P. J. (2007). Model-based IT governance maturity assessments with COBIT. In *ECIS 2007 PROCEEDINGS*. ECIS.

Spamhaus. (2020, 6 8). *Live Botnet Threats Worldwide*. Retrieved June 8, 2020, from Spamhaus: https://www.spamhaustech.com/threat-map/

Standage, T. (2017, October 5). *The crooked timber of humanity*. Retrieved September 18, 2018, from 1843: https://www.1843magazine.com/technology/rewind/the-crooked-timber-of-humanity

Statistica. (2017). *Annual number of data breaches and exposed records in the United States from 2005 to 2018 (in millions)*. Retrieved September 14, 2018, from Statistica: https://www.statista.com/statistics/273550/data-breaches-recorded-in-the-united-states-by-number-of-breaches-and-records-exposed/

Sterling, B. (2011, 6 29). *The dropped drive hack*. Retrieved September 19, 2018, from Wired: https://www.wired.com/2011/06/the-dropped-drive-hack/

Stoll, C. (2005). *The Cuckoo's egg: Tracking a spy through the maze of computer espionage*. Pocket Books.

Strategy and Tactics of Guerilla Warfare. (n.d.). Retrieved 9 9, 2018, from Wikipedia: https://en.wikipedia.org/wiki/Strategy_and_tactics_of_guerrilla_warfare

Sulmeyer, M. (2017, 12 18). *Campaign Planning with Cyber Operations*. Retrieved March 31, 2020, from Georgetown Journal of International Affairs: https://www.georgetownjournalofinternationalaffairs.org/online-edition/2017/12/22/campaign-planning-with-cyber-operations

Sushil Jajodia, V. S. (2016). *Cyber deception*. Springer.

Thomas, T. (2016). *Russian strategic thought and cyber in the armed forces and society: A viewpoint from Kansas. Center for strategic and international studies*. Center for strategic and International Studies.

Thomas Erl, R. P. (2013). *Cloud computing: Concepts, Technology & Architecture (the Pearson Service technology series from Thomas Erl)*. Prentice Hall.

Thompson, J. R., & R. H.-W. (1984). *The cognitive bases of intelligence analysis. U.S. Army, U.S.* Army Research Institute for the Behavioral and Social Sciences. U.S. Army.

Tversky, A., & D. K. (1974). *Judgement under uncertainty: Heuristics and biases* (Vol. 185, p. Science).

U.S. Department of Homeland Security (DHS). (n.d.). *Cybersecurity and Infrastructure Security Agency (CISA)*. Retrieved April 10, 2021, from https://us-cert.cisa.gov/

US Department of Defense. (2008). *Cyberspace operations*. US Department of Defense.

USAF. (2019, 3 15). *The targeting cycle (annex 3-60)*. Retrieved June 22, 2020, from https://www.doctrine.af.mil/Portals/61/documents/Annex_3-60/3-60-D04-Target-Tgt-cycle.pdf

Vogt, P., & F. N. (2007a). Corss-site scripting prevention with dynamic data tainting and static analysis. In *The 2007 Network and Distributed System Security Smposium*. NDSS '07.

Vogt, P., & F. N. (2007b). Cross-site scripting prevention with dynamic data tainting and static analysis. In *The 2007 Network and Distributed System Security Symposium*. NDSS '07.

Whaley, B. (2007). *STRATAGEM - Deception and surprise in war*. Artech House.

Whaley, B., & S. A. (2007). *Textbook of political-military Counterdeception: Basic principles and methods*. National Defense Intelligence College.

Williams, J. (2020, 12 15). *What you need to know about the SolarWinds supply-chain attack*. Retrieved February 15, 2021, from SANS: https://www.sans.org/blog/what-you-need-to-know-about-the-solarwinds-supply-chain-attack/

Zachary Cohen, A. M. (2021, 4 2). *Hunting the hunters: How Russian hackers targeted US cyber first responders in SolarWinds breach*. Retrieved April 2, 2021, from CNN: https://www.cnn.com/2021/04/02/politics/russian-hackers-target-us-cyber-hunters-solarwinds/index.html

Zetter, K. (2007, 11 14). *TOR researcher who exposed embassy e-mail passwords gets raided by swedish FBI and CIA*. Retrieved December 12, 2018, from https://www.wired.com/2007/11/swedish-researc/

Zetter, K. (2009, 11 3). *Mossad hacked Syrian Official's computer before bombing mysterious facility*. Wired.

Zetter, K. (2013, 6 19). *Microsoft launches $100K bug bounty program*. Retrieved March 21, 2020, from wired: https://www.wired.com/2013/06/microsoft-bug-bounty-program/

Zetter, K. (2014). *Countdown to zero day - Stuxnet and the launch of the World's first digital weapon*. Crown.

Zurasky, M. W. (2017). *Methodology to perform cyber lethality assessment*. Old Dominion University.

Chapter 7
Cyber Offense and Targeting

The purpose of this chapter is to provide the reader with a framework to understand cyber targeting. We will start this chapter with a review of well-known kinetic targeting approaches in order to understand the structure of a cyber attack and use this insight to develop a cyber-targeting framework. This cyber targeting framework will help us (1) categorize existing cyber attack approaches in terms of their phase in a kinetic attack process, (2) capture the key information elements required at each step of a cyber target's development, and (3) cost the estimated resources needed for hacker, terrorist, and nation state groups to prosecute an example target.

Cyber Offense and Targeting Example Questions
1. What are key policy and doctrine documents that govern cyber-targeting operations?
2. What are the key processes used in targeting that might be helpful for understanding cyber attackers?
3. What role does technology play in cyber targeting?
4. How is cyber targeting measured, and what are potential improvements in assessing cyber operations?
5. How is cyber targeting different in producing information operations (IO) vs. the more technical confidentiality/integrity/availability (CIA) type of computer-based system effects?

In Chap. 6, we looked at how network defenders actively try to deny access to their systems. Network defense teams use a variety of technologies (e.g., firewalls and encryption), techniques (e.g., moving target defense and denial and deception), and personnel with different skill sets, guided by a defined set of policies and procedures, to build a layered cyber defense.

Security teams continually update network defense layers to ward off the attack. This hardening of network defenses includes training network defense teams to increase their skills in detecting and remediating malware. In addition, security teams update security control processes to account for the latest attack types. For example, in Chap. 3, we looked at SANS 20 and the Australian Essential Eight

© Springer Nature Switzerland AG 2022
J. M. Couretas, *An Introduction to Cyber Analysis and Targeting*,
https://doi.org/10.1007/978-3-030-88559-5_7

(Australian Cyber Security Centre) as examples of process controls that are implemented, by order of effectiveness, to prevent a cyber attacker from penetrating a network. Similarly, security teams use vulnerability advisories (U.S. Department of Homeland Security (DHS)) to identify and replace aging technologies to shore up network defenses.

Attackers adapt to each security measure by tuning their techniques to meet and overcome each new network defense measure. In addition, the combination of multiple system users with constantly changing computer components transforms each network into a multidimensional attack surface that cyber bad actors use to probe and exercise their social engineering and malware techniques.

The daunting scope of cyber terrain requires us to clearly define the key elements that compose a cyber targeting problem. We will therefore start this chapter with the definition of a target, specifically for cyber—

A Target (Merriam-Webster)

1. A goal to be achieved
2. Something or someone fired at or marked for attack
3. Something or someone to be affected by an action or a development

In the context of cyber, targeting can be used in two ways (Bennett & Waltz, 2007):

1. Selection of a cyber object for intelligence, or computer network, exploitation (CNE)
2. Selection of a cyber object for offensive attack to deny (e.g., degrade, disrupt, destroy) or manipulate (e.g., take control of an adversary's system)

The purpose of this chapter is to present a general background on kinetic targeting processes, cyber attack cycles, and their combination, in order to develop a practical cyber targeting framework. We will start with a review of well-known kinetic targeting processes. This review will help us identify how current cyber attack cycles fit into well-known targeting processes. We will then overlay cyber attack cycles onto an accepted kinetic attack process. The result will be a general cyber targeting framework. This cyber targeting framework will then be used to show how to estimate operational resource requirements, at each step of an attack, for a set of malicious cyber actors (MCA).

7.1 Background

Target understanding, along with the associated skills and abilities to deny and manipulate an adversary, is key to successful cyber targeting. Developing any target includes the painstaking steps involved in intelligence gathering, as discussed in Chap. 5. The promise of cyber, however, goes beyond an in-depth intelligence collection exercise. The proposed use of cyber is often co-mingled with force-like effects.

Cyber effects overlap in many people's minds with an almost spooky "action at a distance." This mingling of science fiction with current cyber operations likely stems from the futuristic literature that inspires some of the engineers and scientists responsible for building our computer-based systems.

Science fiction seems to have first hypothesized computer system hacking in the "consensual illusion," written in William Gibson's dystopian novel Neuromancer (Gibson, 1984). Gibson describes how savants hacked their rivals in cyberspace, and also how corporations and states fought each other in that digital realm. Over time, the term "cyberspace" has gradually worked its way into military doctrine.

Originally introduced in Chap. 2, Joint Publication (JP) 3-12 (US Department of Defense, 2008) describes cyberspace operations in terms of physical, logical, and cyber persona layers. These layers are distinct and interrelated elements of the cyber domain (Fig. 7.1).

As shown in Fig. 7.1, each layer of the cyber domain is the ways and means for targeting to achieve denial or manipulation effects. Each layer in Fig. 7.1 also has the potential to be the ends, or final target, in a cyberspace operation. Therefore, we will review existing targeting doctrine before going into cyber-specific examples.

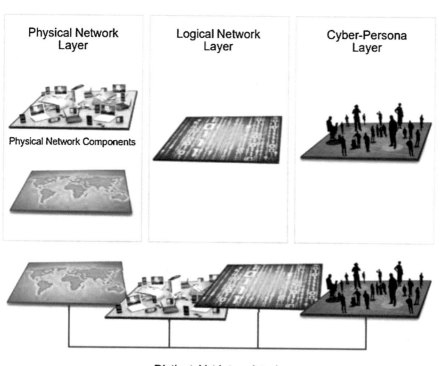

Distinct, Yet Interrelated

Fig. 7.1 Physical, logical, and cyber-persona layers of the cyber domain

7.2   Introduction

Targeting, a mature discipline with existing tool characterizations and application examples, is focused on creating predictably distinct effects through target engagement. In addition, targeting is a process that includes the collaboration of many disciplines, executes through an overall targeting cycle, and seeks to create effects in a systematic manner. This includes decomposing the target, as a component in an adversary's overall system, and estimating effects at each level of target engagement (Fig. 7.2).

As shown in Fig. 7.2, the targeting process is top-down in traversing from the adversary, to the target system, to target system components, and finally to individual target elements. Battle damage assessment (BDA), on the other hand, is used to evaluate effects in reverse. Phase one BDA is used to estimate effects on the individual target element that was affected by the operation. Phase two assesses the behavioral effects of the device that included the target element. And phase three BDA evaluates the overall target system that includes the target element.

The decomposition in Fig. 7.2 can also be applied to cyber. For example, cyber targeting can use the joint targeting process' organizational and process foundation (e.g., "Methodology for Combat Assessment" (Chairman of the Joint Chiefs of Staff (CJCS), 2019)) to develop an approach that uses both the inherent flexibility and dynamics of the cyber terrain to maneuver across the physical, logical, and cyber-persona layers (Fig. 7.1) that compose a cyber target.

In developing our cyber targeting framework, we will start with a review of the kinetic targeting cycle and look at well-known approaches (e.g., JP 3-60). We will

Fig. 7.2 Example target taxonomy for Battle Damage Assessment (BDA) (Chairman of the Joint Chiefs of Staff (CJCS), 2019)

also look at cyber attack frameworks introduced in Chap. 2 (e.g., Lockheed Martin Kill Chain, MITRE ATT@CK) and develop a cyber-specific attack process, the "Cyber Process Evaluator," that overlays cyber considerations onto the kinetic targeting cycle. Using the cyber process evaluator, we will also look at an end-to-end cyber targeting cycle example that evaluates the people, process, and technology resource usage at each step of a cyber operation in order to estimate the associated cost at each step to the attacking organization.

7.2.1   Targeting and Cyber Applications

"Target system analysis is a systematic approach to determine enemy vulnerabilities and exploitable weaknesses. It determines what effects will likely be achieved against target systems and their associated activities." (USAF, 2019)

Target system analysis (TSA) is the incremental development of a target to support a commander's engagement, mission, and campaign-level objectives. Engagement goals are used to describe a specific target, technical details of accessing the target, and estimates of follow-on effects due to engaging the target. Mission goals include multiple engagements to accomplish each mission objective. And campaign goals are the set of missions performed to win an overall campaign. The focus of cyber operations, here, is at the engagement level, where each cyber target is addressed in terms of target system analysis (TSA).

TSA includes evaluating addressable targets based on current access tools and team skill levels. In addition, TSA is used to minimize the likelihood of undesirable consequences and uses a risk/reward evaluation to qualify addressable targets. This includes determining which targets are worth the effort to invest resources in researching the technical, personnel, and associated process vulnerabilities. This analysis also includes developing an exploit for the target's vulnerabilities, building the exploit into a weapon, and executing the mission; described by the six phases in the Joint Publication (JP) 3-60 (USAF, 2019) targeting process:

- Phase 1—Commander's objectives, guidance, and intent
- Phase 2—Target development, validation, nomination, and prioritization
- Phase 3—Capabilities analysis
- Phase 4—Commander's decision and force assignment
- Phase 5—Mission planning and force execution
- Phase 6—Combat assessment

The six phases of the kinetic targeting process span from understanding commander intent to assessing target system effects. This process is easier to follow with a phase diagram that includes the possibility for the inevitable feedback that will occur over the course of the targeting process (Fig. 7.3).

Figure 7.3, in stepping through the six phases of targeting, describes the process of how to understand a target in terms of an organization's (1) mission objectives and (2) technical abilities to perform the mission. In addition, the generality of this

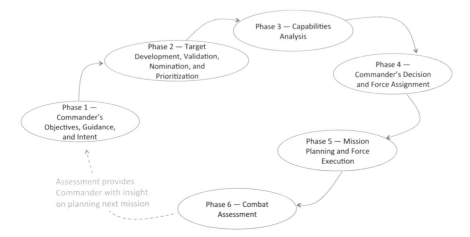

Fig. 7.3 Six phases of targeting (USAF, 2019)

approach, originally developed to methodically understand a target for kinetic applications, is used here to provide an overall framework for cyber targeting.

The goal of Fig. 7.3 is to describe the overall flow of targeting from conceptual development to post-attack analysis. This includes understanding a target's role as a component in an overall system. For example, we want to develop a target in terms of specific effects due to its technical vulnerabilities, as shown in Fig. 7.2. Figure 7.3 is then used to methodically understand the target, operationally, in order to focus effects on controlling specific target behaviors. This detailed target understanding reduces the likelihood of unintended consequences due to collateral damage. As shown, Fig. 7.3 scopes the targeting process based on the commander's objectives, guidance, and intent.

7.2.1.1   Commander's Objectives, Guidance, and Intent

The first phase of targeting, as shown in Fig. 7.3, is coordinating the commander's guidance as it applies to the adversary's systems. Colloquially, the targeteer is trying to find the best "return on investment," in terms of addressable targets that match the commander's current guidance.

As part of a plan, the commander might be interested in relatively short-term effects to accomplish immediate objectives (e.g., technical compromises of vulnerable systems (Chap. 6)). However, the process in Fig. 7.3 is not constrained by time and might be used for an overall campaign. This includes strategic targeting, with long time horizons, for the development and exploitation of a target of interest (e.g., active measures to compromise key people and systems (Chap. 4)). The desired effects and time horizon are key issues when narrowing the candidate list for target development, validation, nomination, and prioritization.

7.2.1.2   Target Development, Validation, Nomination, and Prioritization

Step two in Fig. 7.3 uses the commander's timeframe and desired effects guidance to develop a prioritized list of targets to pursue. This is a look at how candidate targets fit into an adversary's strategic influence system. Target validation includes ensuring that a candidate target fits with both established rules of engagement and the law of armed conflict (LOAC) (USAF, 2019).

Once a set of targets is validated, it is nominated for review by participating organizations. These are the organizations that will be responsible for the work involved in researching the targets, developing the exploits, and weaponizing the asset target interactions (ATIs). In addition, these organizations could be the respective services in a joint forces construct (Joint Staff, 2017). Approval by these parties then results in a prioritization of the targets for requisite capability analysis.

7.2.1.3   Capabilities and Analysis

Phase three of the joint targeting cycle is called capabilities assessment and analysis. This step includes a closer look at the in-house capabilities for achieving the desired effect. Capabilities analysis is composed of four steps: target vulnerability analysis, capabilities assignment, feasibility assessment, and effects estimate.

In Chap. 3, we looked at vulnerability analysis. Each target will be evaluated in terms of its people, process, and technical vulnerabilities (Fig. 7.1) in order to determine the best approach to accomplish targeting objectives. While PASTA (Chap. 3, Sect. 3.3.3) and other vulnerability evaluation approaches provide a clear outline of how to analyze a system for cyber vulnerabilities, gathering the data to apply these approaches remains a challenge.

Target engagement capabilities are assigned to each vulnerability in order to create the candidate asset target interactions (ATIs). The feasibility of each ATI is evaluated in terms of target location (i.e., access) and team capabilities. In addition, part of the feasibility evaluation includes reviewing the effects expected due to prosecuting the target. This step will include considerations on how the target system will react when a target is removed.

Cyber effects estimates are a key part of describing the asset target interaction understanding part of capabilities analysis. These effect estimates will complement the vulnerability and feasibility assessments that facilitate phase four of the targeting cycle, commander's decision, and force assignment.

7.2.1.4   Commander's Decision and Force Assignment

Commander's decision and force assignment leverage information from the previous three phases. Phase four consists of consolidating capability development and target analysis results. Phase four is complete when forces are assigned to missions and associated targets.

The previous three phases were used to develop the target; core knowledge for assigning teams and tools to targets. In addition, a collateral damage estimate (CDE) is used to look at how other systems might be affected by each candidate targeting operation.

In developing the target list, each target is prepared for commander review and apportionment to available teams. A selection among the targets includes a qualitative assessment of the people available, their skills, and how these talents might contribute to the most successful courses of action (COAs) relative to the commander's objectives. Phase four also weighs the combination of the team's skills, tools, and techniques required for target prosecution. This information also helps in developing phase five, mission planning, and operational execution.

7.2.1.5 Mission Planning and Force Execution

Phase five's mission planning and operational execution are where the target lists from phase four get final preparation and the missions are executed. Phase five is also the phase where most of the commercial company's attack cycle steps occur (e.g., Lockheed Martin, FireEye Mandiant, and Cyber Threat Framework (Chap. 2, Table 2.7)).

In developing the first four phases, there will be pieces of the commercial cyber attack cycles touched on. For example, "reconnaissance" from the Lockheed Martin Attack Cycle, for requisite target understanding (i.e., phases two and three), is used to help assign forces in phase four. Once the targets are paired with teams and tools, the remainder of target prosecution occurs in phase five. Evaluating phase five's operations effects occurs in phase six, combat assessment.

7.2.1.6 Combat Assessment

Effects analysis has traditionally been an estimate of what happens when a target is removed or eliminated, from an enemy system under inspection. In the case of cyber, however, the effects analysis question is more challenging. For example, a cyber effect can be either denial or manipulation (US Department of Defense, 2008).

A denial effect is similar to the kinetic question of what happens when a target is removed, similar to being removed by a bomb. In the case of cyber, however, a denial effect can also be perishable, only removing the device for a prescribed period of time. This is the simple case of a target being turned off and then turned back on, as the operation requires.

Cyber can also involve the more challenging question of what happens when a target is manipulated, with the cyber effect changing the target's behavior. Two examples of cyber manipulation include the 2007 Operation Orchard (Clarke & Knake, 2012) attack by the Israelis on a Syrian military installation, where the protecting air defense system was spoofed, and the 2010 STUXNET (Zetter, 2014) industrial control system attack where behavior was changed without the supervising personnel realizing it.

During Operation Orchard, Syrian defense radars were reprogrammed to provide the operator with a steady, unremarkable, view, that looked like nothing was happening while the Israeli's entered Syrian air space, maneuvered to a developing nuclear plant, and destroyed it (Zetter, 2009). STUXNET, more subtle, included the reprogramming of uranium enrichment centrifuge controllers to behave erratically, while the controls looked normal, not displaying the software operation anomaly. This "anomaly" destroyed the industrial controls of the uranium enrichment system and delayed the production of weapons-grade uranium for two to three years (Hayden, 2016).

Describing the targeting process for cyber manipulation, and the subsequent effects analysis, is new to the targeting community. Manipulation effects are very challenging to quantify, usually occurring via qualitative means (e.g., IO operations, Chap. 4), and can result in qualitative effects (e.g., changing opinion). Addressing these outcomes will require both legacy target system analysis (TSA) capabilities and novel modeling and simulation techniques to look at individual target behavior changes that are undergoing a cyber effect.

Phase three of the targeting cycle is where additional quantitative methods are used to describe cyber in a targeting context. For kinetic munitions, validated models are used to describe the ability of a platform (e.g., airplane, missile) to access a target, with kinetic JMEMs used to describe weapon effects on the target (i.e., probability of kill, P_k). However, each of these steps is a challenge to estimate for cyber due to unclear relationships between the teams, targets, and tools in accessing and affecting a target.

Quantitative techniques can be reliably used to estimate kinetic weapon effectiveness and collateral damage (George Cybenko, 2016; Department of Defense, 2002). However, kinetic munition data sets are gathered through multiple test samples, with associated statistical confidence, to assure targeteers of weapon behavior in different circumstances. Getting this level of detail is still a challenge for evaluating cyber effects.

Modeling cyber effects has been attempted via the emulation of kinetic weapons' Joint Munitions Effects Manual (JMEM) approach (Zurasky, 2017; Gallagher, 2008). This approach has promise for suppression effects, similar to the suppression effects provided via conventional munitions. For example, current models can briefly remove a target's core operational capability (i.e., command and control (C2)) to provide a short-term denial effect. Integrity effects or target manipulation, however, span from cognitive (IO) (e.g., Chap. 4) to technical (adversarial control - STUXNET), making the quantification of these effects more challenging. Ideally, the targeting model will both describe and account for how effects propagate through the primary target system and its components.

7.2.2 CARVER Targeting Model Example

Operations assessment, the key goal of the targeting cycle's sixth step, is facilitated by the quick evaluation of the centers of gravity (COG) for a target organization. A center of gravity is a central point in an organization. For cyber, this could be a security operations center (SOC described in Chap. 6) where the organization maintains situational awareness of all of its networks to ensure that an attack has not occurred. Along with Fig. 7.3, several other targeting approaches might be used, depending on the objective.

The CARVER model is a conventional targeting approach and is an acronym that stands for criticality, availability, recuperability, vulnerability, effect, and recognizability (Bradley Greaver, 2017). Developed by the US Special Forces in the 1960s, CARVER is a targeting framework used to facilitate center of gravity (COG) analysis.

CARVER is similar to the MITRE Crown Jewels Analysis discussed in Chap. 6. Each of the CARVER steps is described in Table 7.1.

As shown in Table 7.1, CARVER provides a framework for phases one and two of the standard targeting process (Fig. 7.3) in laying out the key items to be targeted. Building on the COG analysis that CARVER provides, we will look at both the MITRE ATT@CK cycle (Table 2.7) and a more general method to develop a framework that spans the targeting cycle.

7.2.3 Targeting, Attack Cycles, and the Cyber Process Evaluator

Cyber targeting is fortunate to inherit multiple methodologies to structure best practices in the cyber terrain. Joint Publication 3-60, the CARVER matrix and the Lockheed Martin (LM) attack cycle are among the several conceptual models that currently provide guidance to cyber operators (Table 7.2).

As shown in Table 7.2, both CARVER and the LM Attack Cycle fit into the standard targeting steps. However, CARVER is used mostly for the upfront thinking about the targets, and their prioritization, while the LM Attack Cycle helps at the later operational stages of mission planning and force execution. An alternative, more general, framework, the "Cyber Process Evaluator," is used to capture the people, processes and technologies across the targeting cycle (Table 7.3).

Table 7.3 provides an increasingly resolved view of the targeting process. This includes effects determination, center of gravity (COG) identification, mission planning, mission execution, and post-operation assessment. Table 7.3 also adds the "Cyber Process Evaluator," introduced here to capture the people, process, and technology elements for each step of the Joint Publication 3-60 attack cycle. The cyber process evaluator provides an overall framework for combining the policy, process, and technology elements that make up cyber targeting.

Table 7.1. CARVER elements and description

CARVER element	Description
Criticality	This is the target value. How vital is this node to the overall organization? A target is critical when its compromise or destruction has a significant impact on the overall organization's performance. In addition, criticality can be described by many factors, a few of which are: Time—how rapidly will the impact of the target attack affect operations? Quality—what percentage of output, production, or service will be curtailed by target damage? Surrogates—what will be the effect on the output, production, and/or service? Relativity—how many targets are there? What are their locations? How is their relative value determined? What will be affected in the overall system?
Accessibility	How easy is it to reach the target? What are the defenses? Is an insider required to penetrate the organization? Is the target computer on the Internet?
Recuperability	How long will it take for the organization to replace, repeat or bypass the target's compromise? Once the exploit is found by the attacked system, how long will it take the owning organization/system to remediate it and recover operations?
Vulnerability	How much and what kind of knowledge is required to exploit the target? Do I currently have this exploit capability, or do I need to invest in new approaches to access this target? For cyber, an example vulnerability evaluation was described via PASTA (Chap. 3)
Effect	How will the organization be affected by the attack? What kind of reactions should I expect to see from the organization due to a successful attack?
Recognizability	How easy is it to identify the target? How easy is it to recognize that a specific system/network/device is a target and not a security counter measure?

Table 7.2 JP 3-60, CARVER and LM attack cycle—example targeting frameworks

JP 3-60	Phase 1 — Commander's objectives, guidance, and intent	Phase 2 — Target development, validation, nomination, and prioritization	Phase 3 — Capabilities analysis	Phase 4 — Commander's decision and force assignment	Phase 5 — Mission planning and force execution	Phase 6 — Combat assessment
CARVER		CARVER				
LM attack cycle		Reconnaissance			Weaponization, delivery, exploitation, installation, C2, actions on objectives	

Table 7.3 JP 3-60, CARVER, the LM attack cycle, and the cyber process Evaluator

		Phase 1 — Commander's objectives, guidance, and intent	Phase 2 — Target development, validation, nomination, and prioritization	Phase 3 — Capabilities analysis	Phase 4 — Commander's decision and force assignment	Phase 5 — Mission planning and force execution	Phase 6 — Combat assessment
JP 3-60							
CARVER			CARVER				
LM attack cycle			Reconnaissance			Weaponization, delivery, exploitation, installation, C2, actions on objectives	
Cyber process evaluator	People	Define desired effect		Recruit team	Train team	Map target system	
	Process		Determine budget	Appropriate funding	Target recce; obtain access	Emplace device	Post event analysis
	Technology		Determine Technology			Perform mission	

7.3 Target Process Review

While both kinetic and cyber attack cycles have similar process mappings (Tables 7.2 and 7.3), each is also a challenge to measure for effectiveness. Target assessment provides a structured approach for decomposing both physical and cyber targets (e.g., Fig. 7.1).

7.3.1 Target Development and Prioritization

Cyber targeting has unique challenges due to the detailed target understanding required to achieve effects (e.g., Fig. 7.2). One supplement to human understanding for managing the broad scope of cyber targeting is to use knowledge-based systems to support detailed analysis over the course of a target's development. For example, knowledge-based design shows promise in using the limited data available for weapon-target pairing to improve the efficacy of target development at the early stages of the targeting process (Clare Maathuis, 2020; Fox & Hansberger, 2018).

Improving target system analysis (TSA) tools will be helpful in speeding up the cyber targeting process. Fortunately, as discussed in Sect. 7.2, regardless of the type of target addressed, the targeting process is the same. For example, as shown in the CARVER example (Table 7.1), cyber target development and prioritization can be as simple as putting together a matrix of the adversary's center of gravity support infrastructure and determining what is addressable based on current cyber capabilities (Fox & Hansberger, 2018).

7.3.2 Capabilities Analysis and Force Assignment

Capabilities analysis, from Fig. 7.3, includes choosing the right people, processes, and tools/technologies for the targets of interest. While each target–capability pair will be situation specific, an understanding of how the element contributes to the overall system operation (e.g., from Fig. 7.2), combined with measurement of system operation prior to the cyber attack, as discussed in Sect. 7.3.1, will increase target understanding for future effects measurement.

7.3.3 Mission Planning and Force Execution

As shown in Tables 7.2 and 7.3, the Lockheed Martin (LM) attack cycle covers the mission planning and force execution phase of the overall targeting cycle. In addition, MITRE's ATT@CK framework (Chap. 2) focuses on this end game of a cyber mission, providing an extensive body of knowledge (e.g., CARET (MITRE), Chap. 2), for tools, techniques, and known threat actors.

Looking back at Fig. 7.2, we see that the initial phases of target evaluation include the overall system, progressively decomposing the target system until we arrive at the specific element to be addressed. Target decomposition is similar to the analysis that we used in Sect. 3.5, looking for key system vulnerabilities in performing threat analysis. In this case, however, we have the additional challenge of trying to understand an estimated target system weakness well enough to plan on it having an appreciable effect. For example, as shown in Table 7.2, we see that the majority of LM attack cycle activity occurs in the mission planning and execution phase of the targeting process. In addition, Table 7.4 provides the LM attack cycle with example of people, process, and technology use cases and vulnerabilities.

7.3.3.1 People, Process, and Technology Elements across the LM Attack Cycle

An example of vulnerability decomposition across the LM attack cycle is shown in Table 7.4.

The vulnerabilities shown in Table 7.4 occur across an example LM attack cycle, providing examples of known people, process, and technology attack vectors for each threat element. Many of the examples in Table 7.4 are well known. For

Table 7.4. Vulnerability examples across each step of the Lockheed Martin attack cycle (people, process, technology)

LM attack cycle step	Vulnerability examples		
	People	Process	Technology
Reconnaissance	Money, ideology, compromise, ego (MICE) as key staff weaknesses (Scott Stewart, 2009) Social media to identify specific developer skills for a targeted system	UAV monitoring of parking lot activity (Ganesh, 2015); gear used (social media);	NIST vulnerability database—— understand the vulnerabilities in the cyber terrain Vulnerability assessment tools (e.g., Metasploit, Shodan)
Weaponization and delivery	Thumb drive in parking lot (Sterling, 2011)	Solar Winds Supply Chain Attack (Williams, 2020)	
Exploitation	E-mail attachment (spear phishing)		Maintain backdoor in network
Installation			APT 29 (FireEye, 2017a, b)
Command and Control (C2)			Conficker obfuscation of C2 server address (Bowden, 2011)
Actions on objectives		Equifax (Newman, 2017)—machines unprotected (technology) due to delayed maintenance process	

example, MICE for human recruitment (Scott Stewart, 2009) is a widely used conceptual model that is used to describe social engineering based on an individual's personal weaknesses. Other attacks, such as the thumb drive vulnerability (Sterling, 2011), are good examples of how broadly the scope of a cyber operation can reach into simple day-to-day foibles for developing accesses to a cyber target.

In addition to the Lockheed Martin attack cycle, we can look at an attack in terms of the cyber process evaluator, initially discussed in Sect. 7.2.3, as a means to formulate a better defense. Using the cyber process evaluator, we have an opportunity to look at the overall scope of a cyber mission in terms of potential attacker vulnerabilities and determine where we can impose cost and delay on our adversaries to better defend our network.

7.3.3.2 Mission Planning Through Execution Example

A natural place to perform the detailed analysis is during mission planning. This is where we can gather current threat understanding and determine the strengths and weaknesses of a proposed threat, guiding us in our development of cyber defense resources. We can therefore expand on Table 7.3 with activities for each step provides an example attack path for target planning.

The attack path shown in Fig. 7.4 is a set of steps to provide planners with a structured way to perform resource estimates, including time, for different courses of action (COAs). The steps in Fig. 7.4 use the standard policy, process, and technology elements to decompose a mission at the operational level. Understanding threat intelligence for team behavior is key to estimating the time and cost using Fig. 7.4 as a conceptual model. In addition, technical terrain could be in the form of a simple network description in determining the requirements to complete each technical step to be traversed over the course of Fig. 7.4 example attack cycle.

Having the steps laid out, as in Fig. 7.4, provides us with an opportunity to think through the time and cost associated with the successful completion of each step. For example, each step in Fig. 7.4 can be estimated in terms of the time that it takes for a given team to work on a target with the skills and tools at its disposal. Having these nominal time estimates, we can also estimate how the "level of effort" translates into (1) labor cost to work through the task, or, (2) the cost to buy a tool that will execute this operation. In addition, with the steps broken out this discretely, we can also review what additional measures (e.g., denial and deception) might be used to further improve the quality of the operation (i.e., decrease the likelihood of discovery) (Fig. 7.5).

As shown in Fig. 7.5, the time, cost, and effort put into ensuring secure defensive operations are easier to manage as data attributes for each of the steps in Fig. 7.3. Once the attack path is broken out in terms of time, quality, and cost, each of the respective steps can be evaluated, measured, and summed, as shown in Eq. 7.2 (Attack path time).

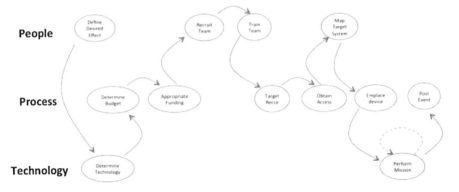

Fig. 7.4 Attack path elements—people, process, and technology

	Independent Variables	Define Effect	Determine Technology	Determine Budget	Appropriate Funding	Recruit Team	Train Team	Target Recce	Obtain Access	Map System	Emplace Device	Perform Mission	Post Event
time	labor / time (hrs)	40	80	20	20	80	160	160	160	80	10	1	160
	time (hrs)	80	20	20	160	40	160	80	20	10	5	160	20
	communications (hrs)	20	40	20	10	40	5	20	20	20	1	1	5
capital	Planning tool(s)	$100,000											
	Traffic Generation											$150,000	
	Attack Tool(s)											$100,000	
	Network Sensors								$50,000				
	System Mapping tool(s)									$100,000			
	Emplacement tool(s) (e.g., rootkit)										$50,000		
	Network C2 System											$100,000	
	After Action Reporting (AAR) System												$50,000
Denial & Deception	Evasive Maneuvers (additive to core mission)												
	alternate team members over mission					160							
	alternative devices & paths									160			
	false communications	40	20	20	10								

Fig. 7.5 Example attack path with time, cost, and quality estimates

$$\text{Total Time} = \sum_i \text{People}_i + \sum_j \text{Process}_j + \sum_k \text{Technology}_k \qquad (7.2)$$

The total time shown in Eq. 7.2 gives the planner a rough idea of what an operation will likely require, in terms of time to completion. Equation 7.3 (Attack path cost) has the additional advantage of helping the planner to partition each step for evaluating individual levels of effort. In addition, Eq. 7.3 uses a similar summation to estimate attack path cost.

$$\text{Total Cost} = \sum_l \text{People}_l + \sum_m \text{Process}_m + \sum_n \text{Technology}_n \qquad (7.3)$$

As shown in Eq. 7.3, we can also estimate the cost imposed on an attacker in terms of the people, processes, and technologies that he will employ in attacking a network. People costs might be relatively straightforward labor costs. Process costs include additional steps that the attacker uses in attempting to breach a network's defenses. And technology cost includes additional tools that the attacker purchases to overcome a network's advanced defensive technologies, as discussed in Chap. 6.

Chapter 6's layered defense (Fig. 6.1) was a defensive architecture managed by a Security Operations Center (SOC) that was designed to impose both process and technology costs on the attacker. For example, the defensive architecture has the potential to force the attacker to do more detailed surveillance for each layer, increasing people and process costs. Similarly, the attacker might need to purchase additional technologies (e.g., denial and deception capabilities) to mask his operations from SOC personnel.

7.3.3.3   Time and Cost Example for Threat Groups

Using Eqs. 7.2 and 7.3, in combination with the attack path description provided in Fig. 7.4, we can now look at simple estimates for what an attack might take, in terms of time and cost, for different example threat groups (Fig. 7.6).

The time/cost estimates shown in Fig. 7.6 describe the imposed resource requirements on an attacker, and her management, to accomplish the operational steps that are required to successfully complete a mission. Nation states, terror groups, and hackers are estimated to have decreasing levels of skill and tool capabilities, respectively, in this simplified description.

Decreased operation time for nation state actors (Fig. 7.6) implies that a nation state actor has better processes, tools, and human resources. The people element of the recent Solarwinds attack on the US government, for example, attributed to Russian cyber actors, was estimated to have over 1000 support personnel (Paul, 2021). And distributing backdoors through the Solarwinds software update system, estimated to serve over 80% of the Fortune 500 (Constantin, 2020), was likely a lengthy and well-funded tool development process.

The goal of Fig. 7.6 is to provide decision makers with a baseline description of how investment in training and technology for cyber defense will impose a cost on attack groups (e.g., access and maneuver). In addition, defense investment might significantly improve the defender's likelihood of providing an adequate defense in denying network access. This is one possible example of quantifying a layered defense, making the cyber terrain difficult to both access, and maneuver, for an attacker with inadequate tools and skillsets.

Fig. 7.6 Operations time vs. cost for different threat groups

Along with showing how a good defense can handle lower-level attackers, Fig.7.6 can also be used to show how proactive security measures (e.g., denial and deception) can be used to impose cost and operations time. These security measures are employed as policies and procedures in the Security Operations Center (Sect. 6.3.3.1) to protect the network and force the attacker to waste resources on paths that have no promise of accessing or exploiting the targeted system.

7.3.4 Post-Operations Assessment

Cyber operations' assessment is challenging due to the multiple sources that are used to provide elements of the evaluation. While physical damage is often key to evaluating conventional weapon effects, cyber is more subtle, and thereby more challenging, in evaluating an operational effect.

Ideally, an attack path is structured as in Table 7.4, leveraging known target elements as described in analyses via Figs. 7.4 and 7.5. This has the potential to provide a before and after snapshot of the target, with a clear understanding of what was done and its relative effectiveness.

7.4 Cyber Targeting Summary

The well-developed frameworks and conceptual models (e.g., CARVER matrix, Lockheed Martin Kill Chain, MITRE ATT@CK) provide us with a great starting point to look at an overall cyber attack cycle in a structured way (Fig. 7.7).

The Joint Publications shown in Fig. 7.7 provide the context and terminology for the targeting process, while the conceptual models clarify the operational behavior of an attack. Revisiting Fig. 7.3, we will use this chapter's developments to fill in elements of the targeting life cycle with more detail (Fig. 7.8).

Using the targeting cycle shown in Fig. 7.8, as a general framework, we were able to better develop time, cost, vulnerability, and other estimates, as provided in this chapter, for what an attack might cost from a simple hacker to a nation state. In addition, the use of cyber targeting will be key in future defensive response and campaign planning.

Cyber Offense and Targeting Questions and Example Answers
1. What are key policy and doctrine documents that govern cyber targeting operations?

    Targeting is generally covered in JP 3-60 (Fig. 7.3) and cyber operations are the subject of JP 3-12, as discussed in Chap. 2. In addition, intelligence is covered by JP 2-0 with JP 3-13 covering the more general information operations. Chapter 2 describes the additional publications that provide guidance concerning cyber (Fig. 2.3).

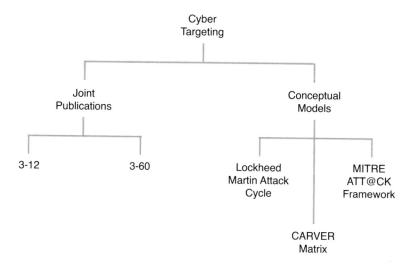

Fig. 7.7 Elements of cyber targeting

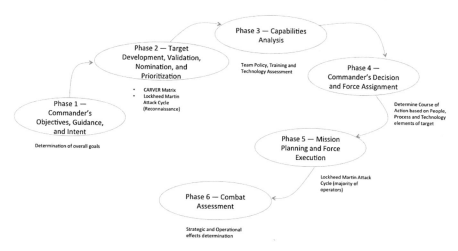

Fig. 7.8 Target cycle with associated cyber analysis capabilities

2. What are the key processes used in targeting that might be helpful in understanding cyber attackers?

As laid out in the policy and doctrine used for cyber targeting (question 1), the six steps in Fig. 7.3 make up the generally accepted targeting process. Within this process, each step might include additional sub-processes for completion. For example, Fig. 7.2 provides a diagram showing how a target is decomposed from a general system to specific elements for increased target resolution.

In Chap. 3 (Sect. 3.5), we looked at different methods to assess system vulnerabilities for cyber defense. Similarly, cyber targeting can leverage some of these defensive methods for system decomposition and potential target description, which we looked at in Fig. 7.2.

An additional challenge, for cyber targeting, is that the mission planner will be expected to provide an effects estimation for the nominated target. This will require a very good understanding of the target environment, as touched on in the ISR discussion of Chap. 5. Once this understanding is available, knowledge-based tools will help in tool/capability understanding based on both SME and empirical data, as discussed in Sect. 7.3.2.

3. How is cyber targeting measured and what are potential improvements in measuring cyber operations?

As discussed in Sect. 7.2.1, cyber targeting includes the initial analysis that leads to a more detailed target system analysis (TSA). This is performed while decomposing the system to understand the specific target element (Fig. 7.2) as discussed in Chap. 5.

Understanding the target and its associated systems provides the targeteer with an ability to answer questions about estimating the level of an effect. For most processes, this requires understanding the baseline process behavior, applying an effect, and then determining how the system's behavior has changed.

4. What role does technology play in cyber targeting?

Technology is important in cyber targeting in order to understand both the tools available to the attacker (e.g., penetration tester) and to understand the defended terrain, as discussed in Chap. 6. These technologies will be especially important to understand during the initial phases of target development (e.g., phase 3 of Fig. 7.3), which include increasing system-level resolution, as shown in Fig. 7.2.

5. How is cyber targeting different in producing IO vs. CIA type effects?

As discussed in Chap. 4, IO effects are essentially the implementation of psychological operations via cyber. These effects are strategic, may take longer to prepare, and can have long-lasting effects on how a target population sees and understands, or frames, issues. This is a challenge to measure. A tangible outcome of this type of activity might be persuading a population to vote differently.

For an IO operation, the effect includes framing an argument differently. For example, increasing the popularity of a candidate so that they can win an election. More generally, as described in China's current IO campaign (i.e., Sect. 4.4.2), a country may use IO to try to shape the information environment. China's current campaign of paying journalists to slant stories so that the rest of the world sees China in a positive light is one example of an IO campaign (i.e., Fig. 4.13).

Cyber, however, includes both IO and the simpler technology-based effects of denying operations (i.e., availability), accessing a "secure" system (i.e., confidentiality), or controlling an adversary system (i.e., "integrity"). While there may be an overlap between confidentiality and IO systems (e.g., extracting e-mails for the 2016 DNC attack), there is little similarity in the technique with integrity attacks and IO. For example, the STUXNET operation provided in

Chap. 5 (e.g., Fig. 5.10) includes tools and processes very different from tools used in a typical IO operation (e.g., Fig. 4.10).

Cyber effects from targeting, however, may provide the same result by using IO or technical means. For example, while IO operations include shifting opinion, more similar to a media operation, and CIA (i.e., confidentiality/integrity/availability) effects will be used to change a device's state, the end result for both might be a vote tally that results in the "right" candidate winning. In addition, both IO operations and CIA effects are strategic. Cyber, therefore, might be used in an application like election security to both (1) use messaging to change the population's mind (IO) or (2) access the system and simply change the vote count (CIA).

Bibliography

Australian Cyber Security Centre. (n.d.). *Essential eight.* Retrieved 4 8, 2021, from Australian Government - Australian Signals Directorate: https://www.cyber.gov.au/acsc/view-all-content/essential-eight

Bennett, M., & Waltz, E. (2007). *Counterdeception principles and applications for national security.* ArTech House.

Bowden, M. (2011). *Worm - the first digital world war.* Atlantic Monthly Press.

Bradley Greaver, L. R. (2017). CARVER 2.0: Integrating the Analytical Hierarchy Process's multi-attribute decision-making weighting scheme for a center of gravity vulnerability analysis for US Special Operations Forces. *Journal of Defense Modeling and Simulation, 15*(1), 111–120.

Chairman of the Joint Chiefs of Staff (CJCS). (2019, 3 8). *Methodology for combat assessment (CJCSI 3162.02).* Retrieved 6 22, 2020, from Joint Chiefs of Staff: https://www.jcs.mil/Portals/36/Documents/Doctrine/training/jts/cjcsi_3162_02.pdf?ver=2019-03-13-092459-350

Clare Maathuis, W. P. (2020, 4 11). *Decision support model for effects estimation and proportionality assessment for targeting in cyber operations.* Retrieved 6 22, 2020, from Science Direct: https://www.sciencedirect.com/science/article/pii/S2214914719309250

Clarke, R. A., & Knake, R. K. (2012). *Cyber war: The next threat to national security and what to do about it.* Ecco.

Constantin, L. (2020, 12 15). *SolarWinds attack explained: And why it was so hard to detect.* Retrieved 12 20, 2020, from CSO Online: https://www.csoonline.com/article/3601508/solarwinds-supply-chain-attack-explained-why-organizations-were-not-prepared.html

Department of Defense. (2002). *Joint Doctrine for Targeting. Department of Defense, Joint Chiefs of Staff Washington, DC.* Department of Defense.

FireEye. (2017a). APT 29. (FireEye, Producer) Retrieved 9 9, 2018, from APT 29: https://www.fireeye.com/current-threats/apt-groups.html#apt29

FireEye. (2017b). *M-TRENDS 2017 - A view from the front lines.* FireEye.

Fox, W. P., & Hansberger, J. (2018). Methodology for targeting analysis for minimal response. *Journal of Defense Modeling and Simulation, 15*(3), 323–336.

Gallagher, M. (2008). *Cyber Analysis Workshop.* MORS.

Ganesh, V. (2015, 12). *Parking lot monitoring system using an autonomous quadrotor UAV.* Retrieved 9 19, 2018, from Clemson University: https://pdfs.semanticscholar.org/9651/9650a1a734cd1feac6f33b7858f476e14f80.pdf

George Cybenko, G. S. (2016). Quantifying covertness in deceptive cyber operations. In V. S. S. Jajodia (Ed.), *Cyber deception: Building the scientific foundation.* Springer.

Gibson, W. (1984). *Neuromancer.* Ace.

Hayden, M. V. (2016). Playing to the edge: American intelligence in the age of terror. Peguin.

Merriam-Webster. (n.d.). *Target (noun, often attributive)*. Retrieved 7 8, 2019, from Merriam-Webster: https://www.merriam-webster.com/dictionary/target

MITRE. (n.d.). *CARET*. Retrieved 9 30, 2018, from CARET: https://car.mitre.org/caret/#/

Newman, L. H. (2017, September 8). *The Equifax breach exposes America's identity crisis*. Retrieved September 7, 2018, from Wired: https://www.wired.com/story/the-equifax-breach-exposes-americas-identity-crisis/

Paul, K. (2021, 2 23). *SolarWinds hack was work of 'at least 1000 engineers', tech executives tell Senate*. Retrieved 2 24, 2021, from Guardian: https://www.theguardian.com/technology/2021/feb/23/solarwinds-hack-senate-hearing-microsoft

Scott Stewart, F. B. (2009, 5 20). *A counterintelligence approach to controlling cartel corruption*. Retrieved 6 22, 2020, from Stratfor: https://worldview.stratfor.com/article/counterintelligence-approach-controlling-cartel-corruption

Sterling, B. (2011, 6 29). *The dropped drive hack*. Retrieved 9 19, 2018, from Wired: https://www.wired.com/2011/06/the-dropped-drive-hack/

U.S. Department of Homeland Security (DHS). (n.d.). *Cybersecurity and Infrastructure Security Agency (CISA)*. Retrieved 4 10, 2021, from https://us-cert.cisa.gov/

US Department of Defense. (2008). *Cyberspace operations*. US Department of Defense, Joint Staff, Arlington.

USAF. (2019, 3 15). *The targeting cycle (Annex 3–60)*. Retrieved 6 22, 2020, from https://www.doctrine.af.mil/Portals/61/documents/Annex_3-60/3-60-D04-Target-Tgt-cycle.pdf

Williams, J. (2020, 12 15). *What you need to know about the solarwinds supply-chain attack*. Retrieved 2 15, 2021, from SANS: https://www.sans.org/blog/what-you-need-to-know-about-the-solarwinds-supply-chain-attack/

Zetter, K. (2009, 11 3). Mossad hacked syrian official's computer before bombing mysterious facility. *Wired*.

Zetter, K. (2014). *Countdown to zero day - Stuxnet and the launch of the world's first digital weapon*. Crown.

Zurasky, M. W. (2017). *Methodology to perform cyber lethality assessment*. Old Dominion University.

Chapter 8
Cyber Systems Design

The purpose of this chapter is to review cyber systems, in terms of their system architecture, operational elements, and processes. This will include a discussion of cyber systems' historical outgrowth from precision targeting developments, along with insight concerning how people and technologies play as elements in the current operational space. In addition, we will look at the composition of a cyber analysis system in terms of both information operations and the collection, storage, and dissemination approach used by Wikileaks.

8.1 Cyber Systems Design Background

In Chap. 1, we described cyber systems as an evolutionary step in precision munitions, whose development started in the Vietnam War. These smart weapons, dating from the late 1960s, were motivated by the Pentagon's dual goal of increasing aerial bombing accuracy in Vietnam while minimizing civilian collateral damage. Smart weapons were more accurate, hitting their desired targets with less sorties based on additional target intelligence before, during, and after each mission. The volume and frequency of these intelligence flows have continued to increase with the progression of munitions in terms of increased accuracy, precision, and standoff. For example, reviewing the progression from dumb bombs to cyber, in Chap. 1 (i.e., Fig. 1.2), we see that persistence is a key outcome for the increasingly software-intensive munitions (Fig. 8.1).

The increased data consumption requirements of cyber are due to extensive ISR use both before and during a mission. For example, as shown in Fig. 8.1, each increment in munition precision is accompanied by upfront design costs (e.g., software development) which are also associated with increased levels of mission data during an operation. Such data flows require improvements in bandwidth, processing power, and network architecture. This need for data and increasing intelligence at the edge led to the development of microcontrollers, mature programming

© Springer Nature Switzerland AG 2022
J. M. Couretas, *An Introduction to Cyber Analysis and Targeting*,
https://doi.org/10.1007/978-3-030-88559-5_8

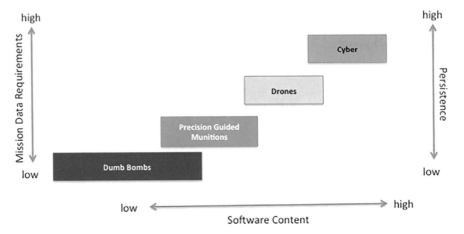

Fig. 8.1 Increased software and more persistent munitions

languages, and fiber optic cables to transport increasingly high levels of data and to manage increasingly complex tasks in order to achieve mission objectives. In addition, this growth in processing power and bandwidth has developed into an exponentially growing cyber ecosystem (Clement, 2019), populated by the current cyber landscape of consumer products (e.g., personal computers, smartphones …) (Warner, 2017), and promises a continued increase via the Internet of Things (IoT), IPv6 and 5G communications.

8.1.1 Intelligence Challenges and Cyber Systems

This ever-expanding technology space of computers, routers, and fiber optic data connections that make up our current networks changed the telecommunications terrain, creating a new environment that challenged traditional intelligence collection. National organizations, therefore, had to change their ways of doing business. For example:

"…In the NSA, the emergence of fibre-optic cables and encryption in the nineties is believed to have led directly to increased funding to support a new emphasis on hacking into computers to gather data which might otherwise be harder to access (Landau, 2014). This activity was undertaken by the Office of Tailored Access Operations (TAO), formed in 1997 (Aid, 2013). 'In the early days TAO used to be just a bunch of hackers!' one member of the department wrote in 2012. 'We did things in a more ad hoc manner … one guy did it all. Now we're more systematic.' TAO now consists of separate teams for each target, one looking at China and North Korea together, another Iran, another Russia, as well as cyber counter-intelligence and counter-terrorism. In the teams are developers to create software and hardware tools alongside analysts who plan operations. A planner would take an operation to a mission director who would assess the risks, and then finally, if approved, hand it to the elite team of hackers in the Remote Operations Center (ROC). In 2004–2005, the hacking team expanded rapidly into a 40,000 square foot office housing 215 personnel able to undertake at least a hundred operations a day. 'What if your job was to exploit a target's computer, collect voice cuts from an adversary's phone system, use a terrorist's web-based email account to infect them with a Trojan horse, and assist the military in locating a high-value terrorist target for capture – all in a day's work? Then you would be working in the Remote Operations Center!' an internal 2006 note explains. ROC's role at the time was described as to collect data, geo-locate individuals, provide real-time support in rendition of 'high-profile terrorists' and manage a global covert infrastructure. Their motto was: 'Your data is our data, your equipment is our equipment – any time, any place, by any legal means'" (Corera, 2016).

Changes in the physical structuring of telecommunications, and the introduction of the Internet, resulted in major changes in how data needed to be collected:

"The scale as well as sophistication of cyber espionage had grown. According to leaked documents, by the end of 2013 the NSA's aspiration was to have 85,000 implants around the world (Der Spiegel, 2013). A system called Turbine promised to allow this procedure to be scaled up to handle 'millions' of implants through automated control. Computer espionage itself could now increasingly be taken out of human hands (Greenwald, 2014)." (Corera, 2016)

The telecommunications revolution, the movement from the plain old telephone system (POTS) to digital systems based on packet communications, therefore changed the people, process, and technology approach required for collection and analysis. Describing these updates in terms of architectures is a natural next step to ensure the sometimes ad hoc linking together of component systems could be described for future design, development, and testing.

8.2   Introduction—Cyber System Architectures

While cyber systems might be looked at as IT, or computer-based, systems, using these technologies for analysis and targeting poses some unique questions:

1. How are cyber systems best described in terms of traditional software architecture?
2. How do system designs help in describing operational architectures used for offensive and defensive cyber operations?
3. What kind of threat descriptions are currently available, and how are they used in cyber systems design?
4. What are examples of designs for cyber analysis and targeting systems, along with their use?

In answering these questions, we will review software and system architecture history, current architecture description languages, and look at the DoD Cybersecurity Analysis and Review (DoDCAR) as an example use of architectures to improve cyber system security.

8.2.1   Cyber and Architecture Background

Information Technology (IT), or cyber, systems, might be considered one of the reasons for current systems architecture to exist. For example, the 1996 Clinger-Cohen Act, supported by the C4ISR Architecture Framework, was followed by the DoDAF in 2003, which continues as an architecture standard to this day. In addition, the commercial sector developed the Unified Modeling Language (UML) in 1995 to manage ever-increasing code complexity. These efforts led to the development of multiple architecture types for system design and development.

One of the challenges addressed by architecture languages, and system description methodologies, is the loosely coupled nature of cyber systems. Prior to software, engineering issues included precision, accuracy, and heat—problems for tightly coupled systems. With the advent of cyber, engineers were faced with systems that not only connected dynamically (e.g., transmission control protocol (TCP)) but included memory. The system complexity, far beyond what a telecommunication switch would provide only a generation or so ago, is still being worked out in the design of secure cyber systems.

8.2.1.1   Architecture Types

An architecture description is common for software-intensive systems (IEEE, 2011). This is due to how architectures are used to combine the various concepts that make up cyber systems. Example architecture types are shown in Table 8.1.

Table 8.1 Example architecture types

Architecture type	Definition
Reference architecture	The reference architecture provides an authoritative source of information about a specific subject area that guides and constrains the instantiations of multiple architectures and solutions (Office of the Assistant Secretary of Defense Networks and Information Integration (OASD/NII), 2010)
Solution architecture	The solution architecture is a structure that portrays the relationships among all the elements of something that answers a problem. It describes the fundamental organization of a system, embodied in its components. This includes how components relate to each other and the environment. The solution architecture provides the principles governing the system's design and evolution (Office of the Assistant Secretary of Defense Networks and Information Integration (OASD/NII), 2010)
Logical architecture	The logical architecture refines, selects, and synthesizes a system's logical elements to provide a framework for verifying that a future system will satisfy its system requirements in all operational scenarios (Zachman)
Physical architecture	The physical architecture is an arrangement of physical elements (system elements and physical interfaces) that provide the solution for a product, service, or enterprise (Zachman)

As described in Table 8.1, a reference architecture includes a template architecture for a given domain. In addition, the reference architecture provides a common vocabulary for discussing implementations, stressing the need for commonality. A solution architecture, on the other hand, focuses on describing a specific solution, or deliverable.

While the reference architecture provides the terms and template for a particular domain, solution architectures are used to describe specific systems. Logical architectures show how a system's components are interconnected, stopping short of specifying implementation technologies. The physical architecture, more specifically, is the layout of a system and its components in a schema, organizing the physical elements of the system.

System descriptions, including the supporting documentation for the architecture types shown in Table 8.1, are also supported by description languages. Architecture description languages (ADLs), designed to document the design of complex systems, now have multiple specialized formats to describe software, hardware, and overall systems.

8.2.1.2 Architecture Description Language (ADL) Background

Early system descriptions include the structured analysis and design technique (SADT) and integrated definition methods (IDEF) to map the inputs, outputs, resources, and controls of an individual component, or overall system. These techniques date from the 1960s to the early 1970s.

While SADT and IDEF addressed overall systems, the Unified Modeling Language (UML) was developed in the 1994–1995 time frame by pioneers Grady Booch, Ivar Jacobson, and James Rumbaugh at Rational Software to manage large, and often disparate, software development projects. UML provided the foundation for both the System Modeling Language (SysML) and the business process modeling notation (BPMN) (Fig. 8.1).

As shown in Fig. 8.2, both the Unified Modeling Language (UML) 2.0 and the Business Process Modeling Notation (BPMN) 1.0 stem from UML 1.1. BPMN are more commonly used to describe the communication flow between nodes that compose a business process. In terms of cyber analysis, MITRE successfully used BPMN in their analyzing mission impacts of cyber actions (AMICA) (Steven Noel, 2015) to evaluate a solution architecture's ability to perform its mission when undergoing attacks on supporting computer systems.

Another branch from UML 2.0 is the System Modeling Language, SysML, designed to support more general system engineering activities in the spirit of SADT and IDEF from a generation ago. SysML is currently supported by multiple tools, shortening the transition from a system design to measurable system characteristics (i.e., executable architecture) (John J. Daly, 2015) (Couretas, 2006).

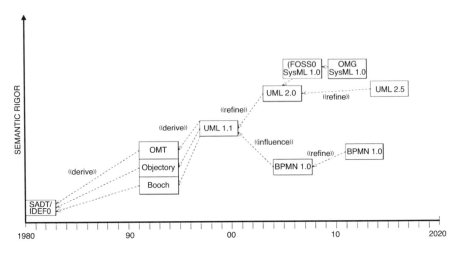

Fig. 8.2 Architecture modeling language evolution (1980–2020) (SysML.org)

8.2.1.3   System Hierarchy Levels

In addition to structured system representations, Bernard P. Zeigler developed the discrete event system specification (DESS) (Zeigler, 1976) as an overall system formalism used for modeling and simulation. DESS is a 5 tuple mathematical object that can be used to represent any type of system and includes a hierarchy of system specifications that helps the modeler clarify the current level of system fidelity.

We can use the levels of system specification to understand how much we know about a target in terms of system analysis (Table 8.2).

The system specification levels in Table 8.2 describe how much we know about a system of interest. This is an increase in resolution, decreasing the amount of abstraction, in moving toward an actual physical understanding of the phenomena. Levels of system specification, as described in Table 8.2, are used to refine system-level models (e.g., for verification, validation, and accreditation (VV&A)).

In addition, as shown in Table 8.2, the often partial observations of a cyber target are represented by the lower level of target identification (e.g., I/O behavior). The I/O behavior estimation is level 1 of the system specification levels. Increasing understanding of the system will require additional information, including the system state and the relation between states based on external inputs. Looking at the system levels, as a representative technique, therefore opens up the analysis for increased levels of resolution via system description.

Levels of system specification provide an objective framework from which to evaluate what we currently know about a target and how to frame what is being performed. For example, in Table 8.2, we describe the levels of system specification (Zeigler, 2019), what we know at each respective level, and how this knowledge fits in a cyber example.

Table 8.2 Levels of system specification

Level	Specification name	What we know at this level	Example: a cyber target
0	Observation frame	How to stimulate the system with inputs; what variables to measure and how to observe them over a time base;	The cyber device has inputs and outputs at the usual operational level, such as data, control, and power. This is the source data for computing the means and standard deviations.
1	I/O behavior	Time-indexed data collected from a source system; consists of input/output pairs	For each control and data input that the cyber device recognizes, the set of possible outputs that the device can produce.
2	I/O function	Knowledge of initial state; given an initial state, every input stimulus produces a unique output.	Assuming knowledge of the cyber device's initial state when starting the process, the unique output response to each control and data input.
3	State transition	How states are affected by inputs; given a state and an input what is the state after the input stimulus is over; what output event is generated by a state.	How the cyber device transits from state to state under control and data inputs and generates outputs from the current state
4	Coupled component	Components and how they are coupled together. The components can be specified at lower levels or can even be structure systems themselves—leading to a hierarchical structure.	A description of a cyber device's I/O data and controls in terms of system behavior and their interactions

The use of structured system descriptions is key to the future simulation of the often high-level system descriptions found in current architecture diagramming techniques. For example, structuring the behavior of a system is just one of the diagrams in the commonly used Department of Defense Architecture Framework (DoDAF).

8.2.1.4 Department of Defense Architecture Framework (DoDAF)

This development of software and system engineering tools took place at the same time that Congress was getting interested in controlling costs for IT systems. The Clinger-Cohen Act (DoD CIO, 1996) was passed in 1996, requiring each US Government acquired IT system to have an accompanying architecture. The C4ISR Architecture Framework (1996) was published to help DoD acquisition professionals meet the intent of the Clinger-Cohen Act. By 2003, The DoD Architecture Framework (DoDAF) version 1.0 was released as a more comprehensive architecture framework; DoDAF is currently at version 2.02 and continues to evolve. While the DoDAF provides a means of description suited to defense acquisition, the

Unified Modeling Language (UML) and System Modeling Language (SysML) are more commonly used to construct cyber systems.

While DoDAF provides all of the products to document an architecture of interest, we are also interested in using these products for system evaluation. This includes reviewing current architecture-based security approaches, studying a contemporary implementation of security architectures (e.g., DoDCAR), and developing an example architecture with a mission thread.

8.2.2   Architectures and Cyber System Evaluation

Architectures are the tools we use to describe the separate parts that make up a complex system. For understanding how architecture is currently used to provide guidance to construct secure cyber systems, there are multiple architecture frameworks. For example, the US National Security Agency's Secure Architecture initiative (NSA), the Microsoft Cybersecurity Reference Architecture (Microsoft), and the NIST Cybersecurity Framework (National Institute of Standards (NIST), 2018) all provide guidance designed to protect computer-based systems from attack.

While the architecture frameworks and products are essential for capturing the detail in software-intensive systems that make up cyber, the majority of questions about cyber have to do with vulnerabilities. For example, in Chap. 3, we looked at example approaches (e.g., PASTA in Sect. 3.3.3) for evaluating system vulnerability, and introduced DoDCAR. We will now take a closer look at DoDCAR as a common defensive cyber architecture framework for better understanding cyber system design applications.

8.2.2.1   DoD Cybersecurity Analysis and Review (DoDCAR)

One of the key architecture-based approaches for cyber defense is the DoD Cybersecurity Analysis and Review (DoDCAR) (Office of the National Manager for NSS) which focuses on matching cybersecurity resources to the cyber threat and mission dependency.

DoDCAR uses threat-based assessments of cybersecurity architecture to provide leadership with an overview of the operational capabilities that are impacted by cybersecurity investment decisions. This enables dependable mission execution based on an analysis-driven, threat-based, process. In addition, DoDCAR leverages existing architecture products (Sect. 8.2.1) to provide standardized data that make it easier to describe and implement the computer security controls (CSCs) discussed in Chap. 2 (Sect. 2.4.2).

A key objective of the DODCAR methodology is to improve the quality of communication surrounding cyber security while leveraging current IT protection standards. For example, DoDCAR uses the National Institute of Standards and Technology (NIST) Cybersecurity Framework (CSF) functions, such as Identify,

Protect, Detect and Respond to talk about how a defender (e.g., Security Operations Center (SOC) in Sect. 6.3.3.1) experiences an attack.

DoDCAR also leverages STIX/TAXII data standardization provided by OASIS (OASIS) for communication cyber observables, events, and threats (Table 8.3).

As shown in Table 8.3, CybOX™, TAXII™, and several other information assurance tools provide a means for cyber professionals to communicate with each other; standard communications for a logical architecture. In addition, STIX™ is a common language for sharing cyber threat information, helping to standardize architecture communications over different cyber threat scenarios and use cases.

In addition to coordinating communications for a system's logical architecture, DoDCAR covers the entire system lifecycle, using DoDAF process artifacts to document the completion of steps from material solution analysis to operations and sustainment (Table 8.4).

As shown in Table 8.4, DoDCAR uses multiple architecture artifacts to describe a system of interest, helping security engineers understand the types of cyber capabilities needed for their system. The use of DoDCAR, especially when paired with more traditional architecture tools, provides a method for communicating architecture, diagrams, and projects. This might be used for technical roadmap development and planning, providing system-level consistency in terminology and measured data for technical and in progress reviews (IPRs). Follow on benefits from a systematic approach include increased agility in responding to new requirements and a clear understanding of current funding for reviews and future justifications.

Tools to implement DoDCAR currently include NextGen (MITRE, 2014), an online source for entering architecture information for assessment. This tool is ideal for evaluating solution architecture information. If available, the systems engineer can directly input physical architecture data to assess the security of that instance.

Table 8.3 Cyber description tools (MITRE)

Title	Description
CybOX™ (Cyber Observable eXpression) (MITRE)	The MITRE website summarizes CybOX™ as "a standardized language for encoding and communicating high-fidelity information about cyber observables." It offers a common structure at the enterprise level that can be used to represent dynamic events and static attributes in the network of interest, together with the associated corrective actions taken
STIX™ (Structured Threat Information eXpression) (MITRE)	The STIX™ framework uses an XML schema to express cyber threat information with a view to enabling the sharing of that information and generating a cyber threat analysis language. It tries to build up the language by using referential relations between tables and nodes, with the goal of creating a standardized way of representing the cyber threat
TAXII™ (Trusted Automated eXchange of Indicator Information) (MITRE)	TAXII™ is a standardized way of defining a set of services and message exchanges for exchanging cyber threat information. It uses XML and is service-oriented with four options (Inbox, Poll, Collection Management and Discovery) and three sharing models (Hub and Spoke, Source/Subscriber and Peer to Peer)

Table 8.4 Example DoDCAR use of architecture artifacts for cyber system security evaluation across an acquisition

DoDCAR process artifacts		Alignment with DAU acquisition phase and associated information														
		Material solution analysis			Technology maturation and risk reduction			Engineering and manufacturing development			Production and deployment			Operations and sustainment		
		ICD	AoA	Draft CDD and TEMP	TEMP	RFP	CDD/TRA	PDR	CDR	CPD	LD	PRR	PPP	PIR	ECPs	EOL
DoDCAR processes	*Threat models* OV-5a OV-5b	Apply framework (OV-5a) to proposed system's threat environment with baseline of mitigation performance (i.e., MOE/KPP)			Identification of test environment, test cases, and technical performance measures (TPMs)			Detailed system specific threat actions (OV-5b) for seventy weighted most probable threat impact actions			Inform Red/Blue team scenarios to most likely threat actions and campaigns. CCORI and CCRI threat scenarios			Support "tuning" of appliances configurations/rules/analytics as adversary behaviors change		
	Architecture models	Cybersecurity performance and affordability parameters. System functions (SV-1, SV-10, CV-2) for Threat mitigation			System tradeoffs and weighted cybersecurity performance and affordability parameters			Detailed system-specific threat mitigation functions for corresponding threat actions (updated OV-5b)			Specific network deployment models, ports and protocols			Threat-based ECP/Tech Refresh designs and functions		
	Scoring model CV-6	Possible combinations of cyber system capabilities for TMRR (initial CV-6)			Cyber effectiveness scores for capabilities in CV-6 for trade-off analyses			Updated scores from detailed design reviews to supplement selecting solutions			Establish capability measure of effectiveness (MOE) feedback loop for deployed systems			Adjusted scoring for threat-based ECP/Tech Refresh		

While DoDCAR shows promise for describing specific cyber defense issues, the application of systems engineering to cyber is still developing. Hibbs (Hibbs, 2019) provides a thoughtful brief on the challenges that systems engineering currently has in being applied to cyber systems. In his 2019 talk, Hibbs said that he believes systems engineering is not currently a key element of cyber today. In addition, Mr. Hibbs said that he believes that this is due to both cultural problems and opinions, which he distilled to six reasons:

- Oversimplifying the complexity of the problem
- Spending excessive funds just discussing the problem
- Commissioning a burdensome documentation effort
- Forcing engineers to perform tasks outside their primary skill sets to accommodate urgent needs
- Meeting the changing needs of the program manager
- Organizing systems engineering efforts is harder than building systems

As Hibbs describes, systems engineering and cyber are still coming together, as an overall approach, for the design, development, and operation of cyber systems. Fortunately, we can use current architecture products to describe solution architectures from historical accounts to better understand how to apply architectures in the future.

8.3   Cyber System Design Example

We introduced an example cyber collection system in Fig. 3.8. This system might also be described as an overall architecture for collecting data (Fig. 8.3).

Figure 8.3 introduces a cyber collection system in terms of a system entity structure (SES), an entity relationship approach developed for managing models that describe a system (Couretas, 2006; Zeigler, 2000). The SES has two operators in

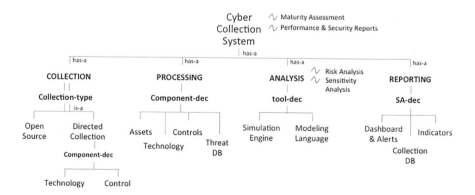

Fig. 8.3 Technical cyber collection system

decomposition (i.e., has-a) and specialization (i.e., is-a). Therefore, each of the collection, processing, analysis, and reporting are components of the cyber collection system. Collection type is specialized to open source and directed collection, where one of them is chosen on the basis of the collection type being performed.

The technical cyber collection system shown in Fig. 8.3 is relatively mechanical in that the key decisions include the type of collection (e.g., open source, directed), relying on preprogrammed processing components and analysis tools to provide reporting that is primarily focused on the collection systems maturity and capability for the activity of interest. A more comprehensive system description includes people and processes involved in the collection.

8.3.1 2016 US Presidential Election Attack (GRU, Guccifer2.0 and Wikileaks)

The general cyber collection system provided in Fig. 8.3 can be scoped down to an architectural replication of the system used during the Russian intelligence services attack on the 2016 US Presidential election. According to the Mueller Report (Mueller, 2019), the Russian intelligence services (i.e., the GRU) leveraged collected data through multiple personas, Guccifer 2.0 being one of the more famous (Poulsen, 2018). The GRU then used DCLeaks and WikiLeaks to disseminate Democratic National Committee (DNC) e-mails to discredit both the candidate and the organization (Fig. 8.4).

In terms of cyber systems design, Fig. 8.4 is a relatively simple process diagram (e.g., DoDAF OV-5) showing the information collection system in operation. The people/organization, the GRU, uses the Guccifer 2.0 persona (Poulsen, 2018) to collect and disseminate over 20,000 Democratic National Committee e-mails, approximately one gigabyte of data, on the web for consumption by interested individuals and mainstream media.

Figure 8.4 might also be looked at as an example process flow for a solution architecture (Table 8.1). The physical architecture, or architecture components of the attack system shown in Fig. 8.4, is shown in Fig. 8.5.

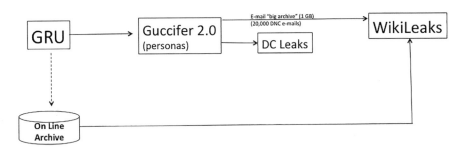

Fig. 8.4 GRU and DCLeaks/WikiLeaks dissemination (Mueller, 2019)

Fig. 8.5 System architecture GRU and DCLeaks/WikiLeaks collection system

As shown in Fig. 8.5, the effectiveness of the GRU/Wikileaks system was likely due to its simplicity. The GRU simply hacked into the DNC, via the persona Guccifer 2.0 (Poulsen, 2018), and dumped the e-mails onto public sites (i.e., Wikileaks, DCLeaks), as described in Chap. 4.

In Fig. 8.5, we also see that the key attributes (i.e., squiggly lines next to "Cyber Collection System") are "trust" and "attention/maneuver." These are example system measurement metrics that show how the system might have been measured in the real world. By releasing the DNC e-mails on the web, there was a degradation of trust between DNC members. For example, there was a formal apology by Debbie Wasserman-Schultz to Bernie Sanders for some of the private comments outed in the e-mails that we noted in Chap. 4.

"Attention/Maneuver" is a bit more abstract in that the e-mail releases caused the DNC to focus on covering for the e-mail releases, at least for a short period, rather than maneuvering toward a successful presidential campaign. Trying to quantify "trust" and "attention/maneuver" points out the complexity in describing cyber systems from an effects standpoint, one of the issues that Hibbs pointed out in Sect. 8.2.2.1.

8.3.2 Wikileaks Operations Example (Costs and Tactics)

While the 2016 US Presidential Election collection architecture (Fig. 8.5) looks simple, this may be due to the limited open-source information that we have available. For example, we have much more documentation concerning the Wikileaks organization. In Chap. 4, we discussed how Wikileaks became a key channel between the collection of data and release to the mainstream media for several high-profile leaks.

Wikileaks originally ran off of one server. With the issue of the "Collateral Murder" video, they subsequently raised approximately $700,000 (Domscheit-Berg, 2011). This allowed Wikileaks to buy state-of-the-art servers, cryptophones,

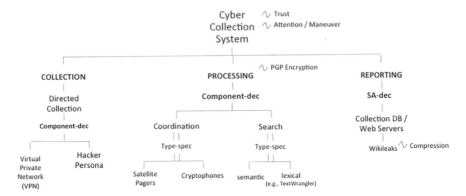

Fig. 8.6 Wikileaks collection system description

and satellite pagers; the tools required to keep the organization secure (Domscheit-Berg, 2011) (Fig. 8.6).

As shown in Fig. 8.6, the multiple components of the Wikileaks system included compression for the voluminous data, encryption (i.e., pretty good privacy (PGP)), search for the exceptional amount of text, and secure communications via crypto-phones (David Leigh, 2011). For example, The Iraq Logs data were archived on a temporary website, compressed in a format called 7z, and then downloaded by the respective news partners (David Leigh, 2011). Due to the free text format of the diplomatic cables, the journalists used "TextWrangler" software to perform word frequency-based searches in developing narratives from the Wikileaks data to reporting for the general public (Fig. 8.7).

As shown in Fig. 8.7, both hackers and foreign intelligence agencies used the Wikileaks site to upload data. While the e-mails were collected and stored, journal-ists still required a search capability to construct digestible stories for the general public. As discussed in Chap. 4, several mainstream media publications (e.g., The Guardian) developed stories based on Wikileaks data.

In this section, we provided an example of cyber collection architecture origi-nally introduced in Chap. 3 (Fig. 3.8). We showed how a version of this architecture was likely used in the 2016 US Presidential Election "outing" attack that leveraged Wikileaks for data distribution. This example effectively included a cyber collection and dissemination mission thread, which used reporting from both the Russian GRU activities (Mueller, 2019) and Wikileaks (Domscheit-Berg, 2011) to provide a cyber system design architecture (Fig. 8.3).

8.4   Summary

Wikileaks is a good example of a cyber system used to exploit a vulnerable cyber system, process the collected data, and provide effects in the real world through an information operation. The collection of key data, stored on secure web servers for

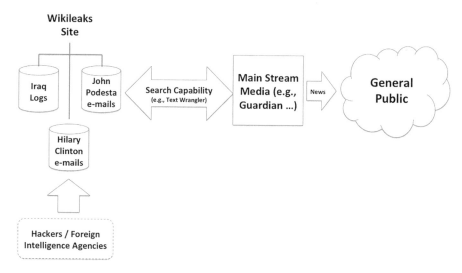

Fig. 8.7 Wikileaks collections and process for main stream media dissemination

follow on dissemination via Main Stream Media, combines cyber targeting, operations, and effects. This example, used by actors that span from hackers to nation states, shows how the Wikileaks example contributes updating legacy active measures to the twenty-first century, through controlled release of information with targeted timing (i.e., e-mails published just before the 2016 Democratic National Convention) and scale (i.e., gigabytes of data) (Mueller, 2019).

In addition, this chapter helps us answer the following questions:

1. How are cyber systems best described in terms of traditional software architecture?

   Information technology (IT)-based systems were mandated by the Clinger-Cohen Act in 1996 to include architectural representation in any US Government procurement. The C4ISR Architecture Framework was the original template to describe these architectures. A few years later (2003) DoDAF was introduced as the means to describe DoD systems, and it has continued to evolve.

   At the same time that government-based architectures were developing, the systems and software community developed their own standards. Process modeling approaches (e.g., SADT and IDEF) provided a foundation for the Unified Modeling Language (UML), used for software engineering, which continues to evolve. In addition, UML was the root for the systems modeling language (SysML) and the business process modeling notation (BPMN).

   In addition, uses of these architectures have been developed specifically for cyber design. One example, DoDCAR, specifies architecture-based products for the communication of cyber defense scenarios. DoDCAR's overall architecture design approach captures cybersecurity performance and affordability parameters, scoring models analyze cost-effectiveness, and possible cyber system capabilities and threat models are described via OV-5 s.

2. How do system designs help in describing operational architectures used for offensive and defensive cyber operations?

   System design methods help in describing the functional elements of systems and their components (Fig. 8.2). In addition, specific system engineering methods (e.g., SysML, BPMN) have been used to describe cyber systems, as solution architectures, for evaluation by simulation.

   Architectures are also useful to provide the terms for a domain (e.g., reference architecture), the implementation (e.g., solution architecture), the system's connections (e.g., logical architecture), and the system components (e.g., physical architecture).

   In addition, architecture approaches can be used for specific scenarios. For example, DoDCAR, originally discussed in Chap. 3, is used specifically to describe a computer-based system, over the course of its acquisition and operational lifecycle, with cyber defense in mind. This includes using architecture products, as design artifacts, to document each step in the system lifecycle.

   And, while DoDCAR is defense focused, we also discussed the Lockheed Martin Attack Cycle (Chap. 1), which provides background for MITRE ATT@CK. MITRE ATT@CK was discussed in Chap. 6 as one way to defeat an adversary during his/her attack.

3. What kind of threat descriptions are currently available, and how are they used in cyber systems design?

   In Chap. 3, we discussed several ways to find a vulnerability and describe it. Architectures leverage these techniques (e.g., PASTA in Sect. 3.3.3) to provide a system structuring for the individual component, for analysis of overall system-level scoring due to the threat, in terms of both fixing the vulnerability and understanding what an exploit could cost.

4. What are examples of designs for cyber analysis and targeting systems, along with their use?

   We introduced a cyber collection system in Fig. 3.8. We use this system as a template to generalize on a collection system architecture. We then reviewed what the collection system from Wikileaks looks like, from open-source reporting, to provide an example cyber system design for producing strategic level effects.

   Wikileaks attacks are some of the more famous for outing US Government communications (i.e., 2010 State Department Cables). In addition, Wikileaks was used to "out" a law firm providing tax havens for Chinese and Russian Government officials (i.e., 2014 Mossack-Fonseca) and 2016 Democratic National Committee (DNC) e-mails.

As discussed in this chapter, architectures and systems engineering provide value for cyber. We saw how BPMN was used by MITRE to evaluate cyber process vulnerability in their AMICA tool/approach (Steven Noel, 2015). Similarly, DoDCAR is currently being used to evaluate cyber vulnerabilities. For example, DoDCAR is implemented in the NextGen tool, which helps users understand their architectural vulnerabilities and possible future compromises.

In this chapter, we also reviewed a "toy" cyber collection system (i.e., Fig. 8.3—based on Fig. 3.8) to determine how we might develop architecture products that describe the 2016 US Presidential election "outing" attack against the Democratic National Committee (DNC) (Fig. 8.4). This architecture included GRU persona and Wikileaks repository components. This extra information provided us with a better understanding of the processing steps used to coordinate a cyber collection operation (Fig. 8.6).

While architecture description helps in abstracting on a real-world example, it remains a challenge to measure a cyber system in terms of overall effects. We looked at how the DNC had trust and attention issues after the attack. However, quantifying this implementation of an active measure, in cyber, remains a challenge.

Bibliography

Acton, J. M. (2017). Cyber weapons and precision guided munitions. In A. L. G. Perkovich (Ed.), *Understanding cyber conflict.*

Aid, M. (2013, October 15). The NSA's new code breakers. *Foreign Policy.*

Akamai. (n.d.). *Global traffic management.* Retrieved 11 14, 2018, from Akamai: https://www.akamai.com/us/en/multimedia/documents/product-brief/global-traffic-management-product-brief.pdf

Alinsky, S. (1989). *Rules for radicals: A practical primer for realistic radicals.* Vintage.

Anderson, N. (2010, 11 7). *How China swallowed 15% of 'net traffic for 18 minutes.* Retrieved 11 13, 2018, from Ars Technica: https://arstechnica.com/information-technology/2010/11/how-china-swallowed-15-of-net-traffic-for-18-minutes/

Andrei Soldatov, I. B. (2015a). *The red web - the struggle between Russia's digital dictators and the new online revolutionaries.* Public Affairs.

Andrei Soldatov, I. B. (2015b). *The red web - the struggle between Russia's digital dictators and the new online revolutionaries.* Public Affairs.

Berman Klein Center. (n.d.). *GhostNet.* Retrieved 4 11, 2019, from GhostNet: https://cyber.harvard.edu/cybersecurity/GhostNet

Bernstein, J. (2017). *Secrecy world - inside the Panama papers investigation of illicit money networks and the global elite.* Henry Holt and Company.

Blank, S. (2013). Russian information warfare as domestic counterinsurgency. *American Foreign Policy Interests, 35*(1), 31–44.

Blank, S. (2017). Cyber war and information war a la Russe. In A. E. G. Perkovich (Ed.), *Understanding cyber conflict.*

Boston Globe. (1998, March 19). *Teen hacker pleads guilty to crippling Massachussetts Airport.* Boston Globe.

Bowden, M. (2011). *Worm - the first digital world war.* Atlantic Monthly Press.

Brenner, B. (2005, 8 31). *Myfip's titan rain connection.* Retrieved 4 10, 2019, from SearchSecurity: https://searchsecurity.techtarget.com/news/1120855/Myfips-Titan-Rain-connection

Brian Grow, M. H. (2011, 4 11). *Special report: In cyberspy vs. cyberspy, China has the edge.* Retrieved 4 10, 2019, from Reuters: https://www.reuters.com/article/us-china-usa-cyberespionage/special-report-in-cyberspy-vs-cyberspy-china-has-the-edge-idUSTRE73D24220110414

Brooks, C. (2011, October 9). *Cybersecurity experts say small buisinesses beware.* Business News Daily.

Bryan Krekel, P. A. (2012, 3 12). *Occupying the information high ground: Chinese capabilities for computer network operations and cyber espionage.* Retrieved 4 10, 2019, from U.S.-China Economic and Security Review Commission: https://www.uscc.gov/Research/occupying-information-high-ground-chinese-capabilities-computer-network-operations-and

Butsenko, A. (2014, October 28). Trolli iz Olgino pereekhali v noviy chetyrekhatazhny office na Savushkina (Trolls from Olgino Moved to a New Four Story Office on Savushkina). DP.ru.

Carlin, J. P. (2018). *Dawn of the code war – America's battle against Russia, China and the rising global cyber threat.* PublicAffairs.

Carr, J. (2012a). *Inside cyber warfare.* O'Reilly.

Carr, J. (2012b). *Inside cyber warfare.* O'Reilly Media.

Carr, J. (n.d.). *Jeffrey Carr.* Retrieved 7 8, 2019, from Wikipedia: https://en.wikipedia.org/wiki/Jeffrey_Carr

Chalfant, M. (2017, 11 20). *Dems call for states to get $400 M election security upgrades.* Retrieved 6 6, 2019, from Hill: http://thehill.com/policy/cybersecurity/361263-dems-say-congress-should-send-400m-to-states-for-election-cyber-upgrades

Clapper, J. R. (2015). *Statement for the record: Worldwide threat assessment of the US intelligence community.* Senate Armed Services Committee.

Clayton, M. (2012, Sep 12). Stealing US business secrets: Experts ID two huge cyber "gangs" in China. *Christian Science Monitor.*

Cleary, G. (2019, 6). *Twitterbots: Anatomy of a propaganda campaign.* Retrieved 7 6, 2019, from Symantec: https://www.symantec.com/blogs/threat-intelligence/twitterbots-propaganda-disinformation

Clement, P. (2019, 7 25). *Internet usage worldwide - statistics & facts.* Retrieved 6 12, 2020, from Statistica: https://www.statista.com/topics/1145/internet-usage-worldwide/

Corera, G. (2016). *Cyber spies - the sSecret history of surveillance, hacking and digital espionage.* Pagasus Books.

Couretas, J. M. (2006). System architectures: legacy tools/methods, DoDAF descriptions and design through system alternative enumeration. *Journal of Defense Modeling and Simulation, 3*(4), 227–237.

CrowdStrike. (2015, 2 6). *CrowdStrike's 2014 global threat intel report: know your adversary and better protect your network.* Retrieved 4 10, 2019, from CrowdStrike: https://www.crowdstrike.com/blog/crowdstrikes-2014-global-threat-report-know-adversary-better-protect-network/

Daly, J. J., & Couretas, J. (2015). *Using conceptual modeling to implement model based systems engineering for program capability analysis and assessment. NDIA systems engineering conference.* NDIA.

Darnay, K. (2014, 1 27). *A look at website lifespans.* Retrieved 7 8, 2019, from Bismarck Tribune: https://bismarcktribune.com/news/columnists/keith-darnay/a-look-at-website-lifespans/article_1d879ae6-851a-11e3-8bd1-0019bb2963f4.html

Data Center Map. (n.d.). *Data center map.* Retrieved 3 13, 2019, from https://www.datacentermap.com/ixps.html

David Leigh, L. H. (2011). *WikiLeaks - inside Julian Assange's war on secrecy.* Public Affairs.

Denning, D. E., & Strawser, B. J. (2017). Active cyber defense - applying air defense to the cyber domain. In A. E. G. Perkovich (Ed.), *Understand cyber conflict - 14 analogies.*

Der Spiegel. (2013, December 30). *Documents reveal top NSA hacking unit.* Der Spiegel.

DHS CISA. (2017, 8 23). *Hidden Cobra – North Korea's DDoS botnet infrastructure.* Retrieved 5 23, 2020, from DHS CISA: https://www.us-cert.gov/ncas/alerts/TA17-164A

DHS CISA. (2018, 12 18). *Hidden Cobra – FASTCash Campaign.* Retrieved 5 23, 2020, from DHS CISA: https://www.us-cert.gov/ncas/alerts/TA18-275A

DHS CISA. (n.d.-a). *Alert (TA18-275A) - Hidden Cobra – FASTCash campaign.* Retrieved 6 11, 2020, from DHS CISA: https://www.us-cert.gov/ncas/alerts/TA18-275A

DHS CISA. (n.d.-b). *Chinese malicious cyber activity.* Retrieved 5 23, 2020, from DHS CISA: https://www.us-cert.gov/china

DHS CISA. (n.d.-c). *North Korean malicious cyber activity.* Retrieved 5 23, 2020, from DHS CISA: https://www.us-cert.gov/northkorea

Dmitry Volchek, D. S. (2015, March 27). *One professional Russian troll tells all.* Radio Free Europe/Radio Liberty.

DoD CIO. (1996, 8). *Clinger Cohen Act of 1996.* Retrieved 6 12, 2020, from DoD CIO: https://dodcio.defense.gov/Portals/0/Documents/ciodesrefvolone.pdf

Domscheit-Berg, D. (2011). *Inside Wikileaks - my time with Julian Assange at the world's most dangerous website.* Crown.

Duggan, M., & Brenner, J. (2012, 2 14). *The demographics of social media users — 2012.* Retrieved 3 13, 2019, from Pew Research: http://www.pewinternet.org/2013/02/14/the-demographics-of-social-media-users-2012/

FireEye. (2017). *M-TRENDS 2017 - A view from the front lines.* FireEye.

Fritz, J. R. (2017). *China's cyber warfare - the evolution of strategic doctrine.* Lexington Books.

Gatlan, S. (2019, 5 6). *Israel bombs building as retaliation for Hamas cyber attack.* Retrieved 6 13, 2019, from Bleeping Computer: https://www.bleepingcomputer.com/news/security/israel-bombs-building-as-retaliation-for-hamas-cyber-attack/

Gavin, F. (2017). Crisis instability and preemption: The 1914 railroad analogy. In A. E. G. Perkovich (Ed.), *Understanding cyber conflict - 14 analogies.*

Geers, K. (Ed.). (2015). *Cyber war in perspective: Russian aggression against Ukraine.* NATO Cooperative Cyber Defence Centre of Excellence.

Georgia Tech Research Institute. (2012). *Cyber threats report 2012.* Georgia Tech information security center, Georgia Tech Cyber Security Summit 2011.

Glenny, M. (2012). *DarkMarket: How hackers became the new mafia.* Vintage.

Goldman, E. O., & Warner, M. (2017). Why a digital pearl harbor makes sense. In A. E. G. Perkovich (Ed.), *Understanding cyber conflict.*

Greaver, B., & Raabe, L. (2018). CARVER 2.0: Integrating the analytical hierarchy process's multi-attribute decision-making weighting scheme for a center of gravity vulnerability analysis for US Special Operations Forces. *Journal of Defense Modeling and Simulation, 15*(1), 111–120.

Greenwald, G. (2014). *No place to hide: Edward Snowden, the NSA and the US Surveillance State.* Hamish Himilton.

Gregory Conti, D. R. (2017a). *On cyber: Towards an operational art for cyber conflict.* Kopidion Press.

Gregory Conti, D. R. (2017b). *On cyber: Towards an operational art for cyber conflict 1st edition.* Kopidion Press.

Heickero, R. (2013). *Emerging cyber threats and Russian views on information warfare and information operations.* Swedish Defense Research Agency.

Heli Tiirmaa-Klaar, J. G.-P. (2014). *Botnets.* Springer.

Henderson, S. (2007). *The dark visitor.* Scott Henderson.

Hibbs, E. (2019, May 15). *Systems engineering in a cyber world - connecting frameworks for program decisions.* Retrieved march 8, 2020, from DISA: https://www.disa.mil/-/media/Files/DISA/News/Events/Symposium-2019/1%2D%2D-Hibbs_Systems-Engineering-in-a-Cyber-World_approved-Final.ashx+&cd=10&hl=en&ct=clnk&gl=us

Hollis, D. (2011, January 11). Cyberwar case study: Georgia 2008. *Small Wars Journal.*

Hsiao, R. (2013, 12 5). *Critical node: Taiwan's cyber defense and Chinese cyber-espionage.* Retrieved 4 10, 2019, from Jamestown foundation: https://jamestown.org/program/critical-node-taiwans-cyber-defense-and-chinese-cyber-espionage/

IDS International. (n.d.). *SMEIR.* Retrieved 1 3, 2019, from https://www.smeir.net/

IEEE. (2011, 12 1). *42010-2011 - ISO/IEC/IEEE systems and software engineering -- Architecture description.* Retrieved 6 13, 2020, from IEEE: https://standards.ieee.org/standard/42010-2011.html

International Telecommunications Union. (2013, May 14–16). Internet Exchange Points (IXPs). *World Telecommunication Policy Forum.*

Internet Exchange Map. (n.d.). *Internet exchange map.* Retrieved 3 13, 2019, from https://www.internetexchangemap.com/

Isikoff, M. (2013, 6 10). *Chinese hacked Obama, McCain campaigns, took internal documents, officials say.*

Jarvis, J. (2011, 3 17). *Revealed: US spy operation that manipulates social media – Military's 'sock puppet' software creates fake online identities to spread pro-American propaganda.* Retrieved 1 3, 2019, from Guardian: https://www.theguardian.com/technology/2011/mar/17/us-spy-operation-social-networks

Jen Weedon, W. N. (2017, 4 27). *Information operations and Facebook.* Retrieved 1 3, 2019, from Facebook: https://fbnewsroomus.files.wordpress.com/2017/04/facebook-and-information-operations-v1.pdf

Keith Collins, S. F. (2018, September 4). *Can you spot the deceptive Facebook post?* Retrieved November 15, 2018, from New York Times: https://www.nytimes.com/interactive/2018/09/04/technology/facebook-influence-campaigns-quiz.html

Koerner, B. I. (2016, 10 23). *Inside the cyberattack that shocked the US government.* Retrieved from Wired: https://www.wired.com/2016/10/inside-cyberattack-shocked-us-government/

Krebs, B. (2014). *Spam nation: The inside story of organized cybercrime - from global epidemic to your front door.* Surcebooks.

Krekel, B. (2009a). *Capability of the people's republic of China to conduct cyber warfare and computer network exploitation.* Retrieved 4 11, 2019, from https://nsarchive2.gwu.edu/NSAEBB/NSAEBB424/docs/Cyber-030.pdf

Krekel, B. (2009b, 10 9). *Capability of the people's republic of China to conduct cyber warfare and computer network exploitation.* Retrieved 4 10, 2019, from The US-China Economic and Security Review Commission: https://nsarchive2.gwu.edu/NSAEBB/NSAEBB424/docs/Cyber-030.pdf

Lambert, N. A. (2017). Brits-Krieg: The strategy of economic warfare. In A. L. G. Perkovich (Ed.), *Understanding cyber conflict - 14 analogies.*

Landau, S. (2014). Under the radar: NSA's efforts to secure private-sector telecommunications infrastructure. *Journal of National Security Law and Policy, 7*, 411.

Maidment, P. (2009, 3 29). *GhostNet in the machine.* Retrieved 4 10, 2019, from Forbes: https://www.forbes.com/2009/03/29/ghostnet-computer-security-internet-technology-ghostnet.html#455f71b7d00e

Mandiant. (2013). *APT1.* Retrieved 10 2019, 4, from Mandiant: https://www.fireeye.com/content/dam/fireeye-www/services/pdfs/mandiant-apt1-report.pdf

Max Boot, M. D. (2013). *Political warfare.* Council on Foreign Relations. Council on Foreign Relations.

Michael Riley, L. D. (2012, 7 26). *Hackers linked to China's army seen from EU to DC.* Retrieved 4 10, 2019, from Bloombert: https://www.bloomberg.com/news/articles/2012-07-26/china-hackers-hit-eu-point-man-and-d-c-with-byzantine-candor

Microsoft. (2011). *How conficker continues to propagate.* Microsoft Security Intelligence Report.

Microsoft. (n.d.). *Microsoft cybersecurity reference architecture.* Retrieved 6 12, 2020, from Microsoft Cybersecurity Reference Architecture: https://virtualizationandstorage.wordpress.com/2020/02/03/microsoft-cybersecurity-reference-architecture-the-microsoft-cybersecurity-reference-architecture/

MITRE. (2014, 10). *NextGen independent assessment and recommendations.* Retrieved 6 15, 2020, from MITRE: https://www.mitre.org/sites/default/files/publications/pr-14-3495-next-gen-independent-assessment.pdf

MITRE. (n.d.-a). Retrieved from https://cybox.mitre.org/language/version2.0/

MITRE. (n.d.-b). Retrieved from https://www.mitre.org/publications/technical-papers/standardizing-cyber-threat-intelligence-information-with-the

MITRE. (n.d.-c). *Lazarus group.* Retrieved 6 11, 2020, from MITRE ATT@CK: https://attack.mitre.org/groups/G0032/

MITRE. (n.d.-d). *The cyber analytics repository (CARET)*. Retrieved 7 8, 2019, from MITRE: https://mitre-attack.github.io/caret/#/

Mueller, R. (2019). *Report on the investigation into Russian interference in the 2016 presidential election*. U.S. Department of Justice. Washington: U.S. Department of Justice.

Nakashima, E. (2018, 10 23). *Pentagon launches first cyber operation to deter Russian interference in midterm elections*. Retrieved 3 12, 2019, from Washington Post: https://www.washingtonpost.com/world/national-security/pentagon-launches-first-cyber-operation-to-deter-russian-interference-in-midterm-elections/2018/10/23/12ec6e7e-d6df-11e8-83a2-d1c3da28d6b6_story.html?utm_term=.8c47d573557b

Nakashima, E. (2019, 2 27). *US disrupted internet access of Russian troll factory on day of 2018 midterms*. Retrieved 3 12, 2019, from Washington post.

National Institute of Standards (NIST). (2018, 4 16). *Cybersecurity framework*. Retrieved 6 12, 2020, from NIST: https://www.nist.gov/cyberframework

NSA. (n.d.). *Secure architecture*. Retrieved 6 12, 2020, from NSA: https://apps.nsa.gov/iaarchive/library/ia-guidance/secure-architecture/index.cfm

Nusca, A. (2010, 3 29). *Malware capital of the world is Shaoxing, China*. Retrieved 4 10, 2019, from ZDNet: https://www.zdnet.com/article/report-malware-capital-of-the-world-is-shaoxing-china/

OASIS. (n.d.). *Advancing open standards for the information society*. Retrieved 2 10, 2018, from OASIS Cyber threat intelligence (CTI) TC: https://www.oasis-open.org/committees/tc_home.php?wg_abbrev=cti

Office of the Assistant Secretary of Defense Networks and Information Integration (OASD/NII). (2010). *Reference architecture description*. Retrieved 6 11, 2020, from https://dodcio.defense.gov/Portals/0/Documents/DIEA/Ref_Archi_Description_Final_v1_18Jun10.pdf

Office of the National Manager for NSS. (n.d.). *DoDCAR*. Retrieved 3 8, 2019, from https://csrc.nist.gov/CSRC/media/Presentations/DODCAR-no-class-markings-Pat-Arvidson/images-media/DODCAR_-no%20class%20markings%20-%20Pat%20Arvidson.pdf

Orr, A. (2018, 11 7). *China re-routed US internet traffic for 2.5 years*. Retrieved 11 13, 2018, from https://www.macobserver.com/link/china-reroute-internet-traffic/

Osnos, E. (2014). *Age of ambition: Chasing fortune, truth, and faith in the new China*. Farrar, Straus and Giroux.

Overy, R. (2014). *The bombers and the bombed: Allied air war over Europe*. Viking.

Pape, R. A. (1996). *Bombing to win: Air power and coercion in war*. Cornell University Press.

Philip Bennett, S. C. (1999, May 25). *NATO Warplanes Jolt Yogoslav Power Grid*. Washington Post.

Popular Mechanics. (2018, March 13). *How long does it take hackers to pull off a massive job like equifax?* Retrieved from Popular Mechanics: https://www.popularmechanics.com/technology/security/a18930168/equifax-hack-time/

Poulsen, K. (2012). *Kingpin: How one hacker took over the billion-dollar cybercrime underground*. Broadway.

Poulsen, K. (2018, 10 25). *'Lone DNC Hacker' Guccifer 2.0 slipped up and revealed he was a Russian intelligence officer*. Retrieved 7 8, 2019, from the daily beast: https://www.thedailybeast.com/exclusive-lone-dnc-hacker-guccifer-20-slipped-up-and-revealed-he-was-a-russian-intelligence-officer

Prosser, M. B. (2006). *Memetics—A growth industry in us military operations (MS Thesis)*. Retrieved 1 3, 2019, from DTIC: https://apps.dtic.mil/dtic/tr/fulltext/u2/a507172.pdf

Raff, A. (2013, 3 5). *Chinese time bomb*. Retrieved 4 10, 2019, from Securlert: http://www.avivraff.com/seculert/test/2013/03/the-chinese-time-bomb.html

Robb, J. (2007, June). The coming urban terror. *City Journal*.

Sanger, D. E. (2017). Cyber, drones and secrecy. In A. E. G. Perkovich (Ed.), *Understanding cyber conflict*.

Sanger, D. E., & Perlroth, N. (2015, 4 15). *Iran is raising sophistication and frequency of cyberattacks, study says*. Retrieved 10 14, 2018, from New York times: https://www.

nytimes.com/2015/04/16/world/middleeast/iran-is-raising-sophistication-and-frequency-of-cyberattacks-study-says.html

SANS. (2016). *Critical security controls*. (SANS, Producer) Retrieved 9 24, 2019, from SANS: https://www.sans.org/media/critical-security-controls/critical-controls-poster-2016.pdf

Savage, C. (2015). *Power wars: Inside Obama's Post - 9/11 Presidency*. Little Brown.

Segura, V. (2009, June 25). *Modeling the economic incentives of DDoS Attacks*. Retrieved from Semantic Scholar: https://pdfs.semanticscholar.org/afdf/d974bc68dc05c48020e-f07a558a61ab94f8a.pdf

Senate Armed Services Committee. (2014). *Inquiry in cyber instrusions affecting US transportation command contractors*. 113th Congress, 2d sess. Washington, DC: Government Printing Office.

Simon Pirani, J. S. (2009, 2 15). *The Russo-Ukrainian gas dispute of January 2009: a comprehensive assessment*. Retrieved 12 11, 2018, from Oxford Institute for Energy Studies: https://www.oxfordenergy.org/wpcms/wp-content/uploads/2010/11/NG27-TheRussoUkrainianGasDisputeofJanuary2009AComprehensiveAssessment-JonathanSternSimonPiraniKatjaYafimava-2009.pdf

Singer, P. W., & Brooking, E. T. (2018). *LikeWar - the weaponization of social media*. Houghton Mifflin.

Smith, P. (2013, 9 24). *Internet exchange point design*. Retrieved 3 13, 2019, from https://www.menog.org/presentations/menog-13/192-MENOG13-IXP-Design.pdf

Snowden, E. (n.d.). Retrieved from https://edwardsnowden.com/2015/01/18/the-roc-nsas-epicenter-for-computer-network-operations/; https://edwardsnowden.com/2015/01/18/interview-with-a-sid-hacker-part-I-how-does-tao-do-its-work/; https://freesnowden.is/2015/01/18/expanding-endpoint-operations

South Front. (2015, May 25). *Cyberkut hacked the site of Ukrainian Ministry of Finance: The country has no money*. Retrieved from South Front: https://southfront.org/cyberkut-hacked-the-site-of-ukrainian-ministry-of-finance-the-country-has-no-money

Spiegel. (2013, Feb 25). *Digital spying burdens German-Chinese relations*. Retrieved 4 10, 2019, from Spiegel: https://www.spiegel.de/international/world/digital-spying-burdens-german-relations-with-beijing-a-885444.html

Steven Noel, J. L. (2015). *Analyzing mission impacts of cyber actions (AMICA)*. NATO.

Stokes, M., & Lin, J. (2011, 11 11). *The Chinese people's liberation army signals intelligence and cyber reconnaissance infrastructure*. Retrieved 4 10, 2019, from Project 2049 Institute: https://project2049.net/2011/11/11/the-chinese-peoples-liberation-army-signals-intelligence-and-cyber-reconnaissance-infrastructure/

Stoll, C. (2005). *The cuckoo's egg: Tracking a spy through the maze of computer espionage*. Pocket Books.

Symantec. (2019). *Internet security threat report*. Retrieved 3 8, 2019, from Symantec: https://www.symantec.com/security-center/threat-report

Symantec. (n.d.). *Symantec*. Retrieved 29 2018, 11, from https://www.symantec.com/

SysML.org. (n.d.). *SysML partners: Creators of the SysML*. Retrieved 6 12, 2020, from SysML.org: https://sysml.org/sysml-partners/

Syverson, P. (n.d.). *Paul Syverson web page*. Retrieved 11 29, 2018, from http://www.syverson.org/

Thomas, T. (2016). *Russian strategic thought and cyber in the armed forces and society: A viewpoint from Kansas. Center for strategic and international studies*. Center for strategic and International Studies.

Timur Chabuk, A. J. (2018, 9 1). *Understanding Russian information operations*. Retrieved 9 9, 2018, from signal (AFCEA): https://www.afcea.org/content/understanding-russian-information-operations

Tkacik, J. J. (2008, 2 8). *Trojan dragons: China's Cyrber threat*. Retrieved 4 10, 2019, from Heritage Foundation: https://www.heritage.org/asia/report/trojan-dragon-chinas-cyber-threat

Twitter. (2017, 7 3). *2016 U.S. Presidential Election - Timeline for attacks*. Retrieved 6 14, 2020, from Twitter: https://twitter.com/symantec/status/881913441946017792

U.S.-China Economic and Security Review Commission. (2011, 8 14). *2011 annual report to congress*. Retrieved 4 10, 2019, from U.S.-China Economic and Security Review Commission: https://www.uscc.gov/content/2011-annual-report-congress

U.S.-China Economic and Security Review Commission. (2014). *USCC annual report*. Retrieved 4 10, 2019, from https://www.uscc.gov/sites/default/files/annual_reports/Complete%20 Report.PDF

US Cyber Consequences Unit (CCU). (2009). *Overview by the US-CCU of the Cyber Campaign against Georgia in August of 2008*. US Cyber Consequences Unit (CCU).

van Eeten, M. (2010). The role of internet service providers in botnet mitigation: an empirical analysis based on spam data. *OECD Science, Technology and Industry Working Papers, 05*.

Vera Zakem, M. K. (2018, 4 1). *Exploring the utility of memes for U.S. government influence campaigns*. Retrieved 1 3, 2019, from Center for Naval Analyses (CNA): https://www.cna.org/ cna_files/pdf/DRM-2018-U-017433-Final.pdf

Warner, M. (2017). Intelligence in cyber - and cyber in intelligence. In A. E. G. Perkovich (Ed.), *Understanding cyber conflict - 14 analogies*. Georgetown University Press.

Watts, B. D., & Keaney, T. A. (1993). *Effects and effectiveness*. US Government Printing Office.

Whaley, B. (2007). *STRATAGEM - deception and surprise in war*. Artech House.

Wilsher, K. (2009, February 7). French fighter planes grounded by computer virus. *The Telegraph*.

Zachman, J. (n.d.). *Conceptual, logical, physical: It is simple*. Retrieved 6 11, 2020, from Zachman international: https://www.zachman.com/ea-articles-reference/58-conceptual-logical-physical-it-is-simple-by-john-a-zachman

Zeigler, B. P. (1976). *Theory of modeling and simulation*. John Wiley and Sons.

Zeigler, B. (2000). *Theory of modeling and simulation*. Academic Press.

Zeigler, B. P. (2019). Introduction to iterative system computational foundations and DEVS. *Winter Simulation Conference*. INFORMS.

Zetter, K. (2011a, April 26). FBI vs. Corefloot botnet: Round 1 goes to the feds. *Wired*.

Zetter, K. (2011b, April 11). With court order, FBI hijacks 'Coreflood' Botnet, Sends Kill Signal. *Wired*.

Zetter, K. (2014). *Countdown to zero day - Stuxnet and the launch of the world's first digital weapon*. Crown.

Zetter, K. (2015, 10 15). *DARPA is developing a search engine for the dark web*. Retrieved 11 15, 2018, from WIRED: https://www.wired.com/2015/02/darpa-memex-dark-web/

Zetter, K. (216, January). *Everything we know about Ukraine's power plant hack*. Wired.

Chapter 9
Measures of Cyber Performance and Effectiveness

The purpose of this chapter is to provide a historical context concerning how cyber has been both employed and measured. This includes unpacking "cyber," as a term used to describe both cyber operations and cyber system security. We will therefore (1) look at cyber operations as a next step in precision guided munition (PGM) development and (2) review cyber security metrics, including measuring effects for non-lethal means. Following the PGM development thread, we will look at cyber operations through the lens of the conventional kinetic joint munitions effectiveness manual (JMEM) work, as applied to cyber, and wrap up with a detailed review of cyber security metrics.

AT9 questions on key performance parameters (KPPs), measures of performance (MOPs), and measures of effectiveness (MOEs) for cyber analysis and targeting

1. How does performance measurement change with cyber versus traditional, kinetic engagement?
2. How do we measure JP 3-12 effects (e.g., deny, manipulate) in current cyber operations?
3. How does cyber Joint Munition Effectiveness Manuals (JMEM) fit in the current measures of cyber performance and effectiveness?
4. How do cyber measures of performance and effectiveness address technical and organizational maturity?

9.1 Background—Information Security, Munitions, and Cyber

In Chap. 8, our cyber systems description included an example architecture approach (e.g., DoDCAR), along with a worked-out example of Glavnoye Razvedyvatelnoye Upravlenie, (Russian: Chief Intelligence Office) (GRU) /Wikileaks cyber collection

© Springer Nature Switzerland AG 2022
J. M. Couretas, *An Introduction to Cyber Analysis and Targeting*,
https://doi.org/10.1007/978-3-030-88559-5_9

system that was used in the 2016 Democratic National Committee (DNC) compromise. In addition, we used a system architecture to describe the components and overall cyber collection system, clarifying the challenges in measuring the overall system (Fig. 9.1).

As shown in Fig. 9.1, the collection and processing system components of a cyber collection architecture are measured in terms of their ability to maintain operational security. The reporting component quantifies file acquisition in terms of the volume of data collected (e.g., terabytes), similar to the Abu Sayyaf collection example discussed in section "2015 Special Forces Raid on ISIS Finance Minister".

Using a cyber collection system to collect and "out" the DNC e-mails affected both internal organization trust and the ability of the DNC to keep focused on its mission of winning the election. However, these organization observations, due to their qualitative nature, are a challenge to prescribe for either an operational plan or a system technical specification. The only thing we can clearly say about a collection operation is how much data was exfiltrated (i.e., reporting component in Fig. 9.1—70 GB for the DNC attack (Whittaker, 2019)).

The 2016 DNC attack included both of the JP 3-12 effects (i.e., denial, manipulation) that we discussed in Chaps. 2 and 7. We also looked at a cyber attack as a threat to be defended against in Chaps. 3 and 6, via cyber risk evaluation frameworks. Measuring cyber, therefore, is a challenge due to the differing perspectives that the operators and information technology (IT) security folks bring to the discussion. For example, operators like to talk in terms of munition effects (i.e., JP 3-12) and IT security personnel talk in terms of threats (e.g., NIST Cybersecurity Framework). These operator and technologist frames of reference, sometimes contrasting, challenge the development of metrics for cyber systems.

Metrics are used to address portions of a system's lifecycle and use. For example, in developing technologies, key performance parameters (KPPs) are useful to see if implementations meet the desired technical objectives. In Fig. 9.1, we saw that

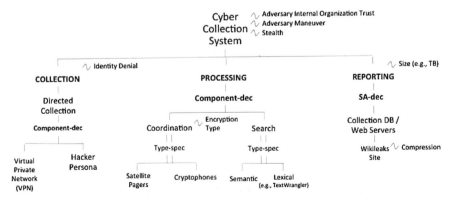

Fig. 9.1 GRU/Wikileaks collection system architecture—collection, processing, and reporting components

maintaining data and communication security was the KPP for the technical collection and processing elements of the system.

Similarly, in using a system, understanding a cyber team's operational maturity might be evaluated via measures of performance (MOPs). For example, in Chap. 6, we looked at a number of security technologies that defensive operators implement in order to maintain security for their system data and communications. In addition, in Sect. 6.4, we looked at how defenders can implement defensive processes (i.e., denial and deception) to challenge an adversary's traversal, should they gain system access. One use of MOPs is to implement technical security components to ensure system confidentiality.

The effectiveness of a cyber actor in prosecuting an individual engagement or mission might be evaluated in terms of MOPs. Measures of Effectiveness (MOEs), usually used at a higher, campaign level, will include an aggregation of these individual missions measured in terms of MOPs.

Modeling and simulation (M&S) professionals discuss of KPPs, MOPs, and MOEs in the context of model use. For example, in developing defense models, it is common to talk about models at the system level, usually an engineered device, in terms of KPPs. Similarly, engagement level modeling might be a one-on-one skirmish; e.g., one missile/one airplane. Following this analogy, a mission will include multiple missiles and airplanes, likely described by both KPPs and MOPs that speak to operator, platform, and weapon effectiveness. And a campaign includes all available assets used to accomplish or defend an objective. A campaign will be described by high level MOEs (e.g., days of war, lives saved/lost).

One way to develop these varying levels of metrics for cyber is to traverse from the system implementation level to the overall campaign (Fig. 9.2).

Fig. 9.2 Systems engineering hierarchy for evaluating KPPs, MOPs, and MOEs

Using Fig. 9.2, we have a straightforward framework for organizing the respective metrics for the campaign, mission, engagement, and system levels that compose a cyber system. While Fig. 9.2 provides a structure for understanding the respective components of a cyber system, and where the metrics fit, it remains a challenge to map contemporary cyber operations, which span from crime to conventional nation state espionage, onto this type of technical hierarchy. In addition, operational cyber system description is complex in that "cyber" can currently inform an operation, be an operation, or simply be a technical tool to enable a conventional munition or delivery capability.

9.1.1   Metrics and Conventional Operations—Viewing Cyber as a Next Step in Precision Munitions

One way to bridge the technologist view of cyber, as an IT system, to an operational capability, is to look at how cyber fits as a next step for military operations. For example, as we discussed in previous chapters (Figs. 1.1, 1.3, and 8.1) cyber might be looked at as the latest arrival in the evolution of precision guided munitions (PGMs).

The use of gravity bombs, and their more recent "smart" variants, PGMs, have the advantage of improved accuracy, requiring fewer aircraft sorties to destroy a target, in order to achieve an "effect." This reduction in aircraft sorties also reduces the time that a pilot is in harms way. An additional upgrade is the combining precision of munitions with the use of unmanned systems, a cruise missile or a drone, which operate at increasingly larger distances (George Perkovich, 2017).

9.1.1.1   Drones, Precision Guided Munitions (PGMs) and Cyber

Perkovich generalizes the connection between drones and precision-guided munitions (PGMs):

> Weaponized drones are an advanced form of PGM. They are especially attractive for targeting single or small numbers of people or other soft targets over long distances with an extremely short time between the decision to fire and the impact on the target. To date, drones have not been used to attack substantial materiel, infrastructure, or military targets, although their payload and lethality is rapidly increasing. (George Perkovich, 2017)

In addition, Sanger (2017) notes the numerous common advantages that both drones and cyber provide:

- Variety of targets
- Low cost of operation and acquisition
- No operator risk
- Long dwell times—potential for greater accuracy
- Real-time control

Cyber has the potential to be more versatile and potentially dangerous than drones. In addition, most adversaries cannot deploy drones over the same long distances that they can for cyber. In addition, the proliferation risk, with cyber, can occur in much shorter time cycles, eliminating any comparative advantage relatively quickly for the attacker.

An additional cost with drones includes skills involved with:

- Piloting
- Image/video processing
- Communicating processed intelligence
- Accurate tasking (e.g., geo locating)

Cyber eliminates each of these costs, except stealthy communication. Cyber also avoids the cost of acquiring and maintaining an airframe, along with the need for a secure airstrip and storage. Implicit in a drone's improvement in safety, standoff and lethality is the development of adequate targeting information, stemming from better intelligence, surveillance, and reconnaissance (ISR). Example metrics for increased bombing precision include:

- Increased accuracy
- Decreased number of sorties
- Decreased pilot exposure

9.1.2 Metrics and Cyber Operations

While the advantages in using PGMs may be intuitive in the case of piloted sortie reduction, abstracting on these metrics can be more challenging in the case of cyber (Acton, 2017). Key Performance Parameters (KPPs), Measures of Performance (MOPs), and Measures of Effectiveness (MOEs) for cyber operations might be standalone, as in our collection system example (Fig. 9.1), or could be a complement to a kinetic operation.

As shown in Fig. 9.2, the metrics have a natural hierarchy, based on the system level (e.g., technical system through campaign employment). In addition, measures of effectiveness (MOEs) are tied to the mission need (Fig. 9.3).

As shown in Fig. 9.3, the mission need drives the overall measure of effectiveness (MOE) with MOPs describing the functional attributes that contribute to the MOE and KPPs describing the technical system performance. For example, in Fig. 9.1, the technical system performance is described with an MOE that is simply a degrading of the target organization performance, whereas the MOP is the volume of data collected, and the KPP is the technical level measures that relate to maintaining a stealthy operation (i.e., encryption level, …).

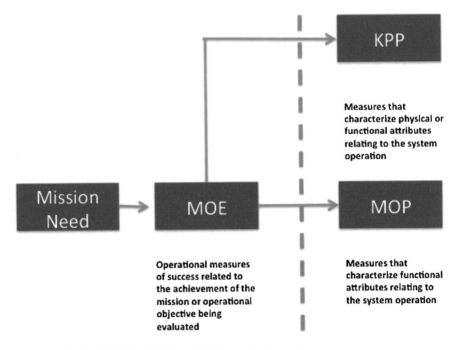

Fig. 9.3 Relationship KPPs, MOPs, and MOEs to mission need

Table 9.1 Increasing munition precision and reduction of piloted sorties

Metrics/capability employed	Dumb bombs	Precision-guided munitions (PGMs)	Drones
Personnel (e.g., Pilot)	N/A	Reduce piloted sorties	Eliminate piloted sorties
MOP		Decreased pilot vulnerability	
KPP		Increased munition accuracy	

9.2 Using the Munition Continuum to Develop Cyber Operational Measurables

The goal of this discussion is to bridge from the somewhat common-sense piloted aircraft metrics to show how cyber might similarly be measured. We will therefore start with a simple example that shows how standard aircraft sorties are improved by increasing munition precision (i.e., PGMs), or completely remote operations (i.e., drones) (Table 9.1).

As shown in Table 9.1, increasing accuracy, a KPP that derives from purely technical developments, contributes to the MOP of decreasing pilot exposure during missions. Aircraft and drones are also used to collect information. Drones increase loiter time due to removing the pilot from the sortie. Additional KPPs include the

speed and altitude at which the drone will fly. Drones also increase red force exposure by increasing the persistence of target surveillance (Table 9.2).

As shown in Table 9.2, looking at the technology spectrum for increasingly
autonomous strike provides us with an ability to look at how aircraft, used for either
targeting or ISR, is measurable, in terms of KPPs (i.e., reduce piloted sorties),
MOPs (i.e., pilot protection, red force exposure).

Cyber can also be looked at as an additional capability in this precision strike
continuum. As a next step after drones, cyber provides both ISR and the ability to
deny or manipulate mission essential equipment for a targeted organization. In fact,
cyber has overlapping capabilities with autonomous platforms when comparing the
use of a drone for jamming, where cyber might be similarly used to disable mission
systems. We can therefore also use this precision strike framework to estimate cyber
effects on the command and control (C2) information systems (Table 9.3).

While the connection between precision, sortie reduction, and pilot safety is
straightforward in the case of improved munitions for bombing sorties, it becomes
more challenging for the stealth and presence requirements for cyber, as shown in
Table 9.3. In addition, cyber MOEs are somewhat qualitative in affecting an organization's internal trust, ability to communicate and maneuver.

As shown in Table 9.3, the measures of cyber become increasingly qualitative in
assessing impact. We started the table with personnel to show the measurable effects
of precision munitions (i.e., reducing piloted sorties), followed by the elimination of
piloted sorties for drones. Drones provide persistent target access, which corresponds to increased red force exposure (e.g., via ISR).

The step from persistent surveillance to a decrease in red force maneuverability
assumes that the red force does not want its actions to be monitored. The next step,
decreasing organization trust, is an insight provided by Warner (2019), where the
perception of loyalty within an organization may be affected by targeting specific
individuals in an organization, in Warner's example with drones. However, cyber
might be used for the same purposes, used to spread rumors about possible collabo-
rators. This type of targeting has the potential for producing stronger impacts when
targeting closed groups that depend on trust to both enter and move up in an organization. Therefore, successful targeting, either kinetically or with IO, of the organization, can result in leadership becoming suspicious about information leaks,
challenging their trust of current members, or reducing their acceptance of new

Table 9.2 Using drones to increase red force exposure

Metrics/capability employed	Dumb bombs	Precision-guided munitions (PGMs)	Drones
Intelligence, surveillance, and reconnaissance (ISR)			Increase red force exposure
Personnel	N/A	Reduce piloted sorties	Eliminate piloted sorties
MOP		Decreased pilot vulnerability	Increased platform persistence, survivability
KPP		Increased munition accuracy	Loiter time, speed, altitude

Table 9.3 Precision munition upgrades (KPPs), ISR improvements (MOPs), and Red Force Maneuverability/Trust (MOEs) across the munitions spectrum

Metrics/capability employed	Dumb bombs	Precision-guided munitions (PGMs)	Drones	Cyber
C2				Reduce red force internal trust
				Reduce red force communications/maneuverability
Intelligence, surveillance and reconnaissance (ISR)				Increase red force exposure
Personnel		Reduce piloted sorties	Eliminate piloted sorties	
MOE	Target organization's ability to communicate and maneuver		Target organization's internal trust Target organization's ability to communicate and maneuver	
MOP	Area coverage rate	Decreased pilot vulnerability Target coverage rate	Increased platform persistence, survivability Target coverage rate	Maintain presence No. information operations
KPP	Random munition accuracy	Increased munition accuracy	Loiter time, speed, altitude	Level of detectability (i.e., stealth) (e.g., encryption)

entrants into the ranks. This potentially decreases both the scale and scope of red force operations. While a challenge to measure, maneuverability and trust are possible IO effects, provided by cyber; candidate MOEs.

While MOEs are a challenge, in cyber, due to developing effects definitions, Table 9.3 shows how our reframing of cyber as a next step in a precision munition continuum provides some initial operational metrics. Looking back at Fig. 9.1, we see example cyber sortie KPPs for collection and processing that are primarily focused on these activities not being detected, with an MOP for reporting being the volume of data collected. Additional MOPs for cyber are related to persistence and access level, prescribing an actor's potential effect level, as discussed in Sect. 7.3 on a target held at risk. MOEs, similarly, describe effects that a threat actor has on a target's maneuverability, internal trust and ability to command and control (C2) red forces (Fig. 9.4).

We see in Fig. 9.4 how precision munition guidance affects measures across the spectrum of systems for piloted aircraft sorties. One of the reasons that we have confidence in these metrics is due to empirical evidence dating back to World War II that provides data for sorties versus munition accuracy. This corpus of data is not available for cyber, making it more challenging to determine good metrics. We handled this challenge by looking at cyber as the latest step in the precision munitions

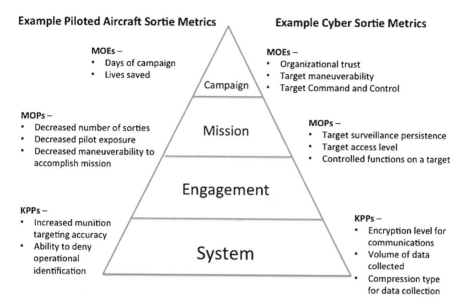

Example Piloted Aircraft Sortie Metrics   **Example Cyber Sortie Metrics**

MOEs –
• Days of campaign
• Lives saved

MOEs –
• Organizational trust
• Target maneuverability
• Target Command and Control

MOPs –
• Decreased number of sorties
• Decreased pilot exposure
• Decreased maneuverability to accomplish mission

MOPs –
• Target surveillance persistence
• Target access level
• Controlled functions on a target

KPPs –
• Increased munition targeting accuracy
• Ability to deny operational identification

KPPs –
• Encryption level for communications
• Volume of data collected
• Compression type for data collection

Campaign

Mission

Engagement

System

Fig. 9.4 Aircraft and cyber sortie examples—KPPs, MOPs, and MOEs

spectrum, helping us estimate metrics for each layer of cyber system use in operations.

In addition, as shown in Fig. 9.4, cyber operations span from IO to traditional collection, thereby making the somewhat loose MOEs for conventional operations seem concrete compared to MOEs for cyber operations (e.g., "trust," "maneuverability," …). Cyber operational MOPs, however, are tied to an activity, helping operators maintain access in order to understand target behavior. Similarly, cyber KPPs help system designers describe and measure the supporting security technologies that enable cyber operational stealth.

Another kinetic analog that provides insight is battle damage assessment (BDA), a measure used every time a bomb is dropped for tactical effect in the conventional world. BDA provides the commander with insight, and analysis, concerning how the target is affected by cyber. This includes the duration, level and sustainability of the effect.

9.3 Cyber and Battle Damage Assessment (BDA)

While MOEs and MOPs provide the commander with an estimate of overall campaign and force-on-force effectiveness, respectively, BDA is more specific, in providing feedback on the operability of an individual target. For example, Battle Damage Assessment is defined as:

… the estimate of damage composed of the physical damage assessment (PDA) and functional damage assessment (FDA), as well as target system assessment, resulting from the
application of lethal or nonlethal military force. (Chairman of the Joint Chiefs of Staff, 2019)

BDA is one method that a commander uses to assess the progression of an air
campaign. One of the advantages of drones is the potential for real-time BDA. This
is also important for cyber operations. For example, Acton (2017) points out that
BDA may be even more important for cyber attacks, in time sensitive military operations, due to cyber effects being temporary, reversible and less observable. In addition, Perkovich points out:

A cyber attacker seeking to disable a set of adversary capabilities in order to conduct other
operations needs to know not only that the degradation of their performance has indeed
occurred but also when they may be fixed. Making such damage assessments is also vital to
figuring if and when malware has spread unintentionally to other targets that might cause
political, diplomatic, economic, or strategic harm to the attacker's position. Knowing the
full extent of damage that has been or may be caused is necessary to inform efforts to neutralize or minimize unintended effects and to manage the consequences, including with
parties not involved in the conflict. (George Perkovich, 2017)

BDA also provides commanders with an assessment of how effective each of her
respective munitions contribute to the overall mission. For conventional munitions,
the Joint Munitions Effectiveness Manuals (JMEMs) have been developed to understand the types of reproducible effects that are available.

9.3.1 Cyber Joint Munitions Effectiveness Manuals (JMEMs)

JMEMs, providing authoritative data concerning weapons effects, are defined as:

… damage / kill probabilities for specific weapons and targets, physical and functional
characteristics of munitions and weapon systems, target vulnerability, obscuration on
weapon effectiveness, and analytical techniques and procedures for assessing munitions
effectiveness. (U.S. Army)

Applying JMEMs to cyber has been an active area of research for over a decade
(Gallagher, 2008). George Cybenko (2016) summarizes the overlaps in traditional
kinetic munitions measurement and information operations through an evaluation
using the Joint Munitions Effectiveness Manuals (JMEMs). For example, in kinetic
warfare, military planners assess munitions through the JMEM process (US Army)
to determine how each weapon's properties contribute to the operation's goal. In
addition, extending JMEM concepts to electronic warfare (EW) (MacEslin, 2006)
and stealth (George Cybenko, 2007) have been proposed.

Similar to our example in Fig. 9.1, George Cybenko (2016) proposes high-level
attributes for a cyber munition to include:

- Covertness
- Precision
- Reusability

- Potential for collateral damage
- Persistence
- Degree of attribution
- Time to deploy on a novel target
- Ease of battle damage assessment
- Reliability

Each of the above attributes has a relative weight in the decision to use cyber versus other means to accomplish operational objectives. Developing work on Cyber JMEMs (Mark Gallagher, 2013) points to interest in providing cyber as a tactical tool in future military operations, similar to present day PGMs and drones.

The ability to assess the effectiveness of a cyber action is as important with cyber as it is with conventional munitions. This is especially true due to the often temporary nature of a cyber action; having the cyber effect in place to complement other activities may be the primary reason for performing the cyber activity in the first place. For example, Israel presumably had a means to verify the effectiveness of their cyber weapon prior to Operation Orchard (Clarke & Knake, 2012), where in 2007 the Israelis spoofed Syrian radars long enough to perform a bombing sortie that destroyed a nuclear facility under construction.

9.3.2 Cyber Operations Lethality and Effectiveness (COLE)

Developing capabilities for missions like the 2007 Operation Orchard example likely included test and evaluation to ensure tool operability prior to the cyber operation. Estimating the effectiveness of a cyber munition has similar challenges to what has been performed with conventional munitions.

As discussed in Chap. 7, standard targeting approaches are just as applicable in cyber as they are for other munitions. This is one of the attributes that leads cyber professionals to look for further analogs between cyber and conventional munitions. A challenge for modern day cyber includes measuring a cyber target's vulnerability (e.g., PASTA in Sect. 3.3.3). The cyber operations lethality and effectiveness (COLE) model (Zurasky, 2017) is designed to provide a reusable approach for cyber target evaluation. To compute an overall probability of kill (P_k), COLE combines the likelihood of success for each step in an attack path, as shown in Fig. 9.5.

As shown in Fig. 9.5, COLE operates along the same lines as the probability of kill (P_k) in kinetic munition evaluation. The elements of the calculation are provided in Table 9.4.

As shown in Table 9.4, COLE is comprehensive in covering each of the elements that a cyber offensive tool developer needs to test and evaluate prior to the execution of a successful cyber attack mission. COLE therefore looks more like a measure of performance (MOP) on the part of the operator, than the KPP of a munition usually associated with a kinetic JMEM.

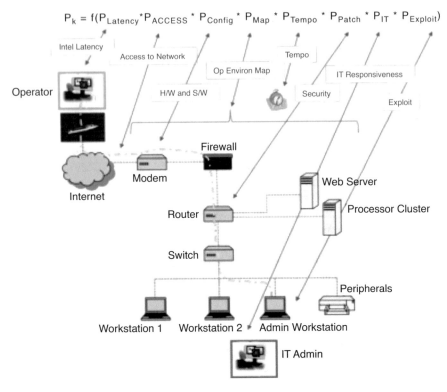

Fig. 9.5 Cyber operation lethality and effectiveness (COLE) kill chain (Zurasky, 2017)

Table 9.4 COLE cyber P_k equation elements

COLE P_k elements	Description
$P_{latency}$	Likelihood of having required supporting intelligence at the right time
P_{access}	Likelihood of knowledge of access points in the targeted system
P_{config}	Likelihood of having needed knowledge of software and hardware configurations in the targeted system
P_{map}	Completeness of network map of targeted system
P_{tempo}	Completeness of understanding operations tempo of targeted system
P_{patch}	Likelihood of a target system IT applying a patch to address software vulnerabilities on the targeted system
P_{IT}	Likelihood of the target system IT being able to detect and respond to a threat actor accessing their system
$P_{exploit}$	Likelihood that current mission exploit will work

COLE is the latest iteration of the former IO and cyber JMEM efforts (Mark Gallagher, 2013) that are intended to be comparable to the systems level parameterization broadly used for kinetic weapon munition effects evaluation. The effect, in this case, is the probability of kill (P_k) that results from the successful delivery chain, and exploit, of a cyber target. In weaponeering, however, accounting for the platform that provides a probability of hit (P_H), with the weapon, or munition, providing a P_k with a given P_H. In addition, other factors might be introduced—probability of detecting the target (P_d), weapon reliability (R_w), system reliability (R_{sys}). Cyber is a challenge with the platform being, for the most part, the operator. And operator effectiveness is often measured by MOPs. Cyber and traditional JMEMs may therefore be operating at different levels of the operational hierarchy as described by Fig. 9.4.

Using COLE to estimate target access is different than the traditional JMEM efforts to provide weapon effects for collateral damage estimates. Cyber is different again in understanding how information, or its compromise, can be used to measurably impact a mission or campaign.

9.3.3 Cyber Effect Significance

While cyber effects currently vary in both their scale and scope, targeting is concerned with achieving a specific effect. Reducing the qualitative nature of cyber is a key analytical challenge, similar to the types of questions often asked of social scientists in making conclusions based on limited data sets about potentially transient phenomena. For example, in Chap. 4, we looked at how cyber is currently a means of implementing active measures, or information operations, to magnify information through social and mainstream media (Fig. 4.7). Measuring the impact and duration of an IO campaign is a real challenge for current, and developing, cyber evaluation techniques (e.g., COLE, Cyber JMEM). Qualitative techniques may therefore be the only overarching tool to assess group sentiment and other soft indicators in order to provide insight on the effectiveness of a particular information operation.

While cyber operations may sometimes look more like an advertising campaign when using information operations as the tool to achieve results (e.g., Fig. 4.10), this leads to a broader insight concerning the importance of intelligence, and mature target understanding, in the planning and execution of cyber operations. For example, a new wrinkle introduced in cyber targeting is the symbiotic relationship between the weapon and the target in cyber operations. For example, the level of intelligence, key to any successful operation, is believed to be even more important in cyber (Dusan Repel, 2015). In fact, it has been argued that the level of intelligence required for cyber weapons is well beyond what is required for physical weaponry:

> Building and deploying Stuxnet required extremely detailed intelligence about the systems it was supposed to compromise, and the same will be true for other dangerous cyber weapons. (Rid, 2012)

In Chap. 5, we saw the extensive preparation involved in the development, deployment, and analysis of each of the active collection techniques (e.g., Duqu,

Flame, SPE) involved in the preparation, engagement, and presence phases of the STUXNET operation. Each of these bots also included a supporting "bot" infra-structure, as discussed in Chap. 6, to command, control, and collect data during this active intelligence operation.

While the effects phase of the STUXNET target, slowing down the Iranian nuclear program by 2–3 years (Hayden, 2016), had the significance to warrant an advanced cyber intelligence operation to achieve the targeted results, the end is not always so clear when planning a cyber operation.

In the less clear cases, one method of scoping an effect's significance is to deter-mine how an operation's cyber effect occurs within the context of the commander's intent for an overall engagement (Phase 1 of Fig. 7.3). Looking at the targeting process this way will require significant upfront resources to ensure that the specific target element (e.g., Fig. 7.2) is understood well enough to develop tools that will provide the effect of interest. Decision support tools are just beginning to appear with promise for addressing targeting as a knowledge-based design problem (Clare Maathuis, 2020).

STUXNET can be looked at as an example of point targeting, as discussed in Chap. 7, that also included a broad area search during the ISR phase of its opera-tions. Similarly, the COLE process, a step-by-step approach for estimating a cyber attack, provides a cyber level KPP for a point target. Area targeting, the use of cyber for broader denial applications, requires a different measurement approach than that used for individual system exploits. For example, denial of service (DoS) attacks are hypothesized by some to be more comparable to chem-bio attacks.

9.3.4   Cyber and Biological Weapons—Cyber Effects Analog

In terms of the challenges for attribution and measuring effectiveness, Koblentz (Koblentz & Mazanec, 2013) believes that cyber weapons are more analogous to biological agents, in terms of their low visibility and the type of effects they can have. For example, detection is challenging for both biological and cyber weapons in terms of their development, penetrations and attacks—for both defensive and offensive purposes. In addition, they can remain concealed even after indications of their existence appeared. For bio weapons:

- Effects may be short, with unclear measurable response, making is a challenge to design a capability to reproducibly induce their perceived effects
- Attribution can also be challenging, in the short term, to determine whether it was an attack or a natural occurrence.

In terms of attribution, and the perishability of effects, there is a clear parallel between bio and cyber effects. Cyber has the advantage of targeting single entities (e.g., Stuxnet's targeting of a specific controller/system combination (Zetter, 2014)), while bio remains primarily an area weapon when delivered by technical means.

9.4 Measuring Non-lethal Capabilities of Cyber

JMEM and chem-bio analogies automatically bring up lethal models for cyber operations that are difficult to substantiate. However, we can take a more moderate path by reviewing the non-lethal capabilities of cyber weapons. For example, non-lethal uses of cyber can be looked at along four areas (Schmidle & Sulmeyer, 2017) (Table 9.5).

As shown in Table 9.5, minimizing collateral damage was one of the key attributes of the STUXNET attack, designed to target a specific Siemens controller configuration (Zetter, 2014). Similarly, incapacitation is possible with cyber in novel ways. Modifying Vice President Cheney's pacemaker is the kind of attack that would have had strategic effects with the possibility of never being detected. And, while stealth and very specific targeting are in the spirit of the PGM argument of Sect. 9.2, cyber can also be reversible, which we also discussed in the MOEs of trust and maneuverability as effects available via either kinetic or cyber targeting.

It remains a challenge to measure the non-lethal capabilities of cyber. While Russian interference in the US elections, as shown in Fig. 9.1, was a challenge to detect, and was only clarified afterward via the Mueller report (Mueller, 2019), the US cost of this "cyber effect" was not quantified. However, Democratic congressman Bennie Thompson of Mississippi requested $400 million, in 2017, for election security upgrades in states across the country (Chalfant, 2017). Congressman Thompson's statement included:

Table 9.5 Examples of cyber uses for non-lethal effects

Non-lethal effect	Description
Minimize collateral damage	Forms of minimization include (1) preventing the spread of computer code beyond the target and (2) minimizing the harm the code causes to non-target systems if it does spread

Incapacitate	An example incapacitation operation is Operation Orchard, executed in 2007, to blind Syrian radars during an Israeli airstrike (Leyden, 2007). Additional examples include
	    2007 Vice President Cheney having the wireless capability of his pacemaker turned off, for fear of getting hacked, while acting as President (Dan Kloeffler, 2013)
	    2014 Sands Casino, in Las Vegas, having its systems taken off line (i.e., availability attack), causing the Casino to temporarily shut down operations (Elgin & Riley, 2014)

Effects reversibility	Non-lethal weapons provide guidance as to categories of reversibility:
	    Non-reversible (e.g., sabotage attacks)
	    Reversible after some period of time
	    Reversible based on operator discretion (e.g., ransomware, denial of service attacks)
	    Reversible with additional time/material/effort to do so (e.g., wiping attacks)

Deter	The two traditional methods of deterrence are cost imposition and denial (Nye, 2016), creating an ongoing discussion as to how to deter an adversary in cyberspace.

> We know that Russia launched an unprecedented assault on our elections in 2016, targeting
> 21 states' voting systems … Congressman Bennie Thompson (Thompson, 2017)

As shown by Congressman Thompson's budget request, the relatively simple collection system in Fig. 9.1 resulted in at least $400 million of spending as an effect of the Russian attacks on the 2016 US Presidential Election. While cyber operations, which include IO and traditional cyber security confidentiality, integrity and availability (CIA), are a challenge to measure, we have multiple cyber security compromise examples over the last few decades. Keeping these computer-based systems secure include measuring cyber performance across confidentiality, integrity and availability (CIA).

9.5 Traditional Measures of Cybersecurity for Performance and Effectiveness

The introduction of IO, while challenging, is within the scope of information related capabilities, as discussed in JP 3-12. In addition, as previously discussed, IO and compromising computer-based systems for data exfiltration, operational denial, or manipulation can all result in the same effect of distrust for the system of interest (Fig. 9.1).

9.5.1 Cyber Security Metrics Discussion

Due to the respective people, process and technology domains that make up a cyber system, describing the "attack surfaces" (Manadhata, 2008) can be very challenging. Therefore, one goal of developing good cyber security metrics is to try to provide the network defender with indicators that will help him detect a security compromise as early as possible.

Cyber security measures therefore possess some basic core properties in order to be valuable and practical to a cyber security analyst (Frei, 2006; SANS Institute, 2006). For example, cyber security metrics might be designed to span from using existing, known, data sets (e.g., CVSS), to estimating the time that it will take for a network to recover from an attack. We can therefore divide our security metrics into core, probability, structurally or time-based (Table 9.6).

As shown in Table 9.6, multiple security metrics are provided to look at the different aspects of vulnerabilities and possible network effects. Choosing the metric will depend on both network defense goals and the types of data available to use the metrics to provide reliable feedback for maintaining secure cyber system operations.

Table 9.6 Example cyber security metrics

Security metric	Examples	
Core	CVSS	Common Vulnerability Scoring System is an open standard for scoring IT security vulnerabilities. It was developed to provide organizations with a mechanism to measure vulnerabilities and prioritize their mitigation. There has been a widespread adoption of the CVSS scoring standard within the information technology community. For example, the US Federal government uses the CVSS standard as the scoring engine for its National Vulnerability database (NVD)
	TVM	Total Vulnerability Measure is an aggregation metric that typically does not use any structure or dependency to quantify the security of the network. TVM is the aggregation of two other metrics called the Existing Vulnerabilities Measure (EVM) and the Aggregated Historical Vulnerability Measure (AHVM)
Probability (attack-graph)	AGP	In Attack Graph-Based Probabilistic (AGP) metric, each node/edge in the graph represents a vulnerability being exploited and is assigned a probability score. The score assigned represents the likelihood of an attacker exploiting the vulnerability given that all pre-requisite conditions are satisfied
	BN	In Bayesian network (BN)-based metrics, the probabilities for the attack graphs are updated based on new evidence and prior probabilities associated with the graph.
Structural (attack-graph)	SP	The Shortest Path (SP) (Ortalo & Y. D., 1999) metric measures the shortest path for an attacker to reach an end goal. The attack graph is used to model the different paths and scenarios an attacker can navigate to reach the goal state which is the state where the security violation occurs
	NP	The Number of Paths (NP) (Ortalo & Y. D., 1999) metric measures the total number of paths an attacker can navigate through an attack graph to reach the final goal, which is the desired state for an attacker
	MPL	Mean of Path Lengths (MPL) metric (Li, 2006) measures the arithmetic mean of the length of all paths an attacker can take from the initial goal to the final goal in an attack graph
Time based (general)	MTTR	Mean Time to Recovery (MTTR) (Jonsson & Olovsson, 1997) measures the expected amount of time it takes to recover a system back to a safe state after being compromised by an attacker
	MTTB	Mean Time to Breach (MTTB) (Jaquith, 2007) represents the total amount of time spent by a red team to break into a system divided by the total number of breaches exploited during that time period
	MTFF	Mean Time to First Failure (MTFF) (Sallhammar & Helvik, 2006) corresponds to the time it takes for a system to enter into a compromised or failed state for the first time from the point the system was first initialized

9.5.2 System Level Cyber Effects

Estimating system metrics, as shown in Table 9.6, is performed by categorizing the effect type with items that can be measured for the cyber attack. For example, we can quantify some well-known attack effects, using the common confidentiality/

integrity/availability CIA Information Assurance (IA) evaluations (Table 9.7).

As shown in Table 9.7, it is possible to observe and quantify an identifiable effect. As discussed in Sect. 7.3.4, one of our goals is post-operations assessment. In Table 9.7, we show "candidate measurables," results from an operation, that provides an initial level of assessment.

9.6 Measures of Cyber Performance and Effectiveness Wrap Up

Maintaining secure cyber systems through good metrics (Table 9.6) is a first step for ensuring that we minimize future denial and manipulation attacks of our systems. For example, our initial look at a cyber attack on the DNC (Fig. 9.1) required a secure cyber system.

While system security is the first step in maintaining the high ground in any cyber operation, we started this discussion with a comparison of cyber to aerial delivery of munitions. Achieving effects in either space requires access, aerial or cyber, which can make secure ingress a challenge. Therefore, we look at cyber as a possible next step in the precision guided munitions (PGM) continuum (Table 9.3).

Table 9.7 Example systems, effects and information assurance categorization

Example system	Effect	Categorization	Example	Candidate measurables
Web services (banks, telecommunications)	Denial of service	Availability	2007 DDoS attack on Estonia (Ottis, 2007)	Total effect time (e.g., down-time, lack of critical service availability) $(\mu = T_{\text{effect}})$
Data service provider	Data extraction	Confidentiality	2017 Office of personnel management (OPM), Equifax (Sanger, The Perfect Weapon - War, Sabotage and Fear in the Cyber Age, 2018)	Effect amount (e.g., number of records exfiltrated) $(\mu = N_{\text{records}})$
Command and control (C2)	Manipulate system to behave in ways advantageous to adversary	Integrity	2008 Russo-Georgian War (Markoff, 2008)	Effect amount (e.g., number of web sites defaced) $(\mu = N_{\text{web\_sites}})$

In this chapter, we found that cyber effects, as a possible next step in PGM has accompanying techniques in the Cyber Operations Lethality and Effectiveness (COLE) model (Fig. 9.5). This is effectively a point targeting example. We also looked at cyber in terms of area effects, described in Sect. 9.3.4, along with stealthier incapacitating techniques provided in Table 9.5.

An additional area targeting example includes Soviet active measures as an IO effect. This is a use of manipulation, a JP 3-12 effect, to describe strategic operations that have been exercised for as long as the conventional military effects (i.e., dropping bombs) have been measured and quantified. A key reason for including Soviet active measures as example cyber operations are that their current cyber integration into Russian all domain operations (e.g., 2008 Georgia, 2014 Ukraine). Measuring the duration and intensity of active measure effects remains a challenge.

While the Joint Munitions Effectiveness Manual (JMEM) is used for conventional munitions, cyber effects have potentially more dimensions to describe. For example, cyber is used to both better understand an adversary and to provide effects via denying or manipulating the use of computers/communications. Even more complicated would be to base the effect (e.g., denial) on the operation's own ISR observations, and for a variable amount of time. Kinetic JMEMs would simply measure whether the denial effect occurred and describe any collateral effects. And, since a kinetic weapon removes its target from the battlefield, no further consideration would be required.

In this chapter, we looked at cyber as a natural progression from PGMs, to drones, to cyber, with increasing target variability at each step in order to provide a conceptual model to start thinking about cyber operations analysis. Example answers to our questions at the beginning of this chapter include:

1. How does performance measurement change with cyber versus traditional, kinetic engagement?

    One of the advantages of conventional performance and effectiveness measurement is simply the duration for which these items have been measured. For example, operations research has been applied to military operations dating back to World War II, with continuous analysis and surveys since that time.

    Munitions effect analysis dates back to the 1960s (US Army), when Joint Munitions Effectiveness Manuals (JMEMs) (Zurasky, 2017) were developed to better understand the effect of a bomb, regardless of how it was delivered. Cyber also has decades of operational history, but mostly in terms of the defensive confidentiality/integrity/availability (CIA) effects.

    More recently, cyber effects have also come to include traditional active measures, applied through cyber space (e.g., 2016 US Presidential Election (Fig. 9.1)). Active measures were used by the Soviet Union against the US and the West over the course of the Cold War (1945–1992) using the media available at the time (e.g., print and radio). It is relatively recent that active measures were convolved with Internet based operations to become a cyber threat.

    As discussed in Chap. 1, Russia has used cyber as part of military operations since the 2007 Estonia DDoS attack, developing and refining cyber based active measures by at least by 2015. Measuring the effects of these cyber active measures is a challenge, as it was for the original active measures, as discussed in Chap. 4. Therefore, in this chapter, we used a precision guided munition (PGM) analog to assess cyber, in terms of technical, human and strategic measurable performance and effectiveness (Sect. 9.1).

2. How do we measure JP 3-12 effects (e.g., deny, manipulate) in current cyber operations?

   We began this chapter with the 2016 DNC cyber attack that included exfiltrating private e-mails and publishing them to "out" key organization members, causing them to change their focus from their primary mission (i.e., winning the election) to performing damage control. In addition, we found in Sect. 9.4, that this operation resulted in a $400 million investment to increase election security.

   Cyber is more nuanced in that we can also have temporary, incapacitating, effects (Sect. 9.4). These might be temporary denial or manipulation to facilitate another mission. For example, Operation Orchard in 2007 (Sect. 9.3.1) is an example where Syrian radars were temporarily spoofed so that bombers could ingress and destroy a nuclear reactor site.

3. How do cyber Joint Munition Effectiveness Manuals (JMEM) fit in the current measures of cyber performance and effectiveness?

   The JMEM, while providing KPP type information on conventional munitions, is a developing idea for cyber (Fig. 9.5). One of the challenges is that the cyber JMEM spans from KPP potentially to MOE, in cases when the exploit is the effect.

4. How do cyber measures of performance and effectiveness address technical and organizational maturity?

   KPPs address the technical maturity of a system, as described in Table 9.3. As we used the term here, MOPs provide a method for describing operator/organizational effectiveness, or maturity, in prosecuting operations.

Bibliography

Acton, J. M. (2017). Cyber weapons and precision guided munitions. In A. L. G. Perkovich (Ed.), *Understanding cyber conflict*. Georgetown.

Andrei Soldatov, I. B. (2015). *The red web - the struggle between Russia's digital dictators and the new online revolutionaries*. Public Affairs.

Biddle, S. (2004). *Military power: Explaining victory and defeat in modern battle*. Princeton University Press.

Bowden, M. (2011). *Worm - the first digital world war*. Atlantic Monthly Press.

Butsenko, A. (2014, October 28). Trolli iz Olgino pereekhali v noviy chetyrekhatazhny office na Savushkina (Trolls from Olgino moved to a new four story office on Savushkina). *DP.ru* .

Chairman of the Joint Chiefs of Staff. (2019, March 8). *Methodology for Combat Assessment (CJCSI 3162.02)*. Retrieved June 19, 2020, from Joint Chiefs of Staff: https://www.jcs.mil/Portals/36/Documents/Doctrine/training/jts/cjcsi_3162_02.pdf?ver=2019-03-13-092459-350

Chalfant, M. (2017, November 20). *Dems call for states to get $400 M election security upgrades*. Retrieved June 6, 2019, from Hill: http://thehill.com/policy/cybersecurity/361263-dems-say-congress-should-send-400m-to-states-for-election-cyber-upgrades

Clare Maathuis, W. P. (2020, April 11). *Decision support model for effects estimation and proportionality assessment for targeting in cyber operations.* Retrieved June 22, 2020, from Science Direct: https://www.sciencedirect.com/science/article/pii/S2214914719309250

Clarke, R. A., & Knake, R. K. (2012). *Cyber war: The next threat to national security and what to do about it.* Ecco.

Cleary, G. (2019, June). *Twitterbots: Anatomy of a propaganda campaign.* Retrieved July 6, 2019, from Symantec: https://www.symantec.com/blogs/threat-intelligence/twitterbots-propaganda-disinformation

Couretas, J. M. (2018). *Introduction to cyber modeling and simulation.* Wiley.

Dan Kloeffler, A. S. (2013, October 19). Dick Cheney feared assassination via medical device hacking: 'I was aware of the danger'. *ABC News.*

David Leigh, L. H. (2011). *WikiLeaks - inside Julian Assange's war on secrecy.* Public Affairs.

Dmitry Volchek, D. S. (2015, March 27). One professional Russian troll tells all. *Radio Free Europe / Radio Liberty.*

Drolet, M. (2018, January 26). *What does stolen data cost [per second]? 58 data records are stolen every second. Guess what the average cost is.* Retrieved September 13, 2018, from CSO Online: https://www.csoonline.com/article/3251606/data-breach/what-does-stolen-data-cost-per-second.html

Dusan Repel, S. H. (2015). The ingredients of cyber weapons. In *10th International Conference on Cyber Warfare and Security (ICCWS15)* (pp. 1–10).

Elgin, B., & Riley, M. (2014, December 11). Now at the Sands Casino: An Iranian hacker in every server. *Bloomberg.*

Frei, S., & U. F. (2006). Why to adopt a security metric? *Quality of Protection, Advances in Information Security, 23* (pp. 1–12).

Gallagher, M. (2008). *Cyber analysis workshop.* MORS. Reston: MORS.

George Cybenko, J. S. (2007). *Quantitative foundations for information operations.*

George Cybenko, G. S. (2016). Quantifying covertness in deceptive cyber operations. In V. S. Sushil Jajodia (Ed.), *Cyber deception: Building the scientific foundation.* Springer.

George Perkovich, A. E. (2017). *Understanding cyber conflict - 14 analogies.* Georgetown.

Greaver, B., M. L. (2018). CARVER 2.0: Integrating the Analytical Hierarchy Process's multi-attribute decision-making weighting scheme for a center of gravity vulnerability analysis for US Special Operations Forces. *Journal of Defense Modeling and Simulation, 15*(1), 111–120.

Hayden, M. V. (2016). *Playing to the edge: American intelligence in the age of terror.* Peguin.

Heckman, K. (2013). *Active cyber network defense with denial and deception.* MITRE. http://goo.gl/Typwi4

Jaquith, A. (2007). *Security metrics: Replacing fear, uncertainty, and doubt.* Addison-Wesley, Pearson Education.

Johnston, P. B., & Sarbahi, A. K. (2016, June 1). The impact of US drone strikes on terrorism in Pakistan. *International Studies Quarterly, 60*(2), 203–219.

Jonsson, E., & Olovsson, T. (1997). A quantitative model of the security intrusion process based on attacker behavior. *IEEE Transactions on Software Engineering.*

Koblentz, G. D., & Mazanec, B. (2013). Viral warfare: The security implications of cyber and biological weapons. *Comparative Strategy, 32*(5), 418–434.

Koerner, B. I. (2016, October 23). *Inside the cyberattack that shocked the US government.* Retrieved from Wired: https://www.wired.com/2016/10/inside-cyberattack-shocked-us-government/

Kohen, I. (2017, August 15). *Cost of insider threats vs. investment in proactive education and technology.* Retrieved September 13, 2018, from CSO Online: https://www.csoonline.com/article/3215888/data-protection/cost-of-insider-threats-vs-investment-in-proactive-education-and-technology.html

Leyden, J. (2007, October 4). Israel suspected of 'hacking' Syrian air defences. *The Register.*

Li, W., & R. V. (2006). Cluster security research involving the modeling of network exploitations using exploitation graphs. *Sixth IEEE International Symposium on Cluster Computing and Grid Workshops.*

Lowrey, A. (2014, December 16). *Sony's very, very expensive hack.* Retrieved September 25, 2018, from Daily Intelligencer: http://nymag.com/daily/intelligencer/2014/12/sonys-very-very-expensive-hack.html

MacEslin, D. (2006). *Methodology for determining EW JMEM.* TECH TALK.

Manadhata, P. K. (2008). *An attack surface metric.* CMU.

Mark Gallagher, M. H. (2013). Cyber joint munitions effectiveness manual (JMEM). *Modeling and Simulation Journal* (pp. 5–14).

Markoff, J. (2008, August 12). Before the gunfire, cyberattacks. *New York Times* .

Mir, A., & Moore, D. (2019). Drones, surveillance, and violence: Theory and evidence from a US drone program. *International Studies Quarterly, 63*(4), 846–862.

Mueller, R. (2019). *Report on the investigation into Russian interference in the 2016 presidential election.* U.S. Department of Justice. U.S. Department of Justice.

Mullin, R. (2014, November 24). *Cost to develop new pharmaceutical drug now exceeds $2.5B.* Retrieved September 13, 2018, from Scientific American: https://www.scientificamerican.com/article/cost-to-develop-new-pharmaceutical-drug-now-exceeds-2-5b/

Nye, J. (2016, April 6). *Foreign Policy Association (blog).* Retrieved June 3, 2018, from Foreign Policy Association (blog): http://foreignpolicyblogs.com/2016/04/06/can-china-be-deterred-in-cyber-space/

Ortalo, R., Deswarte Y., and Kaaniche M. (1999, Oct.). *"Experimenting with quantitative evaluation tools for monitoring operational security," in IEEE Transactions on Software Engineering 25,* (5), pp. 633–650, https://doi.org/10.1109/32.815323.

Ottis, R. (2007). *Analysis of the 2007 cyber attacks against Estonia from the information warfare perspective.* Retrieved September 14, 2019, from Cooperative Cyber Defence Centre of Excellence: https://ccdcoe.org/uploads/2018/10/Ottis2008_AnalysisOf2007FromTheInformationWarfarePerspective.pdf

Popular Mechanics. (2018, March 13). *How long does it take hackers to pull off a massive job like equifax?* Retrieved from Popular Mechanics: https://www.popularmechanics.com/technology/security/a18930168/equifax-hack-time/

Poulsen, K. (2018, October 25). *'Lone DNC Hacker' Guccifer 2.0 slipped up and revealed he was a Russian intelligence officer.* Retrieved July 8, 2019, from The Daily Beast: https://www.thedailybeast.com/exclusive-lone-dnc-hacker-guccifer-20-slipped-up-and-revealed-he-was-a-russian-intelligence-officer

Rid, T. (2012, February 27). *Cyber war: Think again.*

Sallhammar, K., & Helvik, B. (2006). On stochastic modeling for integrated security and dependability evaluation. *Journal of Networks, 1*(5), 31–42.

Sanger, D. E. (2016, April 24). *U.S. cyberattacks target ISIS in a new line of combat.* Retrieved from New York Times: https://www.nytimes.com/2016/04/25/us/politics/us-directs-cyberweapons-at-isis-for-first-time.html

Sanger, D. E. (2017). Cyber, drones and secrecy. In A. E. G. Perkovich (Ed.), *Understanding cyber conflict.* Georgetown.

Sanger, D. E. (2018). *The perfect weapon - war, sabotage and fear in the cyber age.* Crown.

SANS Institute. (2006). *A guide to security metrics.* SANS.

Schmidle, R. E., & Sulmeyer, M. (2017). Nonlethal weapons and cyber capabilities. In A. E. G. Perkovich (Ed.), *Understanding cyber conflict - 14 analogies.* Georgetown University Press.

Segura, V. (2009, June 25). *Modeling the economic incentives of DDoS attacks.* Retrieved from Semantic Scholar: https://pdfs.semanticscholar.org/afdf/d974bc68dc05c48020e-f07a558a61ab94f8a.pdf

Statistica. (2017). *Annual number of data breaches and exposed records in the United States from 2005 to 2018 (in millions).* Retrieved September 14, 2018, from Statistica: https://www.statista.com/statistics/273550/data-breaches-recorded-in-the-united-states-by-number-of-breaches-and-records-exposed/

Thompson, B. G. (2017, November 17). *Election security task force requests funding for state election security, replacing outdated voting machines.* Retrieved June 19, 2020, from https://benniethompson.house.gov/media/press-releases/election-security-task-force-requests-funding-state-election-security-replacing

U.S. Army. (n.d.). *JTCG/ME PO - Joint technical coordinating group for munitions effectiveness program office* . Retrieved June 19, 2020, from https://www.dac.ccdc.army.mil/JTCGMEPO.html

US Army. (n.d.). Joint technical coordinating group for munitions program office.

Warner, M. (2017). Intelligence in cyber - and cyber in intelligence. In A. E. George Perkovich (Ed.), *Understanding cyber conflict - 14 analogies.* Georgetown University Press.

Warner, M. (2019). A matter of trust: Covert action reconsidered. *Studies in Intelligence, 64*(3), 33–41

Whaley, B. (2007a). *Textbook of political-military counterdeception: Basic principles and methods.* National Defense Intelligence College.

Whaley, B. (2007b). *STRATAGEM - deception and surprise in war.* Artech House.

Whittaker, Z. (2019, April 18). *Mueller report sheds new light on how the Russians hacked the DNC and the Clinton campaign.* Retrieved June 17, 2020, from TechCrunch: https://techcrunch.com/2019/04/18/mueller-clinton-arizona-hack/

Zetter, K. (2014). *Countdown to zero day - Stuxnet and the launch of the World's first digital weapon.* Crown.

Zurasky, M. W. (2017). *Methodology to perform cyber lethality assessment.* Old Dominion University. Old Dominion University.

Chapter 10
Cyber Modeling and Simulation for Analysis and Targeting

The purpose of this chapter is to present a general background on cyber modeling and simulation for analysis and targeting. This includes looking at example cyber systems, their structures, and decomposing an example process to estimate the effect of a cyber targeting operation. In addition, we will look at broader cyber modeling and simulation frameworks, along with the use of cyber ranges, to enhance both training and testing in support of cyber analysis and targeting.

Cyber M&S Questions for Analysis and Targeting
1. Why is modeling and simulation (M&S) important for cyber analysis and targeting?
2. What are the examples of cyber models and their uses for analysis and targeting?
3. How does modeling and simulation help with analyzing the parallel operations of cyber systems?
4. How can cyber modeling be expanded to perform vulnerability analysis?
5. What are candidates for cyber effects estimation and how does modeling and simulation improve current analysis approaches?
6. What kind of heuristics help with cyber analysis?

10.1 Background

In Chap. 9, we looked at how key performance parameters (KPPs), measures of performance (MOPs), and measures of effectiveness (MOEs) can be used in the cyber domain. We did this by looking at cyber as the latest development in the evolution of precision guided munitions (PGMs). Implicit in Chap. 9's discussion was that the data for computing KPPs, MOPs, and MOEs was available. This is not always the case.

Developing models is often done when system data is not available. In Chap. 8 (Fig. 8.2), we looked at how modeling frameworks have developed over the last

© Springer Nature Switzerland AG 2022
J. M. Couretas, *An Introduction to Cyber Analysis and Targeting*,
https://doi.org/10.1007/978-3-030-88559-5_10

several decades to help engineers describe systems during design. These models are the only representations that are available for systems not yet built. In addition, these models are the only authoritative system data until the actual system can be built, and its behavior can be measured for its KPPs, MOPs, and MOEs under appropriate conditions.

Modeling is especially important for cyber systems. For example, cyber modeling and simulation spans from system effects estimation to the training of new operators, navigating the objective organizations, technologies, and processes that compose a cyber system. This includes accounting for cyber system complexity, often defined by parallel operations with differing time constants that obfuscate the 1:1 cyber effect-to-component behavior traceability common to more tightly coupled systems. Loosely coupled cyber systems are usually described by the higher-level information assurance (IA) effects (e.g., confidentiality, integrity, availability) that stem from operational anomalies at the application, network or physical layers of the system (Sect. 10.4). Incorporating these IA level effects into training simulations and system availability estimates are good examples of cyber modeling and simulation.

The Cyber Operations Architecture Training System (COATS) (Wells & Bryan, 2015; Morse et al., 2014) incorporates cyber effects into standard military training simulations. In addition, one COATS example synchronously communicates range based cyber effects into a maneuver training exercise (Fig. 10.1).

As shown in Fig. 10.1, the real-time combination of cyber red team behavior with defensive operations integrates cyber into training simulations and provides a clear picture of how well a defensive team did at each step of a cyber attack. While

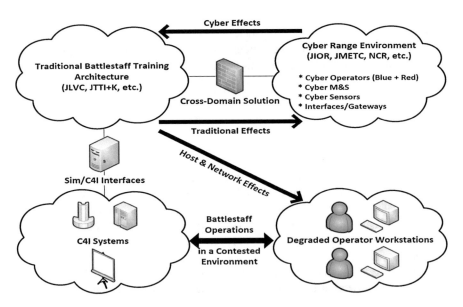

Fig. 10.1 Cyber Operational Architecture Training System (COATS)

COATS focuses on training, Analyzing Mission Impacts due to Cyber Attacks (AMICA) (Noel et al., 2015) is a multi-layered modeling framework for evaluating mission effects due to cyber anomalies (Fig. 10.2).

The system shown in Fig. 10.2 is an example air tasking order system (ATO) production system where the AMICA simulation outputs include throughput changes due to a cyber attack disabling of component computer systems. AMICA's mission system security is measured by KPPs, with the detailed process description, showing ATO throughput vs. system availability changes, measured in MOPs. AMICA is an engagement level model, per our discussion in Chap. 8. MOE type effects from the AMICA level attack would be a reduction in aircraft sorties. In addition, AMICA is one process, of potentially many, individual systems, that can be combined to make up an overall campaign level simulation.

While COATS performs evaluations one effect at a time, AMICA shows the potential for evaluating multiple computer-system level attacks in parallel. AMICA and COATS therefore bring out the simultaneous parallel and sequential elements of modeling cyber systems. While theater-level simulations commonly include simultaneous parallel and sequential campaigns as part of their execution (Wylie, 2014), cyber is complex due to having this combination across component people, process, and technology layers.

10.2 Introduction

Modeling is a natural way to start developing the analysis for a cyber targeting problem. The exercise of modeling helps the targeteer reduce complexity, by clarifying both the effect to be achieved and the target's state change required to achieve it. Table 10.1 describes example modeling tools across the standard target cycle (USAF, 2019).

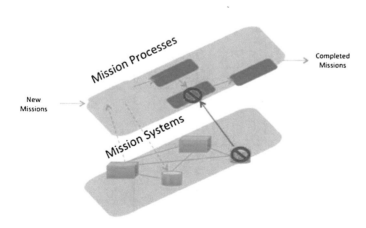

Fig. 10.2 AMICA—mission effects due to cyber system anomalies

The modeling approaches in Table 10.1 span the Joint Publication 3–60 targeting cycle (USAF, 2019) to describe the static and dynamic modeling constructs introduced through this book. In each case, the process is laid out for either the attacking process (Chap. 7) or the target of interest in the following sections.

10.3   Cyber System Description for Analysis and Targeting

As discussed in Sect. 7.3.1, target development and prioritization is the first step in performing analysis for a targeting mission. The challenges at this phase of the operation include determining how a cyber effect contributes to a mission and what kind of effects are actually addressable on the respective targets, and target elements, via cyber. This includes evaluating each "cyber system" in terms of the people, process, and technology elements that compose a network target in order to determine the following:

- Inherent parallel structuring of computer-based cyber systems that use the OSI 7 layer hierarchy
- Sequential nature of a cyber threat and its phase/time behavioral structure

These parallel and series considerations are incorporated into descriptive cyber modeling.

Table 10.1 Modeling approaches and the Joint Publication 3–60 targeting cycle

Joint Publication 3-60 targeting steps	Phase 1 — Commander's objectives, guidance, and intent	Phase 2 — Target development, validation, nomination, and prioritization	Phase 3 — Capabilities analysis	Phase 4 — Commander's decision and force assignment	Phase 5 — Mission planning and force execution	Phase 6 — Combat assessment
Cyber System Vulnerability Estimator (Chap. 7)		X	X			
Cyber Process Models (Sect. 10.4)					X	X
Target Dynamics Modeling for Effects Evaluation (Sect. 10.5)					X	X
Constructive Modeling (Sect. 10.6)					X	X

10.3.1 Parallel/Series Nature of Cyber Systems

While the series element of any process is intuitive, the additional complexity in describing cyber systems stems from the multiple, parallel, operations, that can occur at different time constants and with varying effects. The physical, logical, and social layers common to cyber systems are shown in Fig. 10.3.

As shown in Fig. 10.3, each of the respective layers has parallel components with their own structuring that contribute to its respective cyber process effects. One way to look at these parallel cyber system elements is to use the OSI 7 layer hierarchy.

10.3.1.1 Cyber System with Parallel Layers

Simplifying cyber modeling to sections of the seven-layer OSI stack provides an ability to focus on the system/environment of interest. For example, StealthNet (Torres, 2015), developed by Exata for the Test Resource Management Center (TRMC), uses five layers of the OSI hierarchy as constructive simulation layers to exercise architectures on a cyber range (Fig. 10.4).

The example shown in Fig. 10.4, from the test and evaluation (T&E) community (Bucher, 2012), is also viable as a constructive simulation input for targeting evaluations. Some examples of targeting effects include:

- Loss of communication nodes and lines of communication (e.g., denial of service)
- Loss of fidelity of sources of communication
- Partial loss of information

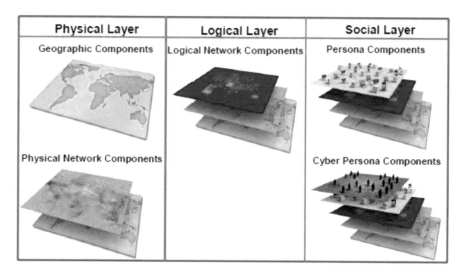

Fig. 10.3 Physical, logical, and social layers of a cyber system

- False or compromised information
- Restrictions in bandwidth

In addition, another view of the OSI hierarchy to evaluate component process functions for modeling and simulation analysis applications is shown in Table 10.2.

The application layer shown in Table 10.2 is the principal target for the confidentiality and integrity attacks that are common in reported cyber attacks. For example, the system data that is the target of many exfiltration and ransomware attacks is stored at the application layer. Similarly, integrity attacks, or changing the behavior of industrial control systems, occur at the application layer. Availability attacks, however, are performed at the networking and physical layers, denying system operations and preventing the use of computer based systems.

In addition, as shown in Table 10.2, clearly differentiating the application, networking, and physical networking layers is used to illustrate the different system considerations over the functional groupings of the layers (Table 10.3).

In Table 10.3, we see where system protection efforts highlighted in Table 10.2 might best be applied. For example, with 75% of the current cyber vulnerabilities at the application layer, this should be a priority in the decision-making process used in planning the acquisition, purchase, and deployment of defensive cyber resources.

As shown in Table 10.2, each section of the OSI 7-layer stack has different identifiable characteristics that might be looked at as unique, and possibly reinforcing, indicators of change, for associated processes that operate in parallel at other layer(s) of the OSI stack. For example, an advanced persistent threat (APT), detectable with unique software tools to protect the application layer, might also show up as a power consumption increase at the physical level due to a bot, or implant's increased run-time requirement. Similarly, as discussed in Chap. 6 (Sect. 6.3.2), increased communications at the network layer might be an indicator of communication back to a command and control (C2) server that is not part of the network of interest (e.g., Conficker maintained periodic communications with obfuscated server addresses for C2 communications (Bowden, 2011)).

As shown in Table 10.2, modeling the network stack in chunks is an example where there is an opportunity to look for reinforcing indicators of suspected malware activity. In addition, while cyber system construction is inherently parallel, due

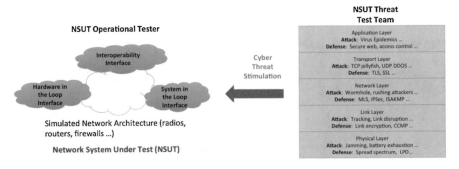

Fig. 10.4 Network emulation—exercising a Network System Under Test (NSUT)

Table 10.2 OSI 7 layer hierarchy—modeling candidates

OSI 7 layer hierarchy	M&S candidates
Application	Statistical Modeling of Software for Vulnerability Detection (e.g., CVSS—Chap. 6); Course of Action Evaluation (e.g., COATS); Persistent Training Environment (e.g., Circadence, Cyber Ranges (Sect. 10.5.3)); Social media platforms
Presentation	Map Networks via Graph Processing (i.e., companies providing this capability include SolarWinds and Blazegraph); Persistent Training Environment (e.g., Circadence, Cyber Ranges (Sect. 10.5.3))
Session	
Transport	
Network	
Data link	
Physical	Use power consumption anomalies for component validity—and malware detection ("Power Finger Printing" (Power Fingerprinting, n.d.))

Table 10.3 Network construction layers and available modeling data

OSI layer(s)	Tool description
Application	Application-level tools, used to verify that a system has a secure design, address an estimated 75% of current cyber vulnerabilities (Wei, 2012)
Presentation through data link	Presentation through data link modeling includes network maps that have the potential to provide a combinatorial description of all the possible paths of a threat vectors
Physical	Physical layer system evaluation is currently being used in Industrial Control System experiments where power consumption anomalies, or "fingerprints," indicate a possible presence of malware (Power Fingerprinting, n.d.)

to operations occurring at each layer of the OSI 7 layer hierarchy, cyber operations usually occur in a series, step-by-step manner, both technically and operationally.

10.3.1.2 Series Processes and Cyber Operations

While the inherently parallel structuring of cyber systems provides both complexity and potentially reinforcing indicators for the cyber analyst, each layer will also include a series process to perform its function. For example, each of the application through data link layers will have its own routine set of sequential operations to maintain its respective functions of application through connection management.

While the OSI 7 technology layers maintain sequential routines, so do attackers as they attempt to penetrate a network. For example, the Cyber Threat Framework (Director of National Intelligence, n.d.) provides a high-level process that describes a cyber operation (Fig. 10.5).

The process described by Fig. 10.5 shows a linear flow, with each phase is reachable from any other phase, adding flexibility in using this approach to describe a cyber operation. In addition, the phases of Fig. 10.5 are where the dynamics of each phase occur. For example, these are where the temporal phases to accomplish each step, and associated resource requirements, are found.

10.4   Cyber Attack Lifecycle Example

One way to look at a cyber targeting cycle is to represent each step of the cyber risk "bow-tie" (Fig. 10.6), which was developed with the ISO 31000 risk standard in mind, to provide a comprehensive system security view of the different controls or countermeasures (Nunes-Vaz et al., 2011, 2014).

As shown, the left side of Fig. 10.6 works to minimize the risk of a cyber attack, while the right side provides the resilience, or consequence management, required to handle a cyber attack currently under way. Figure 10.6 can also be expanded to look at the people, process, and technology elements of cyber defense (Fig. 10.7).

The people, process, and technology elements of cyber defense, at the left side of the "event," show many of the defense tools that we reviewed in Chap. 6. Using the layered defense (Fig. 6.1), the system initially deals with threats via technologies (e.g., firewalls, honeynets …), applying these defenses according to the organization's security policy. As the attacker progresses through the network, the final defense is a team, providing active defense, as the strongest measure of cyber defense.

While the attack in Fig. 10.6 is described by a sequential path, each of the defensive threads (e.g., people, process, and technology) will actually operate in parallel to provide a comprehensive system defense. Each of these layers is also a system vulnerability.

10.4.1   Parallel System Vulnerabilities

The complexity of cyber systems therefore adds requirements concerning the types of capabilities that can be applied to the targeted system. For example, we looked at how an individual vulnerability is converted, via an off the shelf exploit, into a tool in Fig. 3.9. This is the capabilities analysis and force assignment step discussed in Sect. 7.3.2, and can also be used to start the mission planning (Sect. 7.3.3). We can generalize on this exploit development via a look at an overall, parallel, system.

As shown in Fig. 10.7, each of the people, process, and technology components operates in parallel, when configured to defend a cyber system (Fig. 10.8).

In Fig. 10.8 we see the target as a process for the development, and the implementation of both a web facing application and its possible vulnerabilities. For

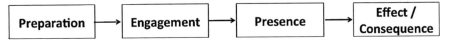

Fig. 10.5 Cyber operational process
Preparation: Pre-execution actions
Engagement: Operational actions
Presence: Maneuver inside target network
Effect/Consequence: Exercise mission

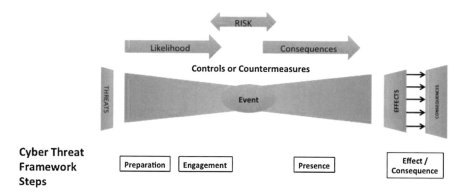

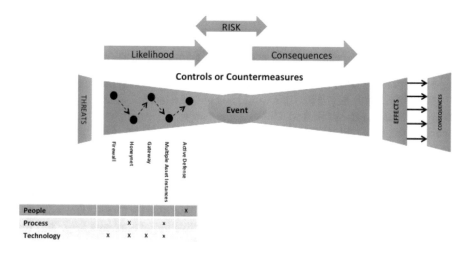

Fig. 10.6 Cyber risk "bow-tie"—prevention, attack, and remediation

Fig. 10.7 People, process, and technology elements of cyber defense

example, the software architect and associated programming team will have vulnerabilities in terms of social engineering. Knowing their interests and hobbies opens these people up to spear phishing exploits for a dedicated cyber attacker. This information might be found on social media.

Similarly, the development process in Fig. 10.8 will have clear steps that the developers use to migrate their applications onto the deployment platform. In the case of an Internet facing application, a web presence may provide a persistent target, with known exploits. While each of the process steps for deploying the system on the web will likely have vulnerabilities that can be exploited via social or technical means, once the application is deployed, there will be a clear set of technical vulnerabilities and exploits, as discussed in Chap. 3, for an attacker to use.

Each of the vulnerabilities are operating on their respective planes, or in parallel, in a cyber system, as shown in Fig. 10.8. Therefore, an expression for the overall vulnerability in a parallel cyber system can described by Eq. 10.1 (Cyber system vulnerability equation (parallel)).

$$V_T \cong 1 - \prod_{i=1}^{n} (1 - V_i) \tag{10.1}$$

The vulnerability estimates provided by elements of the people, process, and technology elements from Figs. 10.6 and 10.7 are input into Eq. 10.1 in order to estimate a system's overall vulnerability.

While we are accounting for an aggregate vulnerability of the target in Eq. 10.1, this is an early step in the targeting process, providing a shorthand method of describing the vulnerabilities for the major elements (e.g., people, process, technology) that account for the target's operation. Equation 10.1 is different from the Cyber Operations Lethality Effectiveness (COLE) equation (Zurasky, 2017), described in Sect. 9.3.3, in that we are estimating some of the qualitative elements of a target, while still including the technical element. Qualitative elements of the target include using social engineering to exploit the people involved in the design and development of the software (e.g., software design staff). COLE could be a component in Eq. 10.1 by providing the technical level vulnerability/exploit.

In addition, an analytic like this would be applied when information is still developing on the target of interest. This is stage two of the targeting cycle (Fig. 7.3), where a target is still under inspection for possible candidacy. Moving a target to stage three of the targeting cycle, capabilities evaluation, will require additional information for better understanding of how the target system works, in terms of the people, the processes, and the specific technologies.

Equation 10.1 provided initial insight concerning target vulnerability, as used in stage two of the targeting cycle (Fig. 7.3). In addition, Eq. 10.1 opened up a discussion on how best to approach each of the respective target elements (Fig. 7.2) for their vulnerability, and possible exploit, evaluation.

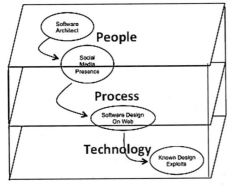

V_{people} : system vulnerability introduced by the people developing and using the system

$V_{process}$: system vulnerability introduced by commonly used processes

$V_{technology}$: system vulnerability introduced by technology components in the system – technology can be further decomposed into application, networking and physical

Fig. 10.8 Vulnerabilities across the people, process, and technology planes

10.5 Target System Description

Stage three of the targeting cycle, where modeling and simulation helps with target analysis may also help in the initial behavior description of the target. A target behavior model becomes important is step six, combat assessment, for assessing effects due to the cyber attack, where we want to measure how our strategy worked. For example, as shown in Eq. 10.1, a cyber target includes the people, the processes, and the technologies that make up the overall system of interest. This could be defensive, where the target is a malicious cyber actor that needs to be neutralized in a defended network, or offensive, where a red team is working to capture a key node. In either case, we will use our developing target analysis approach to better understand performance against a designated target.

For example, each of the people, the process, and the technology components of a cyber system has natural cycles that both affect our cyber terrain and might be used in mission/target planning (Sect. 7.3.2). Some examples include:

- Daily work schedules (e.g., backups)
- Weekly patch updates
- Vulnerability updates (weeks–months)
- IT refresh (months–years)
- Human resource changes (months–years)

Describing a system in terms of its timing, or cycles, provides the analyst with a clear understanding of target behavior, and a baseline for estimating effects. A first step in this direction is to look at a target system as a discrete event system.

10.5.1 Target System as a Discrete Event System

Process cycles that occur in the day-to-day operations of a cyber system can be described for each cyber target in terms of a discrete event system, which includes states and times for each system phase (Fig. 10.3).

$$M_1 = \left(S_1^1, t_1^1\right), \left(S_2^1, t_2^1\right), \ldots \left(S_n^1, t_n^1\right)$$

(10.2)

- S_n^1: nth phase for system 1
- t_n^1: time in nth phase

Using Eq. 10.2 (Target system and temporal states) as a timing description of the behavior for a cyber system, we can compute the average time that it takes for a target system to complete its cycle, traversing a memorized set of states to perform a key organizational function. Summing the time that it takes to complete one process cycle is called the system's makespan (Eq. 10.3 (Process state set completion time (makespan))).

$$T_{total} = \sum_n t_n \qquad\qquad (10.3)$$

The total time estimate shown in Eq. 10.3 provides potentially unique identifying information for an attacker of interest. For example, processes might be identifiable in the hours, days, weeks, or months required for backups, patch management, or upgrades, respectively. The use of process description is a refinement of the timing information introduced in Chap. 7, which focused on gross resource application (i.e., people, process, technology) involved in prosecuting a target.

10.5.1.1 Time Difference and Operations Example

Targeting is often thought of in terms of attacking a static target; a computer system in this case. Part of this thinking comes from the weaponeering legacy that we talked about in Chap. 8, where a JMEM (Sect. 9.3.1) is provided for a given weapon type, in terms of the amount of damage that the weapon will have on a target. Cyber, as discussed in Sect. 10.4.1, is fluid in accounting for the people, process, and technologies of a target of interest. While a target could be a static system just waiting to be attacked, it is more likely that the target will be dynamic and morphing through its different phases while an attacker tries to figure out how to manage a successful engagement. For example, we discussed moving target defenses in Chap. 6, along with describing the different steps that an attacker goes through during target prosecution in Chap. 7. Therefore, a "standard" cyber target will likely be a dynamic system, whether it's a node being defended from a red team penetration test or a malicious cyber actor that we are targeting (e.g., layer 2 in Fig. 6.1) to deny them access to our network.

The goal here is to play the role of an active defender, using the cyber threat framework (CTF) to monitor and assess the behavior of an attacker in a network where we assume that all of the attacker's moves are observable. In addition, we will assume easy translation of attacker behavior observations, or set of dynamic states that a target traverses, to the CTF phases of preparation, engagement, presence, and effect phases of our layered defense (Fig. 6.1). This includes an assumption that a homomorphism relation holds in mapping observed target states to prescribed CTF phases. In formal modeling terms, each of our four CTF phases will ideally have enough observations for a distinct I/O function relation (Zeigler, 2019), as described by the levels of system specification (Table 8.2).

In this hypothetical example, we will use the CTF's four phase model to estimate identifiable attacker behaviors; using the time that an attacker spends in a given state, or set of states, to evaluate both our current defense strategy and any active targeting of the attacker that we can manage during the engagement. This is useful for understanding how the attacker is navigating the current cyber terrain, and which defenses are providing challenges. For example, if an attacker seems to be spending all of his time in preparation, he is having a hard time getting through the current set of defenses—the current network defense strategy is working. Similarly, if the attacker is firmly in the effects phase, he has complete access to the target, and the target is held at risk. Candidate examples include:

- Preparation time > (engagement/presence/effect) time → new defenses
- Effect time > (preparation/engagement/presence) time → complete access

Individual phase times, associated with identifiable behaviors, provide analytical information on the underlying process. For example, long preparation times show that the network defenses are a challenge to overcome, and keep the perpetrator outside of the defended network. The other extreme, where the system is always in the "effects" phase, shows that the perpetrator has complete system access. In addition, using the target phases shown in Eq. 10.1, along with their associated times, also provides the analyst with a tool for labeling a system with an individual identity; phase/time deviations then provide data to estimate potential effects from a given operation.

Looking back at Chap. 9, we have both key performance parameters (KPPs) and measures of performance (MOPs) for the example collection system. For example, we've denied the attacker the KPP of stealth (i.e., 0%) in assuming that his behaviors are always observable. In addition, our perfect observability provides for attack state estimation, in terms of having a clear picture for each step of the attacker's traversal through the network, an MOP of 100% on our part.

The MOP for the attacker, however, is more challenging, in that we will likely do a calculation that includes our estimates of attacker skill and tool capability in the context of the cyber terrain defense that we have provided, for each CTF phase of the operations. For example, "preparation" will be determined by what the attacker can understand about our network prior to "engagement" or attaining "presence." These will be specific, tool, and skill-related assessments that we can also address through live-virtual-constructive observations (Sect. 10.6) in monitoring actual operators in different cyber terrains.

In addition to red team performance operation over the course of an operation, we are also interested in what modeling and simulation can provide in terms of estimating both the types and significance of cyber effects.

10.5.2   Target State Differences for Effects Estimation

Revisiting Eq. 10.2, we can use system state differences to estimate the level of effect. For example, in computing the potential difference in system behavior due to a cyber operation, we can compare state/time snapshots of the system of interest both before and after the operation is complete (Eq. 10.4 (Next step target system)).

$$M_2 = S^2, t^2 \equiv \left(S_1^2, t_1^2\right), \left(S_2^2, t_2^2\right), \dots \left(S_n^2, t_n^2\right) \tag{10.4}$$

One example of calculating the time difference between the two systems, M_1 and M_2, is done by a simple sum of squares evaluation (Eq. 10.5 (Cyber effect estimate—sum of squares over known states)).

$$\text{Effect}_{\text{estimate}} \equiv \sqrt{\sum_n \left(t_n^2 - t_n^1\right)^2} \tag{10.5}$$

One caveat for using representative statistics on a semi-observable system is that we may still be operating within the tolerance of the system itself, rather than viewing an anomalous effect induced by cyber means. The goal, showing the level of system deviation induced by cyber means, is still useful, however, because in some cases we want a temporary effect that is reversible when a complimentary engagement/mission is complete—structural, and potentially destructive, effects may be detrimental to the overall cyber targeting goal. We can also look at the mean time in each state for both gross differences and effect sizes.

10.5.2.1   Effect Sizes

As defined in Joint Publication (JP) 3–12 (US Department of Defense, 2008), cyber effects can be strategic, operational, or tactical. Also, cyber effects are usually in one of two forms:

- Deny (e.g., degrade, disrupt, destroy)
- Manipulate (e.g., take control of an adversary's system)

In addition, cyber is characterized by both duration and reversibility, with reversibility being "new" in conventional targeting terms, as kinetic fires usually provide an irreversible effect.

Effects, either kinetic or cyber, change the state of the object targeted. This might be denial of a key public service, as shown in Sect. 7.2.2 CARVER targeting model, or simply a reversible change in the target's behavior. Measuring the amount of the effect is achieved through statistical understanding of the "as is" and "to be" of the phenomena under inspection.

While some effects may be relative to the mission/campaign being supported, we want to look at individual cyber effects as measurable entities. For example, effect size should be something measurable, as a reflection of target understanding. Chap. 4's Information Operations showed how active measures can be used to sway opinion, potentially moving the number of people believing a certain message from a minority to the majority.

Determining effect sizes is work that has been addressed in the social sciences by comparing the behavior between two observations. This could be two samplings, a before and after for the application of a cyber effect, for example. Looking at the average difference between collected distributions for the two observations, to produce an effect size, is provided by Cohen's d (1988) (Eq. 10.6 (Effect determination)).

$$\theta = \frac{\mu_1 - \mu_2}{\sigma} \tag{10.6}$$

- θ: effect size, standardized mean difference between two populations
- μ_1: mean for one population
- μ_2: mean for other population
- σ: standard deviation based on either or both populations

In practice, population values for Eq. 10.6 are usually not known and need to be estimated from sample statistics, based on specimen, or target, observations. Using effect size evaluations for cyber includes sampling from both the initial system and the system after the cyber effect. These two systems provide the data for estimating the differences in system behavior differences that quantify the effect.

10.5.2.2   Effect Types

Effects are a challenge to quantify, in terms of both their development cost and their persistent value. For example, a 2014 study estimated the cost to bring a prescription drug, or active ingredient, with a provable medical effect, to market at approximately $2.5 Billion (Mullin, 2014). Little research, however, exists concerning how much an active ingredient improves the overall health of society and subsequent savings in health care. This kind of calculation is clearly challenged by unquantifiable measures of medical care costs solely dependent on the condition treated, increased lifespan due to the treatment, or the even more nebulous quality of years lived after receiving the medication. We have similar challenges in quantifying the social good that derives from a cyber security capability that protects our on-line data and transactions.

Similar to health care, cyber effects are similarly challenging to quantify. The "deny" and "manipulate" cyber effects, from Joint Publication 3–12, provide a basis for describing the change of state that a cyber-targeting operation has on a system. For example, we used a collection example in Chaps. 8 and 9 to show how a cyber system might be modeled, with its results quantified by the amount of data collected. This collection example leveraged well known, "big data," cyber ISR collections, from Chap. 5, to look at one measure of performance (MOP) for a cyber operation.

10.5.2.3   Attack as Moving Target Example

Cyber defenders have an advantage in that they control the sensors, and observability, of a target within their visible area of operations. In addition, as discussed in Sect. 10.5.1.1, we an observer an attacker's behavior for each of the CTF phases to get a clear picture of how we are doing to prolong his stay in a phase, or even deny his moving to his next objective in our defended network. In addition, with enough observation, we can observe a mean time in a state both before and after the effect is applied. For each of the states in the system, we can then calculate effect (Eq. 10.7 (Cohen's d—effect calculation for each system phase)).

$$d_1 = \frac{\mu_{t_1^1} - \mu_{t_1^2}}{\sigma_1} \tag{10.7}$$

We can use Eq. 10.7 to compute the difference for an attacker's mean time in each of the respective phases for a hypothetical example (Table 10.4).

As shown in Table 10.4, two example attackers, with very different mean preparation times, show up strongly with a Cohen d of 0.4 for the preparation phase. For the defender, this effect size is a favorable indicator for a security investment. An attacker with this kind of result may have a skill, or ISR, deficit, causing challenges in getting onto and maneuvering through the network. Differences in the engagement and presence phases, at 0.22, show some difference, while a $d = 0.11$ for the effect phase shows little difference. In addition, Eq. 10.5 and 10.6, while effects estimates for each CTF step, can also be scoped down to specific measurables for any phase in the attack process.

10.5.3   Static Cyber System Description

As discussed, it is possible to describe broadly scoped systems in terms of their means and standard deviations for the common IA CIA effects. In addition, in Table 10.4, we reviewed examples of clear effects estimation via behavioral changes in terms for the standard CTF phases of a target system. We also captured the effects of a dynamic system by using both the mean time for each state, and its standard deviation. This included a standard phase structuring of the attack process that is assumed to be general enough to describe any threat.

An alternative approach for estimating a threat is to look at its complexity, and thereby "guess" at the level of effort it will take to gain access. Figure 10.9 shows an example system entity structure (SES) (Zeigler et al., 2000), or tree structuring of a target.

As shown in Fig. 10.9, an example target consists of the process and technology elements, each with addressable components that require some level of effort to understand. For example, under "process," each of the component technologies includes three alternatives, with a cross product $3 \times 3 \times 3$ equaling 27. Similarly, the cross product of the number of technology alternatives is 243. Multiplying the

Table 10.4 Cohen's d with two example processes

Phase	μ_1	σ_1	μ_2	σ_2	$\mu_1-\mu_2$	σ_{ave}	Cohen's d
Preparation	10	2	30	3	1	2.50	0.40
Engagement	6	1.5	5	1.2	0.3	1.35	0.22
Presence	4	1	3	0.8	0.2	0.90	0.22
Effect	1	0.1	1	0.09	0.01	0.10	0.11

process and technology enumerations together results in over 6500 potentially unique system compositions that make up the target of interest.

Enumerating the total number of possible target combinations, for addressable vulnerabilities, is an additional approach for the analyst to perform a rough order of magnitude to estimate the level of effort for accessing a target of interest.

Using these operational effects, in terms of enumeration, provides an initial estimate for the possible effort required to prosecute the target. In addition, this approach provides a baseline for describing the system behaviors that we can build upon with constructive modeling to elaborate on the system elements and develop more detailed performance estimates.

10.6 Cyber Modeling and Simulation Environments

Static target definition and analytical approaches for cyber systems analysis are used to provide first order analysis for cyber systems of interest. We initially introduced analytical approaches, for cyber system vulnerability assessment, in Chap. 3 (e.g., PASTA). In this section we will review tools for more detailed analysis via constructive modeling, live-virtual-constructive environments and cyber ranges.

10.6.1 Constructive Modeling Environments

As shown in Fig. 3.5, a threat analysis process (i.e., V diagram), there is usually a clearly specified context for building a model, and a representation for its environment. This includes emulating traffic, estimating processing capabilities and using

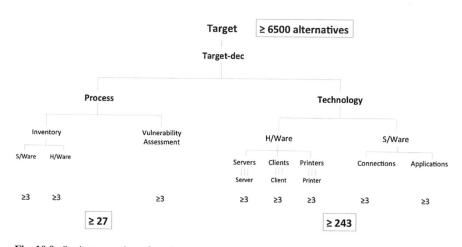

Fig. 10.9 Static structuring of a cyber target

sensors to measure the system's response to external events. One approach is to represent the generator, processor, and transducer components of a model (Fig. 10.10).

The general construct shown in Fig. 10.10 describes the environment (generator), the system under test (processor), and the collection of tested behavior (transducer). Multiple frameworks currently exist for cyber modelling (Table 10.5).

As shown in Table 10.5, Scalable's EXata is one of the few tools identified to date that provides a representation of the impact of an event in cyberspace on both the informational and operational capabilities of a mission. Similar to COATS (Wells & Bryan, 2015), Scalable's Network Defense Trainer (NDT) creates linkages between a cyber training environment and the classical domain training exercises.

OPNET is a de facto standard for network modeling, with DISA leveraging OPNET for implementation of libraries for specific JCSS components, which includes more specific entities for client modeling and simulation. In addition, extending to cyber, the OPNET-based SELEX ES NetComm Simulation Environment (NCSE) provides users with an ability to model and simulate operational network assets. By implementing a "System-in-the-Loop" capability, users can establish a "Live-Constructive" connection and allow real hardware or applications and the simulation environment to interact as a common operational picture. NCSE incorporates communications effects with the rest of the simulation to generate an enhanced awareness of the impact a cyber attack might have on the scenario.

Both the Scalable NDT and SELEX' NCSE are training platforms to improve the situational awareness of the respective cyber and mission operators. These training platforms are usually targeted toward training mission operators to know when to call the cyber professionals, who maintain specialized tools.

Fig. 10.10 Generator, processor, and transducer (GPT) components of a constructive model (Zeigler et al., 2000)

Table 10.5 Example constructive cyber modeling and simulation environments

Tool	Provider	Uses
Network Risk Assessment Tool (NRAT) (Whiteman, 2016)	US Government Off the Shelf Technology (GOTS) that provides an accredited method for cyber defense	Assessing risk to the information services and supported operational missions/objectives from potential cyber attacks
Scalable Networks' Exata Environment	Scalable Networks	Originally developed for the Joint Tactical Radio System (JTRS), EXata was extended for cyber via StealthNet (Torres, 2015) by the Test Resource Management Center (TRMC) Scalable Network Defense Trainer (NDT)—a European variant introduced primarily for defense modelling within NATO countries
OPNET	A suite of protocols and technologies to design, model, and analyze communication networks	OPNET provides a plug and play environment for developing network models. Examples of cyber simulators built on OPNET include DISA's Joint Communication Simulation System (JCSS) builds on OPNET to create a specific component library SELEX ES NetComm Simulation Environment (NCSE) is the Selex ES Modeling and Simulation environment
NS2/3/4	Open source tool (Wikipedia, n.d.)	Open source network simulator (NS) originally funded by DARPA through the VINT project

10.6.2 Live-Virtual-Constructive (LVC) Cyber Training Tools

Cyber models are one part of the training system used to develop mission and cyber professionals. Figure 10.11 (Stine, 2012) provides a notional interaction between the network, cyberspace, and mission operations that help inform how cyber effects are included in standard training simulations.

The information flows in Fig. 10.11 provides just one view of an "as is" architecture, parts of which are emulated by cyber simulators in training future cyber defenders. The objective here is to emulate a real-world system, like that shown in Fig. 10.4, with the right mix of Live-Virtual-Constructive (LVC) assets (Fig. 10.12).

As shown in Fig. 10.12, each of the LVC modes has different associated skill acquisition goals. One example, the air domain, includes extensive use of simulation for initial familiarization through skill development. Achieving the realism common to LVC for the air domain will require either producing simulations of realistic fidelity (constructive modelling) or consistently providing live injects (live/

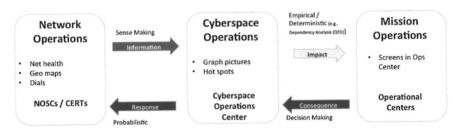

Fig. 10.11 Network, cyberspace, and mission operations—information flows and events

virtual modeling) into the cyber training. A few considerations for achieving realistic fidelity and timeliness in building cyber M&S for defense training are shown in Table 10.6.

The techniques shown in Table 10.6 will contribute to cyber defense training simulators to keep pace with the fast moving nature of the cyber domain. M&S, therefore, can be used to quickly evaluate cyber operator skills against current cyber threats (Chap. 3). Another approach is to do "live" testing on cyber ranges, exercising with realistic malware effects.

10.6.3 Cyber Ranges

A recent literature review describes the specialization of cyber ranges (Yamin et al., 2020). Examples include the multiple cyber ranges and test beds are currently being used for testing and training (Davis & Magrath, 2013), with more coming on-line each month; often for educational purposes. Each of these ranges includes a modeling platform for representing cyber systems of interest. As discussed, OPNET and Scalable's Exata are key names in producing cyber M&S development platforms. For example, the Joint Communication Simulation System (JCSS), which leverages OPNET, is commonly used for long haul communication evaluation. Similarly, CORONA (Norman & Davis, 2013) uses OPNET for T&E experiments, which usually include a cyber range.

The speed and combinatorial nature of the evolving cyber threat demands a more flexible modeling and simulation (M&S) that uses cyber ranges for both training and system test and evaluation. One approach for reviewing the base-line processes is to conduct cyber-range events. This includes the logical range construct (Damodaran & Smith, 2015) that provides event environments to be constructed in a location independent manner, leveraging range specific tools (e.g., traffic generation).

Cyber-range events vary in complexity and their objectives, covering a broad spectrum of event types. For example, some events are conducted for training cyber defense forces, and others are conducted for developmental testing (DT) or operational testing (OT). Events may also be conducted for experimentation with technology or tactics, or to assess mission readiness. An early example of a cyber range

Fig. 10.12 Live-virtual-constructive (LVC) and skills development

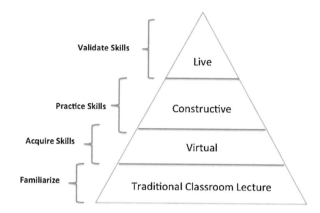

Table 10.6 LVC contributions to cyber training realism

LVC element	Description
Live	Injecting effects from operators into the simulation
Virtual	Injecting effects from ranges into the training simulation (e.g., COATS)
Constructive	Use of simulators to inject cyber effects into tactical exercise

used for academic studies is the DETER test bed (2004). In addition, the DoD Enterprise Cyber Range Environment (DECRE (DOT&E, 2013)) is a federated set of ranges within DoD, comprised of the National Cyber Range (NCR), the DoD Cyber Security Range, the Joint Information Operations Range (JIOR), and the Joint Staff J6's C4 Assessments Division (C4AD) (Table 10.7).

The cyber ranges shown in Table 10.7 are used to support cyber-range events such as Cyber Flag (Hansen, 2008). In addition, the ranges in Table 10.7 can be combined into an overall "logical range" (Damodaran & Smith, 2015), where each range's respective specialty (e.g., traffic generation) is leveraged virtually to provide a location agnostic testing/training environment.

10.7 Summary

Depending on the targeting application, there is currently a mosaic of tools across the people, policy, process, and technology threads that compose a cyber target. Cyber ranges are a good candidate for evaluating actual systems. In addition, there are application, networking, and physical layer simulators for estimating follow on effects from a cyber event.

Process modeling and the description of specific effects due to cyber are a first step in the developing of valid component models for estimating target effects. Describing the technical elements (e.g., application, networking, physical) and

Table 10.7 Example cyber ranges

Cyber range	Management	Function
National Cyber Range (NCR)	Joint Mission Environment Test Capability (JMETC)	General cyber system evaluation capability for large scale scenario testing (Pathmanathan, 2013)—transitioned from DARPA in October, 2012.
DoD Advanced Cyber Range Environment (ACRE) (MANTECH, 2018)	Defense Information Systems Agency (DISA)	Evaluation of bandwidth effects on operations
C4AD	Test Resource Management Center (TRMC) (Ferguson, 2014)	Interoperability assessments, technology integration and persistent C4 environment
Joint IO Range (JIOR)	Joint Staff J7	Nation-wide network of 68 nodes for live, virtual, and constructive operations across the full spectrum of security classifications (National Defense Authorization Act (NDAA) for Fiscal Year 2013, 2012)

entity-types (i.e., people, process, technology) that compose the actual device determines both how a phenomenon is modeled and its expected validation level. These are important first steps in developing cyber targeting effects simulations that scale, for both high fidelity training and to answer the critical strategic and tactical questions.

In addition, we presented modeling and simulation tools in terms of their fit across the targeting process (Fig. 7.3). For example, each of the elements listed in Fig. 10.13 includes a step number, designating where in the targeting cycle that the modeling or assessment capability will be used.

As shown in Fig. 10.13, up front analytics (e.g., Eq. 10.1 and COLE) are used as tools for doing initial analysis of targets, and their elements. Nominating targets, step three in the overall targeting cycle (Fig. 7.3) can include more advanced emulations and simulations via LVC (Sect. 10.6.2) and cyber ranges (Sect. 10.6.3).

The next major role for modeling and simulation in the targeting cycle is step six. For example, assessing target effects are performed by (1) direct inspection of what occurred (e.g., % downtime, number of files extracted, repostings of IO messaging) or (2) behavioral anomalies in a described system of interest (e.g., Cohen's d).

Cyber M&S Questions for Analysis and Targeting

1. Why is modeling and simulation (M&S) important for cyber analysis and targeting?

   Cyber systems are often complex and require gross assumptions concerning both their construction and their operation. Models provide a way to clearly represent the overall system, and its individual components. Originally introduced in Chap. 8 (Fig. 8.2), modeling has grown in tandem with the design, development and deployment of computer-based systems.

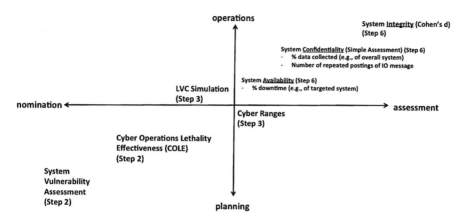

Fig. 10.13 Modeling and simulation applications for cyber analysis and targeting

As the model resolution increases, the network defender can more clearly forecast the effect of a given cyber action.

2. What are examples of cyber models and their uses for analysis and targeting?

Cyber M&S has been used effectively in both exercises and for constructive simulation. For example, in Sect. 10.1 we looked at COATS, a simulation that includes the injection of cyber effects to increase operator awareness and improve their ability to respond.

In addition, in Sect. 10.1 we also looked at AMICA, a constructive model that propagates computer-system compromises (i.e., measured in KPPs) to an Air Tasking Order (ATO) compromises (i.e., measured in MOPs) that eventually lead to aircraft sortie reductions (i.e., measured in MOEs).

3. How does modeling and simulation help with analyzing the parallel operation of cyber systems?

As discussed in the AMICA example, computer systems have multiple processes operating in parallel, where multiple compromises are possible simultaneously.

In addition, as shown in Fig. 10.3, a cyber system has physical, logical, and persona systems, an inherently parallel structure, where models make for a handy way to account for all the possible vulnerabilities, and exploits, that could attend simultaneously. In fact, one accounting for a possible simultaneity in modeled cyber system exploit events is called superdense time (Nutaro, 2020).

4. How can cyber modeling be expanded to perform vulnerability analysis?

In this chapter, we looked at both computer system vulnerability (Figs. 10.2, 10.4, and 10.11) and operational vulnerability (Figs. 10.1, 10.6, 10.7, 10.8, 10.9, and 10.10). While the technical system, due to multiple, simultaneous, processes, is a challenge to model, including the interleaved layers (e.g., Fig. 10.3) can be even more challenging for vulnerability estimates of operational systems.

5. What are candidates for cyber effects estimation and how does modeling and simulation improve current analysis approaches?

In Chap. 9, we talked about KPPs, MOPs, and MOEs to describe cyber security, operations, and effects, respectively. In addition, in we looked as "measurables" from well-known cyber attacks.

Modeling each of the respective use cases can start with the mean of the files extracted (i.e., OPM breach) or the duration of the denial of service (i.e., Operation Ababil). These "models," consisting of a mean, and standard deviation, if multiple samples are available, provide the analyst with a way to organize attacks by type (e.g., availability, confidentiality, integrity) and amount (e.g., duration of service denial, number of records exfiltrated, number of systems controlled by attacker).

6. What kind of heuristics help with cyber analysis?

We used Cohen's d as an example approach to help quantify the difference between two means of statistical samples in order to estimate a behavior change in a process. For example, this might be a change in a delay time associated with the process completing an activity. Cohen's d lends itself to observations, with target access, so as to extract samples of reasonable size, for comparison, both before and after applying the cyber effect.

In addition, we used Cohen's d to look at example effects across the CTF phases for a generalized cyber attack (Sect. 10.5). We found that even this simple approach, using means and standard distributions, example timing data for each of the respective attack phases, can tell us a lot about the effectiveness of defenses and whether a target is owned by an adversary.

We also generalized on the approach (Sect. 10.5.1) to show how individual timing differences for a cyber-system process can help us with discerning specific cyber effects due to measurable "before" and "after" behavioral anomalies.

Bibliography

Arnwine, M. (2015). Developing the infrastructure and methodologies for cyber security. *18th Annual Systems Engineering Conference.* NDIA.

Bowden, M. (2011). *Worm - the first digital world war.* Atlantic Monthly Press.

Bucher, N. (2012). Simulation and emulation in support of operational networks: "ALWAYS ON". In *NDIA 15th annual systems engineering conference.* NDIA.

Center for Computational Analysis of Social and Organizational Systems (CASOS). Center for computational analysis of social and organizational systems (CASOS). http://www.casos.cs.cmu.edu/. Accessed 14 Oct 2018.

Chabuk, T., & Jonas, A. (2018, September 1). Understanding Russian information operations. *Signal (AFCEA).* https://www.afcea.org/content/understanding-russian-information-operations. Accessed 9 Sept 2018.

Cohen, J. (1988). *Statistical power analysis for the behavioral sciences.* Lawrence Earlbaum Associates.

Couretas, J. M. (2018). *Introduction to cyber modeling and simulation.* Wiley.

Damodaran, S. K., & Smith, K. (2015). *CRIS cyber-range lexicon.* Cyber Range Interoperability Standards Working Group.

Davis, J., & Magrath, S. (2013). *A survey of cyber ranges and testbeds.* Defence Science and Technology Organisation.

Defense Science Board. (2013). *Resilient military systems and the advanced cyber threat* (p. 146). Office of the Under Secretary of Defense for Acquisition, Technology and Logistics.

Department of Defense. (2002). *Joint doctrine for targeting* (p. 100). Doctrine, Joint Chiefs of Staff Washington, DC, Department of Defense, Arlington: Department of Defense.

Department of Defense. (2012). Joint publication 1–13.4 military deception. Department of Defense.

Dietrich, N. (2011). Development and the deployment of Cosage 2.0. *WinterSim., 8.*

Director of National Intelligence. (n.d.). Cyber threat framework. https://www.dni.gov/index.php/cyber-threat-framework. Accessed 20 Oct 2018.

DISA. Joint Communication Simulation System (JCSS). *Joint Communication Simulation System (JCSS).* https://www.disa.mil/Mission-Support/Enterprise-Engineering/JCSS. Accessed 10 Oct 2018.

DoD Operational Test and Evaluation (DOT&E). (2015). Cyber security. .

DOT&E. (2013). Test and evaluation resources. http://www.dote.osd.mil/pub/reports/FY2013/pdf/other/2013teresources.pdf. Accessed 11 Feb 2018.

Evans, D., Nguyen-Tuong, A., & Knight, J. (2011). Effectiveness of moving target defenses. In A. Ghosh, V. Swarup, C. Wang, & S. W. S. Jajodia (Eds.), *Moving target defense.* Springer.

Ferguson, C. (2014). Distributed cyber T&E. In *NDIA Annual T&E Conference,* Washington, p. 26.

Frank, A. B. (2017). Toward computational net assessment. *Journal of Defense Modeling and Simulation (Sage), 14*(1), 79–94.

Frei, S., Fiedler, U., May, M., & Plattner, B. (2006). Why to adopt a security metric? *Quality of Protection, Advances in Information Security, 23.*

Greenberg, A. (2018, February 12). 'Olympic destroyer' malware hit Pyeongchang ahead of opening ceremony. *Wired.* https://www.wired.com/story/olympic-destroyer-malware-pyeongchang-opening-ceremony/. Accessed 14 Oct 2018.

Hansen, A. P. (2008). *Cyber flag - a realistic cyberspace training construct.* Wright Patterson Air Force Base: AFIT.

Hayden, M. V. (2016). *Playing to the edge: American intelligence in the age of terror.* Penguin.

Jaquith, A. (2007). *Security metrics: Replacing fear, uncertainty, and doubt.* Addison-Wesley, Pearson Education.

Johnson, R. E. (1995). A stochastic version of the concepts evaluation model (CEM). *Naval Research Logistics,* 233–246.

Joint Chiefs of Staff. (2011). *CJCS instructions.* www.jcs.mil/Library/CJCS-Instructions/udt_46626_param_orderby/Info/udt_46626_param_direction/ascending. Accessed 11 Feb 2018.

Jonsson, E., & Olovsson, T. (1997). A quantitative model of the security intrusion process based on attacker behavior. *IEEE Transactions on Software Engineering, 4* (23), 235–245. https://doi.org/10.1109/32.588541.

Lewis, J. (2011, July 11). Cyber attacks, real or imagined, and cyber war. *Center for Strategic and International Studies (CSIS).* https://www.csis.org/analysis/cyber-attacks-real-or-imagined-and-cyber-war. Accessed 10 Oct 2018.

Li, W., & Vaughn, R. (2006). Cluster security research involving the modeling of network exploitations using exploitation graphs. In *Sixth IEEE International Symposium on Cluster Computing and Grid Workshops.*

Maathuis, C., Peters, W., & Van Den Berg, J. (2020, April 11). Decision support model for effects estimation and proportionality assessment for targeting in cyber operations. *Science Direct.* https://www.sciencedirect.com/science/article/pii/S2214914719309250. Accessed 22 June 2020.

Manadhata, P. K. (2008). *An attack surface metric.* CMU.

MANTECH. (2018). Cybersecurity. http://www.mantech.com/solutions/Cyber%20Security/Pages/default.aspx. Accessed 11 Feb 2018.

Morse, K., Wells, W., & Drake, D. (2014). Realizing the Cyber Operational Architecture Training System (COATS) through standards. *SIW.* Orlando: SISO.

Mullin, R. (2014, November 24). Cost to develop new pharmaceutical drug now exceeds $2.5B. *Scientific American.* https://www.scientificamerican.com/article/cost-to-develop-new-pharmaceutical-drug-now-exceeds-2-5b/. Accessed 13 Sept 2018.

Noel, S., Ludwig, J., Jain, P., Johnson, D., Thomas, R., & McFarland, J. (2015). Analyzing mission impacts of cyber actions (AMICA).

Norman, R., & Davis, C. (2013). Cyber operations research and network analysis (corona) enables rapidly reconfigurable cyberspace test and experimentation. *M&S Journal,* 15–24.

Nunes-Vaz, R., Lord, S., & Ciuk, J. (2011). A more rigorous framework for security-in-depth. *Journal of Applied Security Research, 23,* 372–393.

Nunes-Vaz, R., Lord, S., & Bilusich, D. (2014). From strategic security risks to national capability priorities. *Security Challenges, 10*(3), 23–49.

Nutaro, J. (2020). Toward a theory of superdense time in simulation models. *ACM, 30*(3).

Ortalo, R., Deswarte, Y., & Kaaniche, M. (1999). Experimenting with quantitative. *IEEE Transactions on Software Engineering, 25,* 633–650.

Pathmanathan, A. (2013). 30th Annual International Test and Evaluation Symposium (ITEA). *11,* (14). http://www.itea.org/~iteaorg/images/pdf/conferences/2013_Annual/Panel_2_Pathmanathan.pdf. Accessed 1 Jun 2015.

Perlroth, N., & Hardy, Q. (2013, January 8). Bank hacking was the work of Iranians, officials say. *The New York Times.* https://www.nytimes.com/2013/01/09/technology/online-banking-attacks-were-work-of-iran-us-officials-say.html. Accessed 10 Oct 2018.

Power Fingerprinting. (n.d.). Power fingerprinting. *Power Fingerprinting.* https://www.pfpcyber.com/. Accessed 20 Oct 2018.

Repel, D., & Hersee, S. (2015). The ingredients of cyber weapons. In *10th international conference on cyber warfare and security (ICCWS15)* (pp. 1–10).

Rid, T. (2012, February 27). Cyber war: think again..

Romanosky, S. (2016). Examining the costs and causes of cyber incidents. *Journal of Cybersecurity, 2*(2).

Sallhammar, K., Helvik, B., & Knapskog, S. (2006). On stochastic modeling for integrated security and dependability evaluation. *Journal of Networks, 1*(5), 31–42.

Sanger, D. E., & Perlroth, N. (2015, April 15). Iran is raising sophistication and frequency of cyber-attacks, study says. *New York Times.* https://www.nytimes.com/2015/04/16/world/middleeast/iran-is-raising-sophistication-and-frequency-of-cyberattacks-study-says.html. Accessed 14 Oct 2018.

SANS Institute. (2006). *A guide to security metrics.* SANS.

Stine, K. (2012). Inside NIST's cybersecurity strategy. *Washington Technology.*

Taylor, J. G. (2000). Hierarchy of models approach for aggregated attrition. In *Winter simulation conference.*

"The DETER Testbed: Overview." *The DETER Testbed: Overview.* 25 Aug 2004. http://www.isi.edu/deter/docs/testbed.overview.pdf. Accessed 6 May 2015.

Torres, G. (2015). *Test & evaluation/science & technology net-centric systems test (NST) focus area overview* (p. 37). Center for Systems and Software Engineering, USC, Pt Mugu: Center for Systems and Software Engineering.

United States of America. (2018). *United States of America vs. internet research agency LLC* (p. 37). INDICTMENT, Department of Justice.

US Department of Defense. (2008). *Cyberspace operations* (p. 104). Joint Staff, US Department of Defense.

USAF. (2019, March 15). "THE TARGETING CYCLE (Annex 3-60)." Curtis E. LeMay Center for Doctrine Development and Education. https://www.doctrine.af.mil/Portals/61/documents/Annex_3-60/3-60-D04-Target-Tgt-cycle.pdf. Accessed 22 June 2020.

Velez, T. U., & Morana, M. (2015). *Risk centric threat Modeling: Process for attack simulation and threat analysis.* Wiley.

Verizon. (2018). 2018 data breach investigations report. *Verizon.* http://www.verizonenterprise.com/industry/public_sector/docs/2018_dbir_public_sector.pdf. Accessed 17 Oct 2018.

Waltz, E. (2000). *Information warfare: Principles and operations.* ArTech House.

Wei, O. K. (2012, November 14). Securing your web application against security vulnerabilities. *IBM.* http://www-07.ibm.com/sg/smarterbusiness/meettheexperts/includes/downloads/Securing_Your_Web_0910_eve.pdf. Accessed 9 Oct 2018.

Wells, D., & Bryan, D. (2015). Cyber operational architecture training system – cyber for all. In *Interservice/industry training, simulation, and education conference (I/ITSEC)* (p. 9). NDIA.

Whiteman, B.(2016, March 15). IATAC. *IATAC.* https://www.csiac.org/wp-content/uploads/2016/02/Vol11_No1.pdf. Accessed 20 Oct 2018.

Wikipedia. (n.d.). Ns_simulator. https://en.wikipedia.org/wiki/Ns_(simulator). Accessed 26 Oct 2018.

Wylie, J. C. (2014). *Military strategy: A general theory of power control.* Naval Institute Press.

Yamin, M. M., Katt, B., & Gkioulos, V. (2020). Cyber ranges and security testbeds: Scenarios, functions, tools and architecture. *Computers & Security, 1.*

Yildrim, U. Z. (1999). *Extending the state-of-the-art for the COMAN/ATCAL methodology* (p. 161). NPS.

Zeigler, B. P. (1976). *Theory of modeling and simulaton.* Wiley.

Zeigler, B. P. (2019). Introduction to iterative system computational foundations and DEVS. *Winter Simulation Conference.* INFORMS.

Zeigler, B. P., Praehofer, H., & Kim, T. G. (2000). *Theory of modeling and simulation.* Academic Press.

Zurasky, M. W. (2017). Methodology to perform cyber lethality assessment. Doctoral Dissertation, Old Dominion University, Old Dominion University, p. 60.

Chapter 11
Cyber Case Studies

The purpose of this chapter is to review the use cases that we referenced and developed, throughout the chapters, to describe cyberspace analysis and targeting. These use cases span the policy, process and technology exemplars that we used to illustrate key cyber analysis and targeting concepts, organized in the order that they might be employed in an analysis and targeting exercise.

11.1 Introduction: Cyber Use Cases for Analysis and Targeting

Throughout the previous chapters, our goal was to illustrate cyber analysis and targeting applications with use cases, or analogies, to provide conceptual structures that guide us in areas where data are still developing for the new field of cyberspace operations. These examples span from policy frameworks (Chap. 2), to system vulnerability analyses (Chap. 3), to analysis of information operations (Chap. 4), and to the description of a cyber target as a discrete event system for effects evaluation (Chap. 10).

We also provided a general targeting process, in Chap. 7, that includes quantifiable steps that a decision maker will use in nominating, prosecuting, and assessing a cyber targeting mission. These targeting steps often require more detailed knowledge of both cyber systems (Chap. 5) and their environments (Chaps. 6 and 8) to provide meaningful answers as to how to successfully complete each step. Our review of use cases will therefore follow the targeting process described in Chap. 7 (Fig. 7.2) to guide our steps through this book's exemplars to nominate, prosecute, and assess a cyber target.

© Springer Nature Switzerland AG 2022
J. M. Couretas, *An Introduction to Cyber Analysis and Targeting*,
https://doi.org/10.1007/978-3-030-88559-5_11

11.2 Cyberspace Mission Analysis

A key goal of cyber analysis and targeting will be to provide effects, via cyber means, that meet a commander's objectives, guidance, and intent. In the previous chapters, we looked at each of the various elements of cyber analysis and targeting that help to structure the types of questions that cyberspace effects can answer. These effects should help a commander meet her engagement, mission, and campaign objectives. This might include questions on when to apply cyber, against what kind of targets, and for how long, in order to achieve short, intermediate and long-term operational objectives. In addition, we will review the application of cyber means within current analysis and policy frameworks.

11.2.1 Cyber Analysis and Policy Frameworks

One of the current challenges in discussing cyber is that it is ill-defined. For example, cyberspace has sometimes been called a murky, "space in between," area between diplomacy and kinetic operations. For example, in 2014, Eric Rosenbach, then an Assistant Secretary of Defense, said it well—.

> "The place where I think [cyber operations] will be most helpful to senior policy makers is what I call "'the space between.' What is the space between? … you have diplomacy, economic sanctions … and then you have military action. In between there's this space, right? In cyber, there are a lot of things that you can do in that space (in) between that can help us accomplish the national interest" (Maurer, 2014).

Cyberspace, as an operational domain, provides policymakers with a maneuver space previously unavailable. However, due to the ill-defined nature of cyber as a means of applying the standard levers of government power, it is a challenge to determine how cyber fits in the spectrum from diplomatic to kinetic action. For example, in Chap. 2, we used "the line," a boundary on the spectrum that spans from diplomacy to kinetic warfighting, to portray a way to measure when a cyber attack might merit a kinetic response (i.e., Schmitt Criteria) (Fig. 11.1).

Fig. 11.1 "The Line"—spectrum of conflict

"The Line," in Fig. 11.1, helps provide context for future analysis and targeting examples. This may even include updating considerations on how to respond to information operations (IO). For example, the area and point targeting discussed in Chap. 4 produced dozens of riots during the 2016 U.S. Presidential election (Mueller, 2019), all safely commanded and controlled from cyberspace.

Similarly, understanding the relative positioning of cyber in the spectrum of hostilities helps clarify why we used the autonomous system analogy when developing cyber system key performance parameters (KPPs), measures of performance (MOPs), and measures of effectiveness (MOEs) (Chap. 9). Using the munitions model, where cyberspace operations is simply the latest iteration in a smart bomb continuum, can change how a strike, like those provided in the Mueller Report (Mueller, 2019), might be classified. A revisit of the Tallinn Manual (Schmitt, 2013), and formally approving an internationally accepted document that clearly codifies how cyber attacks fit on the diplomatic/military engagement spectrum, is a possible next step in clarifying the use of cyber.

11.3 Target Identification

The Line in Fig. 11.1 is an initial guide as to where a cyber effect may fit in the spectrum of conflict. In addition, the Line may also provide guidance concerning target identification, providing a guideline for how provocative the access and control of an adversary's cyber targeted device might be, in terms of both device location and the type of effects that will derive from a proposed mission. Understanding target sensitivity will therefore help clarify the best choices concerning a target type for the desired effect.

With commander guidance clarified in terms of the types of desired effects, the next step, determining and developing targets, includes multiple tools that will help to determine which targets are the most promising, and whether cyber is a means to achieve those ends.

11.3.1 CARVER (Criticality, Availability, Recuperability, Vulnerability, Effect, and Recognizability)

The CARVER model (Bradley Greaver, 2017) is a conventional targeting approach. In addition, CARVER is an acronym that stands for Criticality, Availability, Recuperability, Vulnerability, Effect, and Recognizability. Developed by U.S. Special Forces in the 1960s, CARVER is a targeting framework to facilitate center of gravity (COG) analysis (Table 11.1).

As shown in Table 11.1, the CARVER framework facilitates the discussion on how cyberspace effects help a mission commander achieve her mission objectives.

Table 11.1 CARVER elements and description

CARVER element	Description
Criticality	This is the target value. How vital is this node to the overall organization? A target is critical when its compromise or destruction has a significant impact on the overall organization's performance. In addition, criticality can be described by many factors, a few of which are– Time—How rapidly will the impact of the target attack affect operations? Quality—What percentage of output, production, or service will be curtailed by target damage? Surrogates—What will be the effect on the output, production, and/or service? Relativity—How many targets are there? What are their locations? How is their relative value determined? What will be effected in the overall system?
Accessibility	How easy is it to reach the target? What are the defenses? Is an insider required to penetrate the organization? Is the target computer on the internet?
Recuperability	How long will it take for the organization to replace, repeat, or by pass the target's compromise? Once the exploit is found, how long will it take for the system to remediate it and recover operations?
Vulnerability	How much and what kind of knowledge is required to exploit the target? Do I currently have this capability, or do I need to invest in new approaches to access this target?
Effect	How will the organization be affected by the attack? What kind of reactions should I expect to see from the organization due to a successful attack?
Recognizability	How easy is it to identify the target? How easy is it to recognize that a specific system/network/device is a target and not a security counter measure?

In addition, CARVER addresses additional characteristics that help to determine the most appropriate target for a mission. For example, the discussion of criticality is important to clarify for the overall mission or campaign plan. CARVER, therefore, helps with the upfront considerations of the appropriate type targets to meet the commander's objectives.

In addition, CARVER, though dated, includes "recuperability" in its concise analysis of targeting. Recuperability speaks directly to the perishable effects of a cyber attack that challenge current JMEM comparisons (Chap. 9). A key product of CARVER includes the delivery of candidate targets, based on the mission objectives, for further analysis. Analyzing targets, for their vulnerabilities and possible compromise, will include a review of the target and its environment.

11.3.2   Cyber System Vulnerability Estimation and Tool Development

One of the key elements of CARVER is its use in evaluating a target's vulnerability. Vulnerability, as used here, describes the ease of exploiting the target. This might include scoring vulnerabilities, as discussed in the overview of DREAD, STRIDE, and CVSS (Chap. 3).

In Chap. 3, we also discussed how some of the early cyber threat models, which include the Common Vulnerability Scoring System (CVSS), are used to identify and score system vulnerabilities. CVSS, still widely used, was developed by Carnegie Mellon University (CMU) and is a more rigorous approach to vulnerability evaluation. Provided by the National Institute of Standards (NIST), CVSS is described as follows:

"The Common Vulnerability Scoring System (CVSS) provides an open framework for communicating the characteristics and impacts of IT vulnerabilities. CVSS consists of three groups: Base, Temporal and Environmental. Each group produces a numeric score ranging from 0 to 10, and a Vector, a compressed textual representation that reflects the values used to derive the score. The Base group represents the intrinsic qualities of a vulnerability. The Temporal group reflects the characteristics of a vulnerability that change over time. The Environmental group represents the characteristics of a vulnerability that are unique to any user's environment. CVSS enables IT managers, vulnerability bulletin providers, security vendors, application vendors and researchers to all benefit by adopting this common language of scoring IT vulnerabilities." (Peter Mell, 2007)

While several other frameworks have been developed, the more famous DREAD, STRIDE, and CVSS provide some of the initial approaches for classifying and estimating potential vulnerability damages. Finding system vulnerabilities remains a challenge. In addition, the process for attack simulation and threat analysis (PASTA) (Sect. 3.3.3) provides a knowledge-based method to find vulnerabilities and estimate the level of threat that they actually pose to the system. This includes an in-depth analysis of the technologies that compose a system, their popularity/availability, and the types of transactions that provide an external attack surface to bad actors.

While we reviewed DREAD, STRIDE, and CVSS, the vulnerability analysis of PASTA provided the context for why using an architecture framework (e.g., DoDCAR), to account for each of the system components and its current security level. In addition to using DoDCAR to manage the security maturity level for a defended system, we reviewed the identification of vulnerabilities for both remediation and possible exploitation (e.g., penetration testing). For example, software vulnerabilities (CVE and CWE) might be scored (CVSS and CWSS) and then developed into malware (e.g., Metasploit discussion) via a software development process, to produce a tool for exploiting a specific vulnerability (Fig. 11.2).

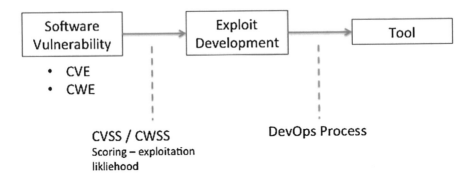

Fig. 11.2 Software vulnerability to exploit development—attack tool process

As shown in Fig. 11.2, each of the steps used to facilitate the communication of cyber threats might also be used for attack tool development, and these steps are already built into "white hat" hacker tools like Rapid7's Metasploit (Rapid7).

11.4   Capabilities Analysis

Analyzing the respective capabilities in a targeting scenario includes understanding the cyber system terrain and system metrics that will be used for evaluating system performance.

11.4.1   Cyber Security Technologies

In Chap. 6, we covered cyber system terrain from a network defenders' point of view. Looking at a network as end points, connections, and key nodes for the application of requisite security technologies helps provide an overview of a network's defense. The technologies covered compliment the processes and skilled people that make up the security operations center (SOC), where key policies are enforced and maintained by the description of the system in terms of network architecture.

11.4.2   Cyber System Architectures

System architectures provide a means to organize component technologies and processes that make up a cyber system. This is especially important for cyber systems, as both the scope and scale can easily exceed what an individual, or a network security team, can keep in memory. System architectures (e.g., DoDCAR), therefore, provide artifacts to account for the terms that describe the security system's context and use, its components, its connections and, possibly, its actual implementation. While architectures provide for a detailed description of system's components and their overall structuring, it remains a challenge to understand the system in terms of its performance and effectiveness.

11.4.3 Cyber System Metrics: Key Performance Parameters (KPPs), Measures of Performance (MOPs), and Measures of Effectiveness (MOEs)

Using architecture, and its artifacts, is helpful for organizing a cyber system. Understanding the behavior of the system, however, will require an understanding that includes key performance parameters (KPPs), metrics that speak to the engineering, or system level behavior of a device. Of course, the KPPs should reflect the system's original requirements. However, some cyber systems, due to the fast-changing operational environment and technologies that compose them, might be a challenge to trace back to a requirements document.

Measures of performance (MOPs), different from system level KPPs, include operational considerations, factoring in human performance. MOPs are more challenging to measure, as operator training and familiarity with scenarios may bear on the results. Similarly, MOPs are measured through engagements and missions, with the assumption that the system for which KPPs are measured will behave consistently.

While MOPs reflect a human in the loop, measures of effectiveness (MOEs) are even more challenging to define, in terms of determining whether an effect has been achieved. This is a reflection of the cyber means employed (e.g., Sect. 11.3.2) combined with a commander's abilities concerning the overall management of the mission/campaign. In addition, as discussed in Chap. 9, the use of cyber to achieve hard-to-measure effects like "trust" makes it even more challenging to put a traditional number scale onto cyber MOEs.

While mission level MOEs are a challenge to quantify, the process that led up to describing effectiveness is very helpful for the commander to make decisions concerning the assigning of forces. For example, understanding her force's capabilities, in terms of MOPs, is very helpful to determine which targets are in fact addressable. Similarly, having KPPs for the cyber system to be employed will help a commander understand her strengths and weaknesses. For example, COLE (Fig. 9.5) can help with probability estimates for hypothetical target success scenarios. This thinking will build on conclusions from earlier analysis, including the products provided by CARVER (Sect. 11.3.1).

11.5 Mission Planning and Force Execution

The majority of cyber operations occur around mission planning and force execution. This is reflected by the Lockheed Martin attack cycle, where weaponization, delivery, exploitation, installation, command and control (C2), and actions on objectives occur, as shown in Table 7.2. In addition, as discussed in Chap. 7, cyber effects span the confidentiality, integrity, and availability (CIA) spectrum of information assurance. This is a challenge for mission planners who are incorporating "cyber" into mission planning.

Cyber includes both IO and the simpler technology-based effects of denying operations (i.e., availability), accessing a "secure" system (i.e., confidentiality) or controlling an adversary system (i.e., "integrity"). While there may be an overlap between confidentiality and IO systems (e.g., extracting e-mails for the 2016 DNC attack), there is little similarity in the techniques used for integrity attacks and IO. The STUXNET example, provided in Chap. 5 (e.g., Fig. 5.10), includes tools and processes very different from tools used in a typical IO operation (e.g., Fig. 4.10).

Cyber effects from targeting, however, may provide the same result for either computer network attack (CNA) or IO operations. For example, while IO operations include shifting opinion, more similar to a media operation, and confidentiality/integrity/availability (CIA) effects will be used to change a device's state, the end result for either might be a vote tally that results in the "right" candidate winning. In addition, both IO operations and CIA effects are strategic. Cyber, therefore, might be used in an application like election security to both (1) modify the messaging to change the population's mind or (2) compromise the system and simply change the vote count.

11.5.1 Effect Likelihood—JMEMs and COLE

An additional challenge is that cyber also operates as a pseudo-kinetic actor, challenging analysis and targeting to assess effects that can take multiple forms over a range of time periods. Successful kinetic effects simply remove their targets from the battlefield. Cyber effects, however, span the compromise of information (i.e., confidentiality), the misuse of information (i.e., integrity), or the denial of information (i.e., availability). Quantifying any of these effects is still an art, with one source of inspiration coming from legacy joint munition effectiveness manuals (JMEMs) that used to describe the performance of conventional bombs and guided munitions.

Due to the decoupling of ISR from kinetic strike, conventional munitions lend themselves to a more detached, technical assessment, with methods and techniques developed over decades. The Joint Munitions Effects Manual (JMEM) (US Army), for example, is used to document an explosive munitions' effects in engineering level detail.

As discussed in Chap. 7, standard targeting approaches are just as applicable in cyber as they are for other munitions. This is one of the attributes that leads cyber professionals to look for further analogs between cyber and conventional munitions. For example, one challenge for modern day cyber is measuring a cyber target's vulnerability. The cyber operations lethality and effectiveness (COLE) model (Zurasky, 2017) is designed to provide a reusable approach for cyber target evaluation. To compute an overall probability of kill (P_k), COLE combines the likelihood of success for each step in an attack path (Fig. 11.3).

As shown in Fig. 11.3, COLE operates along the same lines as the probability of kill (P_k) in kinetic munition evaluation. The elements of the calculation are provided in Table 11.2.

As shown in Table 11.2, COLE is comprehensive in covering each of the elements that a cyber attacker needs to test and evaluate prior to executing a successful cyber attack mission. COLE therefore looks more like a measure of performance (MOP), on the part of the operator, than the KPP of a munition usually associated with a kinetic JMEM.

COLE is the latest iteration of the former IO and cyber JMEM efforts (Mark Gallagher, 2013), designed to be comparable to the systems level parameterization broadly used for kinetic weapons. The effect, in this case, is the probability of kill (P_k) that results from the successful delivery chain, and exploit, of a cyber target. In weaponeering, however, accounting for the platform that provides a probability of hit (P_H), with the weapon, or munition, that provides a P_k with a given P_H. In addition, other factors might be introduced—probability of detecting the target (P_d), weapon reliability (R_w), system reliability (R_{sys}), etc. Cyber is a challenge with the platform being, for the most part, the operator. And operator effectiveness is often measured by MOPs. Cyber and traditional JMEMs may therefore be operating at different levels of the operational hierarchy as described by the modeling and simulation hierarchy (Fig. 9.4).

Table 11.2 COLE cyber P_k equation elements

COLE P_k elements	Description
$P_{latency}$	Likelihood of having required supporting intelligence at the right time
P_{access}	Likelihood of knowledge of access points in the targeted system
P_{config}	Likelihood of having needed knowledge of software and hardware configurations in the targeted system
P_{map}	Completeness of network map of targeted system
P_{tempo}	Completeness of understanding operations tempo of targeted system
P_{patch}	Likelihood of a target system IT applying a patch to address software vulnerabilities on the targeted system
P_{IT}	Likelihood of the target system IT being able to detect and respond to a threat actor accessing their system
$P_{exploit}$	Likelihood that current mission exploit will work

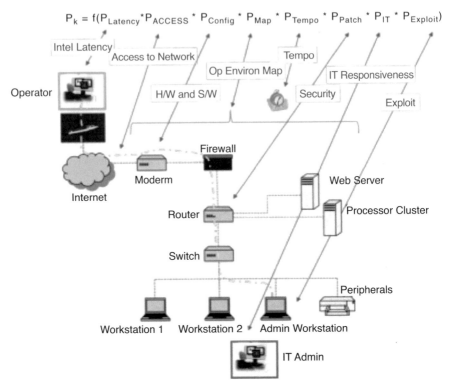

$$P_k = f(P_{Latency} * P_{ACCESS} * P_{Config} * P_{Map} * P_{Tempo} * P_{Patch} * P_{IT} * P_{Exploit})$$

Fig. 11.3 Cyber Operation Lethality and Effectiveness (COLE) Kill Chain (Zurasky, 2017)

While COLE provides a step-by-step approach for a cyber attack, designed to provide a cyber level KPP, this is a measure of the point targeting discussed in Chap. 7. Area targeting, the use of cyber for broader denial applications, may require a different measurement approach than that used for individual system exploits.

11.5.2   Information Operations (IO)—Cyber Targeting via Social Media

The vulnerability analysis and penetration testing in Chap. 3 focused on the technical elements of a computer system. In Chap. 4, we reviewed Soviet era active measures, the disinformation cycle and an adaptation of area/point targeting for messaging via social media. For example, transforming any falsehood into disinformation relies on the use of seemingly legitimate sources to validate a message. This message is then continuously communicated in order to anchor a message into the public's mind as a truth (Fig. 11.4).

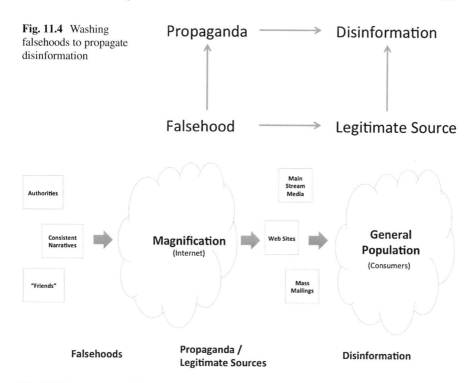

Fig. 11.4 Washing falsehoods to propagate disinformation

Fig. 11.5 Internet magnification of messaging

As shown in Fig. 11.4, disinformation comes from the repetition of a falsehood through propaganda or from a seemingly legitimate source. The Internet can be used to provide implied validity through repetition and the resending of false messages by seemingly authoritative sources. Bad actors can therefore use the Internet to create a barrage of information, making it a challenge to find the truth (Fig. 11.5).

Figure 11.5, an operational example of Pacepa's (Ion Mihai Pacepa, 2013) conceptual model for disinformation (Fig. 11.4), provides an example of the magnification possible for a small group of influencers. As shown, the information operation leverages commercial media, through valid messages, to magnify the number of recipients to be influenced.

While social engineering (Hadnagy, 2018) is a well-known part of any cyber campaign, Fig. 11.6 connects the respective elements of a traditional marketing (Kotler, 2017), or influence, campaign and its cyber components.

As shown in Fig. 11.6, the crossover between a traditional marketing campaign's sequence of Plan/Build/Send/Effect and the Cyber Threat Frameworks' (CTF) (Director of National Intelligence) Preparation/Engagement/Presence/Effect.

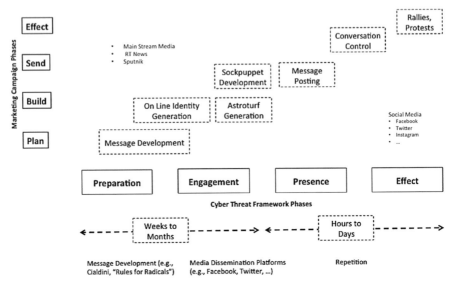

Fig. 11.6 Lifecycle of an influence operation

11.6   Mission Assessment

While measuring cyber effects might be challenged by the scope of potential outcomes, some possibly unexpected, the process that we are following helps focus our efforts on measuring collections, behavior anomalies, or target systems being down. The complexity of cyber systems, including logical, physical and persona aspects, adds additional uncertainty regarding effects estimation. However, Cohen's d, as discussed in Chap. 10, provides a starting point for measuring cyber effects from an observable target.

11.6.1   Effect Size—Cohen's d

While some effects may be relative to the mission/campaign being supported, we want to look at individual cyber effects as measurable entities. For example, effect size should be something measurable, as a reflection of target understanding. Chap. 4's Information Operations showed how active measures can be used to sway opinion, potentially moving the number of people believing a certain message from a minority to the majority.

Determining effect sizes is work that has been addressed in the social sciences by comparing the behavior between two observations. This could be a comparison of two samplings, a before and after for the application of a cyber effect. Cohen's d

(Cohen, 1988) looks at the average difference between collected distributions for two observations to produce an effect size (Eq. 11.1).

$$\theta = \frac{\mu_1 - \mu_2}{\sigma}$$ (11.1)

Effect determination

θ: effect size, standardized mean difference between two populations.

μ_1: mean for one population.

μ_2: mean for other population.

σ: standard deviation based on either or both populations.

In practice, population values for Eq. 11.1 are usually not known and need to be estimated from sample statistics, based on specimen, or target, observations. Sampling system behavior both before and after a cyber attack are therefore required to quantify the effect.

11.6.2   Measure of Effectiveness (MOE)

While observing the before and after of a cyber event may be a challenge for performing effects estimation, we can also use measures of effectiveness (MOE), from Chap. 9, to evaluate an overall operation. One of the advantages of using MOEs is that they might be collected by other means (e.g., open-source collection (Chap. 5)) to assess a mission or campaign. In addition, an MOE that can be derived from an assessment that uses the original mission context as a reference point.

An example MOE might include rolling up a set of collections to create a "big data" cyber intelligence system, as discussed in Chap. 5, that results in searches with meaningful data that show connections and entities not available with individual data set analysis. Similarly, assessing effects like organizational trust or maneuver may require collection from multiple sources in order to prove that the cyber campaign was effective in affecting an organization in terms of time/cost.

Qualitative cyber effects have an analog in the 1999 NATO bombing of Serbia (Hosmer, 2001). For example, the general bombing of military installations was having limited effects in slowing down Serbian operations. However, bombing the private assets of the friends of Premier Milosevic resulted in the Serbian government coming to the negotiating table in short order. This targeting analysis required strategic assessment, understanding how the levers of power actually worked in the Milosevic regime, in order to provide an effective targeting scheme that delivered the desired effects.

11.7   Summary

Targeting is a challenge regardless of the domain. For example, understanding what is important to the target, the opinion of Milosevic's friends in the 1999 NATO Serbian bombing example can be a challenge to discern when beginning a campaign. Similar issues in understanding the overall context for a targeting mission also occur in the case of cyber targeting for determining the key target elements associated with important addressable processes and measurable mission and campaign level effects.

Effect estimation remains a challenge for cyberspace, as the observations before and after the application of a cyber mission are required to make an effective comparison, in terms of target process behavioral change. We can also look at measures of effectiveness (MOEs), effects on a target as a result of an overall cyber campaign. In Chap. 8, for example, we talked about how a cyber campaign might undermine organizational trust, an attribute that is hard to quantify.

Quantifying cyber effects is a challenge due to, sometimes, qualitative nature of the phenomena being observed. For example, describing operation execution is provided by evaluating the system's key performance parameters (KPPs) and operator measures of performance (MOPs). In Chap. 9, we looked at system level KPPs in terms of stealthy operation. This ensures that the system can maintain a low profile while operating as a cyber collection system. Similarly, we measured the amount of data collected during an engagement as an example MOP for a hypothetical collection mission.

We can also develop MOPs for cyber-based messaging campaigns. As shown in Fig. 11.6, messaging campaigns look a lot like marketing campaigns. Getting out an effective message, similar to a good advertising "jingle," is a challenge to develop and get in the right context so that the audience immediately gets it, possibly using it in everyday conversation as accepted knowledge. The Internet can make the process easier by lowering the cost of developing a message and using social media to propagate the message in more pliable channels (Fig. 11.4).

While social media message dissemination corresponds to advertising, constructing cyber access and maneuver tools corresponds to software development. For example, as shown in Fig. 11.2, using estimated target vulnerabilities to develop exploits, and design them into a tool, is key to the process that a penetration tester might use to get technical access to our systems. A commercial example of this is Metasploit, a tool that uses published vulnerability information to develop exploits and package them for commercial use. Metasploit includes these exploits in a tool for penetration testers to use at the click of a button. In addition, penetration testers also have tools like Shodan (Shodan) that show Internet facing nodes associated with industrial control systems (ICS). Tool development for accessing networks, even the sensitive controllers for our critical infrastructure, are widely available to penetration testers through openly published sources.

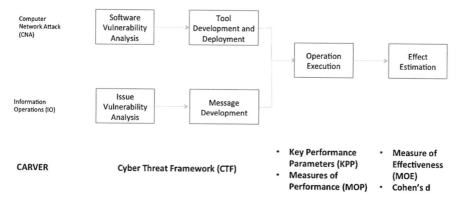

Fig. 11.7 Chapter use case examples and the analysis and targeting process

Determining which target systems to focus tool development efforts on, however, results from system vulnerability analysis based on current targeting objectives. Both software system and organizational issue vulnerabilities result from early target selection, based on the objectives and goals of a cyber operation. We reviewed CARVER (Table 11.1) as an approach for determining key nodes based on a set of time proven assessment criteria. In addition, we discussed how a cyber response fits within current policy frameworks (e.g., "The Line" (Fig. 11.1)). Developing a cyber response, based on current policy, process and technology formulations is provided in Fig. 11.7.

As shown in Fig. 11.7, CARVER provides initial guidance concerning the type of node, or target, that will most likely meet the intent of a cyber operation. If a computer network attack (CNA) makes the most sense, a software/system analysis will help determine the types of vulnerabilities that will be used for tool development and deployment. Similarly, if information operations (IO) are more likely to provide the effect, the issues to be discussed will be analyzed, with subsequent message development. Executing the operation is slightly different, in that the technical operation will consist of accessing the target node (e.g., Fig. 11.3), and the IO operation will look more like a marketing campaign (e.g., Fig. 11.6). Estimating the effects, however, may be similar, in the cases of relatively abstract measures of effectiveness (e.g., decreasing organization internal trust). Each of these examples is provided in Table 11.3.

As shown in Table 11.3, in the development of approaches for cyber analysis and targeting, we have reviewed and built upon, existing policies, process, and processes that leverage existing cyber doctrine, static and dynamic modeling approaches, and theory concerning the application of cyber as a munition.

Table 11.3 Policy, process, and technology use cases and exemplars for cyber analysis and targeting

	Use case/exemplar
Policy	"The line"—Threshold between diplomacy and kinetic action (Chap. 2)
Process	Cyber threat framework (Chap. 3) DoDCAR for cyber system architecture construction (Chap. 3) System's "V-curve" for risk identification and remediation (Chap. 3) IO area/point targeting and legacy active measures (Chap. 4) Cyber implemented active measures (Chap. 4) Six marketing principles for social engineering (i.e., Cialdini) (Chap. 4) Internet magnification of messaging (Chap. 4) Cyber process evaluator (Chap. 7) Architecture description of cyber systems (Chap. 8) Measures of cyber system effectiveness (Chap. 9) Series/parallel description of cyber processes (Chap. 10)
Technology	Tool development (Chap. 3) Cyber "big Data" systems (Chap. 5) Cyber collection system examples (Chap. 8 and 9) Cyber as a next step in the precision guided munition spectrum (e.g., for KPP, MOP, and MOE development) (Chap. 9)

Bibliography

U.S.-CHINA ECONOMIC and SECURITY REVIEW COMMISSION. (2011, 8 14). *2011 ANNUAL REPORT TO CONGRESS.* Retrieved April 10, 2019, from U.S.-CHINA ECONOMIC and SECURITY REVIEW COMMISSION: https://www.uscc.gov/content/2011-annual-report-congress

9/11 Commission. (2004). *The 9/11 Commission Report.* Retrieved August 4, 2019, from https://www.9-11commission.gov/report/911Report.pdf

Acton, J. M. (2017). Cyber weapons and precision guided munitions. In A. L. G. Perkovich (Ed.), *Understanding cyber conflict.* Georgetown.

Aid, M. (2013, October 15). The NSA's new code breakers. Foreign Policy .

Akamai. (n.d.). *Global Traffic Management.* Retrieved November 14, 2018, from Akamai: https://www.akamai.com/us/en/multimedia/documents/product-brief/global-traffic-management-product-brief.pdf

Alinsky, S. (1989). *Rules for radicals: A practical primer for realistic radicals.* Vintage.

Anderson, N. (2010, 11 7). *How China swallowed 15% of 'Net traffic for 18 minutes.* Retrieved November 13, 2018, from Ars Technica: https://arstechnica.com/information-technology/2010/11/how-china-swallowed-15-of-net-traffic-for-18-minutes/

Andrei Soldatov, I. B. (2015a). *The red web - the struggle between Russia's digital dictators and the new online revolutionaries.* Public Affairs.

Andrei Soldatov, I. B. (2015b). *The red web - the struggle between Russia's digital dictators and the new online revolutionaries.* Public Affairs.

Angus King, M. G. (2020, 3). *United States of America cyberspace solarium Commission.* Retrieved May 9, 2020, from The Cyberspace Solarium Commission: https://www.solarium.gov/

Arquilla, J. (2017). From Pearl Harbor to the "harbor lights". In A. E. George Perkovich, Understanding cyber conflict - 14 analogies. : Georgetown.

Australian Government. (2018). *Cyber security strategy.* Retrieved March 10, 2019, from Cyber security strategy: https://cybersecuritystrategy.homeaffairs.gov.au/executive-summary-0

Whaley, B., & S. A. (2007). *Textbook of political-military Counterdeception: Basic principles and methods.* National Defense Intelligence College.

Barnes, J. E. (2019, 8 18). U.S. cyberattack hurt Iran's ability to target oil tankers, officials say. New York times.

Watts, B. D., & T. A. (1993). *Effects and effectiveness*. US Government Printing Office.

BBC. (2016, October 27). *18 revelations from Wikileaks' hacked Clinton emails*. Retrieved August 21, 2018, from BBC: https://www.bbc.com/news/world-us-canada-37639370

BBC News. (2018, 3 28). *BBC News*. Retrieved July 6, 2019, from Islamic state and the crisis in Iraq and Syria in maps: https://www.bbc.com/news/world-middle-east-27838034

Ben Collins, G. R. (2018, 3 1). *Leaked: Secret Documents From Russia's Election Trolls*. Retrieved September 9, 2018, from Daily Beast: https://www.thedailybeast.com/exclusive-secret-documents-from-russias-election-trolls-leak?ref=scroll

Bennett, C. (1995). *How Yugoslavia's destroyers harnessed the media*. Retrieved August 27, 2018, from PBS frontline: https://www.pbs.org/wgbh/pages/frontline/shows/karadzic/bosnia/media.html

Berman Klein Center. (n.d.). *GhostNet*. Retrieved April 11, 2019, from GhostNet: https://cyber.harvard.edu/cybersecurity/GhostNet

Bernstein, J. (2017). *Secrecy world - inside the Panama papers investigation of illicit money networks and the global elite*. Henry Holt and Company.

Biddle, S. (2004). *Military power: Explaining victory and defeat in modern Battle*. Princeton University Press.

Blank, S. (2017). Cyber war and information war a la Russe. In A. E. G. Perkovich (Ed.), *Understanding cyber conflicT*. Georgetown.

Blank, S. (2013). Russian information warfare as domestic counterinsurgency. *American Foreign Policy Interests, 35*(1), 31–44.

Borland, J. (2019, 6 5). *As Europe Went to the Polls, Cyber Election Efforts Paid Off*. Retrieved June 28, 2020, from Symantec: https://symantec-enterprise-blogs.security.com/blogs/election-security/europe-went-polls-cyber-election-efforts-paid?om_ext_cid=biz_social3_EMEA_EMEA-All_facebook_EMEA-All,election-security,FY20-Q1,Blog,EMEA,org.

Boston Globe. (1998, March 19). *Teen Hacker Pleads Guilty to Crippling Massachussetts Airport*. Boston Globe .

Bowden, M. (2011). *Worm - the first digital world war*. Atlantic Monthly Press.

Bradley Greaver, L. R. (2017). CARVER 2.0: Integrating the analytical hierarchy Process's multi-attribute decision-making weighting scheme for a center of gravity vulnerability analysis for US special operations forces. *Journal of Defense Modeling and Simulation, 15*(1), 111–120.

Brenner, B. (2005, 8 31). *Myfip's titan rain connection*. Retrieved April 10, 2019, from SearchSecurity: https://searchsecurity.techtarget.com/news/1120855/Myfips-Titan-Rain-connection

Brian Grow, M. H. (2011, 4 11). *Special report: In cyberspy vs. cyberspy, China has the edge*. Retrieved April 10, 2019, from Reuters: https://www.reuters.com/article/us-china-usa-cyberespionage/special-report-in-cyberspy-vs-cyberspy-china-has-the-edge-idUSTRE73D24220110414

Brooks, C. (2011, October 9). Cybersecurity experts say small buisinesses beware. Business News Daily.

Bryan Krekel, P. A. (2012, 3 12). *Occupying the information high ground: Chinese capabilities for computer network operations and Cyber espionage*. Retrieved April 10, 2019, from U.S.-CHINA ECONOMIC and SECURITY REVIEW COMMISSION: https://www.uscc.gov/Research/occupying-information-high-ground-chinese-capabilities-computer-network-operations-and

Burgess, M. (2016, 4 6). *Panama papers: The security flaws at the heart of Mossack Fonseca*. Retrieved June 25, 2020, from Wired.

Soderbergh, S., Burns, S. Z., & Soderbergh, S. (2019). *The laundromat [motion picture]*. USA.

Butsenko, A. (2014, October 28). Trolli iz Olgino pereekhali v noviy chetyrekhatazhny office na Savushkina (trolls from Olgino moved to a new four story office on Savushkina). DP.ru .

Carlin, J. P. (2018). *Dawn of the code war - America's Battle against Russia, China and the rising global Cyber*. PublicAffairs.

Carr, J. (2012a). *Inside Cyber warfare*. O'Reilly.

Carr, J. (2012b). *Inside Cyber warfare*. O'Reilly Media.

Carr, J. (2012c). *Inside Cyber warfare: Mapping the Cyber underworld*. O'Reilly Media.

Carr, J. (n.d.). *Jeffrey Carr*. Retrieved July 8, 2019, from Wikipedia: https://en.wikipedia.org/wiki/Jeffrey_Carr

Theohary, C. A.,& J. R. (2011, 3 8). *Terrorist Use of the Internet: Information Operations in Cyberspace*. Retrieved August 22, 2018, from Congressional Research Service: https://digital.library.unt.edu/ark:/67531/metadc103142/m1/1/high_res_d/R41674_2011Mar08.pdf

CBS News. (2003, 8 19). *SoBig virus may be spam scam*. Retrieved June 27, 2020, from CBS News: https://www.cbsnews.com/news/sobig-virus-may-be-spam-scam/

Center for Computational Analysis of Social and Organizational Systems (CASOS). (n.d.). *Center for Computational Analysis of Social and Organizational Systems (CASOS)*. Retrieved October 14, 2018, from http://www.casos.cs.cmu.edu/

Chalfant, M. (2017, 11 20). *Dems call for states to get $400 M election security upgrades*. Retrieved June 6, 2019, from Hill: http://thehill.com/policy/cybersecurity/361263-dems-say-congress-should-send-400m-to-states-for-election-cyber-upgrades

Chaos Computer Club. (n.d.). *Chaos Computer Club*. Retrieved from https://www.ccc.de/en

Chapman, J. (1909). *Doctors of the church*. Robert Appleton Company.

Chinascope. (2019). *The Passionate Time of Chinese Hackers*. Retrieved July 7, 2019, from Chinascope: http://chinascope.org/archives/6520

Cialdini, R. B. (2007). *Influence - the psychology of persuasion*. Collins Business.

Clapper, J. R. (2015). *Statement for the record: Worldwide threat assessment of the US intelligence community. Senate armed services Committee*. Senate Armed Services Committee.

Clayton, M. (2012, Sep 12). Stealing US business secrets: Experts ID two huge cyber "gangs" in China. Christian Science Monitor .

Cleary, G. (2019, 6). *Twitterbots: Anatomy of a propaganda campaign*. Retrieved July 6, 2019, from Symantec: https://www.symantec.com/blogs/threat-intelligence/twitterbots-propaganda-disinformation

Clement, P. (2019, 7 25). *Internet usage worldwide - statistics & facts*. Retrieved June 12, 2020, from Statista: https://www.statista.com/topics/1145/internet-usage-worldwide/

Cohen, J. (1988). *Statistical power analysis for the behavioral sciences*. Lawrence Earlbaum Associates.

Corera, G. (2016). *Cyber Spies - the sSecret history of surveillance, hacking and digital espionage*. Pagasus Books.

Council on Foreign Relations. (2005). Titan rain. Council on Foreign Relations.

Couretas, J. (2017). *A developing science of Cyber security - an opportunity for model based engineering and design. NDIA systems engineering conference (p. 55)*. NDIA.

Couretas, J. M. (2018). *Introduction to Cyber Modeling and simulation*. John Wiley and Sons.

CrowdStrike. (n.d.). Retrieved November 29, 2018, from https://www.crowdstrike.com/

Crowdstrike. (2016, 6 15). *Bears in the Midst: Intrusion into the Democratic National Committee*. Retrieved September 21, 2019, from Crowdstrike: https://www.crowdstrike.com/blog/bears-midst-intrusion-democratic-national-committee/

CrowdStrike. (2015, 2 6). *CrowdStrike's 2014 Global Threat Intel Report: Know Your Adversary and Better Protect Your Network*. Retrieved April 10, 2019, from CrowdStrike: https://www.crowdstrike.com/blog/crowdstrikes-2014-global-threat-report-know-adversary-better-protect-network/

Darnay, K. (2014, 1 27). *A look at website lifespans*. Retrieved July 8, 2019, from Bismarck tribune: https://bismarcktribune.com/news/columnists/keith-darnay/a-look-at-website-lifespans/article_1d879ae6-851a-11e3-8bd1-0019bb2963f4.html

Data Center Map. (n.d.). *Data Center Map*. Retrieved March 13, 2019, from https://www.datacentermap.com/ixps.html

Sanger, D. E., E. S. (2017, 6 12). U.S. *Cyberweapons, used against Iran and North Korea, are a disappointment against ISIS*. Retrieved August 27, 2018, from New York times: https://www. nytimes.com/2017/06/12/world/middleeast/isis-cyber.html

Sanger, D. E., J. R. (2018, July 15). *Tracing Guccifer 2.0's many tentacles in the 2016 election*. Retrieved September 9, 2018, from New York times: https://www.nytimes.com/2018/07/15/us/ politics/guccifer-russia-mueller.html

Sanger, D. E., N. P. (2015, 4 15). *Iran is raising sophistication and frequency of cyberattacks, study says*. Retrieved October 14, 2018, from New York times: https://www.nytimes.com/2015/04/16/ world/middleeast/iran-is-raising-sophistication-and-frequency-of-cyberattacks-study- says.html

David Leigh, L. H. (2011). *WikiLeaks - inside Julian Assange's war on secrecy*. Public Affairs.

Davis, J. (2007, 8 21). Hackers take down the Most wired country in Europe. Retrieved July 7, 2019, from WIRED: https://www.wired.com/2007/08/ff-estonia/

Denning, D. (2017, 10 5). *Cyberwar: How Chinese hackers became a major threat to the U.S.* Retrieved 6 26, 2020, from Newsweek: https://www.newsweek.com/ chinese-hackers-cyberwar-us-cybersecurity-threat-678378

Department of Defense (DoD). (2018). *DoD Cybersecurity Strategy*. Retrieved March 10, 2019, from DoD Cybersecurity Strategy: https://media.defense.gov/2018/Sep/18/2002041658/-1/-1/1/ CYBER_STRATEGY_SUMMARY_FINAL.PDF

Department of Defense. (2012). *Joint Publication 1–13.4 Military Deception*. Department of Defense.

Department of Homeland Security (DHS). (2018, 5 15). *DHS Cybersecurity Strategy*. Retrieved March 10, 2019, from DHS Cybersecurity Strategy: https://www.dhs.gov/sites/default/files/ publications/DHS-Cybersecurity-Strategy_1.pdf

Department of Justice. (2014, 5 19). *U.S. Charges Five Chinese Military Hackers For Cyber Espionage Against U.S. Corporations And A Labor Organization For Commercial Advantage*. Retrieved June 25, 2020, from https://www.justice.gov/usao-wdpa/pr/ us-charges-five-chinese-military-hackers-cyber-espionage-against-us-corporations-and

Spiegel, D. (2013a, December 30). Documents reveal top NSA hacking unit. Der Spiegel .

DHS CISA. (n.d.-a). *Alert (TA18-275A) - HIDDEN COBRA – FASTCash campaign*. Retrieved June 11, 2020, from DHS CISA: https://www.us-cert.gov/ncas/alerts/TA18-275A

DHS CISA. (n.d.-b). *Chinese malicious Cyber activity*. Retrieved May 23, 2020, from DHS CISA: https://www.us-cert.gov/china

DHS CISA. (2016, 2 25). *Cyber-attack against Ukrainian critical infrastructure (ICS Alert (IR-ALERT-H-16-056-01))*. Retrieved June 27, 2020, from ICS Alert (IR-ALERT-H-16-056-01): https://www.us-cert.gov/ics/alerts/IR-ALERT-H-16-056-01

DHS CISA. (2018, 12 18). *HIDDEN COBRA – FASTCash campaign*. Retrieved May 23, 2020, from DHS CISA: https://www.us-cert.gov/ncas/alerts/TA18-275A

DHS CISA. (2017, 8 23). *HIDDEN COBRA – North Korea's DDoS botnet infrastructure*. Retrieved May 23, 2020, from DHS CISA: https://www.us-cert.gov/ncas/alerts/TA17-164A

DHS CISA. (n.d.-c). *North Korean malicious Cyber activity*. Retrieved May 23, 2020, from DHS CISA: https://www.us-cert.gov/northkorea

DHS Science and Technology (S&T) Directorate. (n.d.). *Telephony Denial of Service*. Retrieved June 27, 2020, from DHS Science and Technology (S&T) Directorate: https://www.dhs.gov/ sites/default/files/publications/508_FactSheet_DDoSD_TDoS%20One%20Pager-Final_ June%202016_0.pdf

Director of National Intelligence. (n.d.-a). *Cyber Threat Framework*. Retrieved October 20, 2018, from https://www.dni.gov/index.php/cyber-threat-framework

Director of National Intelligence. (n.d.-b). *Cyber Threat Framework*. Retrieved February 15, 2019, from Cyber Threat Framework: https://www.dni.gov/index.php/cyber-threat-framework

Diresta, R. (2018, 3 8). *How ISIS and Russia won friends and manufactured Crowds*. Retrieved July 7, 2019, from wired: https://www.wired.com/story/isis-russia-manufacture-crowds/

Djabatey, E. (2019, 10 17). *Reassessing U.S. Cyber operations against Iran and the use of force*. Retrieved October 20, 2019, from just security: https://www.justsecurity.org/66628/reassessing-u-s-cyber-operations-against-iran-and-the-use-of-force/

Dmitry Volchek, D. S. (2015, March 27). One professional Russian troll tells all. Radio Free Europe / Radio Liberty .

DoD CIO. (1996, 8). *Clinger Cohen Act of 1996*. Retrieved June 12, 2020, from DoD CIO: https://dodcio.defense.gov/Portals/0/Documents/ciodesrefvolone.pdf

DoD. (2001, 10). *DoDD 3600.01 Information Operations*. Retrieved March 10, 2019, from Homeland Security Digital Library: https://www.hsdl.org/?abstract&did=439849

Doman, C. (2016, 7 6). *The first Cyber espionage attacks: How operation moonlight maze made history*. Retrieved August 4, 2019, from medium: https://medium.com/@chris_doman/the-first-sophistiated-cyber-attacks-how-operation-moonlight-maze-made-history-2adb12cc43f7

Domscheit-Berg, D. (2011). *Inside Wikileaks - my time with Julian Assange at the World's Most dangerous website*. Crown.

Denning, D. E., & B. J. (2017). Active cyber defense - applying air defense to the cyber domain. In A. E. G. Perkovich (Ed.), *Understand Cyber conflict - 14 analogies*. Georgetown.

Dragos. (n.d.). *Dragos*. Retrieved November 29 2018, from https://dragos.com/

Dustin Volz, J. F. (2016, March 24). *U.S. indicts Iranians for hacking dozens of banks*, New York dam: Reuters.

Economist. (2007, 7 12). *A world wide web of terror*. Retrieved August 11, 2019, from Economist: https://www.economist.com/briefing/2007/07/12/a-world-wide-web-of-terror

Eeten, M. V. (2010). *The role of internet service providers in botnet mitigation: an empirical analysis based on spam data*. OECD Science, Technology and Industry Working Papers , 05.

Egloff, F. (2017). Cybersecurity and the age of privateering. In A. E. G. Perkovich (Ed.), *Understanding Cyber conflict - 14 analogies*. Georgetown.

Goldman, E. M., & M. W. (2017). Why a digital Pearl Harbor makes sense. In A. E. G. Perkovich (Ed.), *Understanding Cyber conflict*. Georgetown.

England, R. (2019, 8 13). *UN claims North Korea hacks stole $2 billion to fund its nuclear program*. Retrieved August 18, 2019, from Engadget: https://www.engadget.com/2019/08/13/un-claims-north-korea-hacks-stole-2-billion-to-fund-its-nuclear/

Hutchins, E. M., M. J. (2012). *Intelligence-driven computer network Defense informed by analysis of adversary campaigns and intrusion kill chains*. Retrieved August 6, 2019, from Lockheed Martin: https://www.lockheedmartin.com/content/dam/lockheed-martin/rms/documents/cyber/LM-White-Paper-Intel-Driven-Defense.pdf

European Union. (2018). *General Data protection regulation (GDPR)*. Retrieved June 27, 2020, from GDPR: https://gdpr-info.eu/

Feldman, S. (2019, 2 25). *Russia has the fastest hackers*. Retrieved September 3, 2019, from Statistica: https://www.statista.com/chart/17151/government-hack-speed/

Ferguson, N. (2018). *The square and the tower - networks and power, from the freemasons to Facebook*. Penguin.

FireEye. (n.d.-a). *APT 29*. (FireEye, Producer) Retrieved September 9, 2018, from APT 29: https://www.fireeye.com/current-threats/apt-groups.html#apt29

FireEye. (2017a). *APT 29*. (FireEye, Producer) Retrieved September 9, 2018, from APT 29: https://www.fireeye.com/current-threats/apt-groups.html#apt29

FireEye. (2014). *APT28: A WINDOW INTO RUSSIA'S CYBER ESPIONAGE OPERATIONS?* Retrieved September 9, 2018, from FireEye: https://www.fireeye.com/content/dam/fireeye-www/global/en/current-threats/pdfs/rpt-apt28.pdf

FireEye. (n.d.-b). *FireEye*. Retrieved November 29, 2018, from https://www.fireeye.com/

FireEye. (2017b). *M-TRENDS 2017 - A view from the Front lines*. FireEye.

FireEye. (2017c). *M-TRENDS 2017 - A view from the Front lines*. FireEye.

Frank, A. B. (2017). Toward computational net assessment. *Journal of Defense Modeling and Simulation, 14*(1).

Fritz, J. R. (2017). *China's Cyber warfare - the evolution of strategic doctrine*. Lexington Books.

Gallagher, M. (2008). *Cyber analysis workshop. MORS.* MORS.

Gatlan, S. (2019, 5 6). *Israel bombs building as retaliation for Hamas Cyber attack.* Retrieved June 13, 2019, from bleeping computer: https://www.bleepingcomputer.com/news/security/israel-bombs-building-as-retaliation-for-hamas-cyber-attack/

Gavin, F. (2017). Crisis instability and Preemption: The 1914 railroad analogy. In A. E. G. Perkovich (Ed.), *Understanding Cyber conflict - 14 analogies.* Georgetown.

George Cybenko, G. S. (2016). Quantifying covertness in deceptive Cyber operations. In V. S. S. Jajodia (Ed.), *Cyber deception: Building the scientific foundation.* Springer.

George Cybenko, J. S. (2007). *Quantitative foundations for information operations.*

George Perkovich, A. E. (2017). *Understanding Cyber conflict - 14 analogies.* Georgetown.

George Washington University. (n.d.). *National Security Archive.* Retrieved January 30, 2020, from https://nsarchive.gwu.edu/news/cyber-vault/2018-11-07/presidential-orders

Georgia Tech Research Institute. (2012). Cyber threats report 2012. *Georgia Tech information security center, Georgia Tech Cyber Security Summit, 2011.*

Glenny, M. (2012). *DarkMarket: How hackers Bacame the new mafia.* Vintage.

Global Security. (2005). *Global security.* Retrieved from Letter from al-Zawahiri to al-Zarqawi: https://www.globalsecurity.org/security/library/report/2005/zawahiri-zarqawi-letter_9jul2005.htm

Government of Canada. (2018). *National Cyber Security Strategy.* Retrieved March 10, 2019, from National Cyber Security Strategy: https://www.publicsafety.gc.ca/cnt/rsrcs/pblctns/ntnl-cbr-scrt-strtg/index-en.aspx

Graff, G. (2018, 10 31). *CHINA'S 5 STEPS FOR RECRUITING SPIES.* Retrieved July 7, 2019, from Wired: https://www.wired.com/story/china-spy-recruitment-us/

Greenberg, A. (2017, 6 22). *Experts suspect Russia is using Ukraine as A cyberwar testing ground.* Retrieved June 27, 2020, from NPR: https://www.npr.org/2017/06/22/533951389/experts-suspect-russia-is-using-ukraine-as-a-cyberwar-testing-ground

Greenberg, A. (2019a). *Sandworm: A new era of cyberwar and the hunt for the Kremlin's Most dangerous hackers.* Doubleday.

Greenberg, A. (2019b). *Sandworm: A new era of cyberwar and the hunt for the Kremlin's Most dangerous hackers.* Doubleday.

Greenwald, G. (2014). *No place to Hide: Edward Snowden, the NSA and the US surveillance state.* Hamish Himilton.

Gregory Conti, D. R. (2017a). *On Cyber: Towards an operational art for Cyber conflict.* Kopidion Press.

Gregory Conti, D. R. (2017b). *On Cyber: Towards an operational art for Cyber conflict 1st edition.* Kopidion Press.

Koblentz, G. D., & B. M. (2013). Viral warfare: The security implications of Cyber and biological weapons. *Comparative Strategy, 32*(5).

Hadnagy, C. (2018). Social engineering: The science of human hacking.

Hancock, D. (2003, 8 21). *Virus disrupts train signals.* Retrieved July 7, 2019, from CBS News: https://www.cbsnews.com/news/virus-disrupts-train-signals/

Harris, G. (2018, 3 4). *State Dept. was granted $120 million to fight Russian meddling. It has spent $0.* Image. Retrieved September 9, 2018, from New York times: https://www.nytimes.com/2018/03/04/world/europe/state-department-russia-global-engagement-center.html

Hayden, M. V. (2016). *Playing to the edge: American intelligence in the age of terror.* Peguin.

Heickero, R. (2013). *Emerging Cyber threats and Russian views on information warfare and information operations.* Swedish Defense Research Agency.

Heli Tiirmaa-Klaar, J. G.-P. (2014). *Botnets.* Springer.

Henderson, S. (2007). *The dark visitor.* Scott Henderson.

Hibbs, E. (2019, May 15). *Systems engineering in a Cyber world - connecting frameworks for program decisions.* Retrieved March 8, 2020, from DISA: https://www.disa.mil/-/media/Files/DISA/News/Events/Symposium-2019/1%2D%2D-Hibbs_Systems-Engineering-in-a-Cyber-World_approved-Final.ashx+&cd=10&hl=en&ct=clnk&gl=us

HM Government. (2016). *National Security Strategy 2016–2021*. Retrieved March 10, 2019, from National Security Strategy 2016–2021: https://assets.publishing.service.gov.uk/government/uploads/system/uploads/attachment_data/file/567242/national_cyber_security_strategy_2016.pdf

Hollis, D. (2011, January 11). Cyberwar case study: Georgia 2008. Small Wars Journal.

Hosmer, S. T. (2001). Why Milosevic decided to settle when he did. RAND.

Hsiao, R. (2013, 12 5). *Critical node: Taiwan's Cyber Defense and Chinese Cyber-espionage.* Retrieved April 10, 2019, from Jamestown foundation: https://jamestown.org/program/critical-node-taiwans-cyber-defense-and-chinese-cyber-espionage/

IDS International. (n.d.). *SMEIR*. Retrieved January 3, 2019, from https://www.smeir.net/

IEEE. (2011, 12 1). *42010–2011 - ISO/IEC/IEEE Systems and software engineering -- Architecture description*. Retrieved June 13, 2020, from IEEE: https://standards.ieee.org/standard/42010-2011.html

International Telecommunications Union. (2013, May 14–16). *Internet Exchange Points (IXPs)*. World Telecommunication Policy Forum.

Internet Exchange Map. (n.d.). *Internet Exchange Map*. Retrieved March 13, 2019, from https://www.internetexchangemap.com/

Ion Mihai Pacepa, R. J. (2013). *Disinformation - former spy chief reveals secret strategies for undermining freedom, attacking religion and promoting terrorism*. WND.

Isikoff, M. (2013, 6 10). Chinese hacked Obama, McCain Campaigns, took internal documents, officials say.

Jarvis, J. (2011, 3 17). *Revealed: US spy operation that manipulates social media - Military's 'sock puppet' software creates fake online identities to spread pro-American propaganda*. Retrieved January 3, 2019, from Guardian: https://www.theguardian.com/technology/2011/mar/17/us-spy-operation-social-networks

Jen Weedon, W. N. (2017, 4 27). *Information operations and Facebook*. Retrieved January 3, 2019, from Facebook: https://fbnewsroomus.files.wordpress.com/2017/04/facebook-and-information-operations-v1.pdf

Daly, J. J., & J. C. (2015). *Using conceptual Modeling to implement model based systems engineering for program capability analysis and assessment. NDIA systems engineering conference.* NDIA.

John Wu, D. I. (2013). *Introduction to computer networks and cybersecurity*. CRC Press.

Johnston, P. B., & Sarbahi, A. K. (2016). The impact of US drone strikes on terrorism in Pakistan. *International Studies Quarterly, 60*(2), 203–219.

Joint Staff. (2016). *CJCSI 3370.01B target development standards*. Joint staff.

Joint Staff. (2018, 6 8). *Joint publication 3-12 cyberspace operations*. Retrieved September 16, 2019, from joint publications: https://www.jcs.mil/Portals/36/Documents/Doctrine/pubs/jp3_12.pdf

Katelyn Polantz, S. C. (2018, 7 14). *12 Russians indicted in Mueller investigation*. Retrieved September 9, 2018, from CNN: https://www.cnn.com/2018/07/13/politics/russia-investigation-indictments/index.html

Keith Collins, S. F. (2018, September 4). *Can You Spot the Deceptive Facebook Post?* Retrieved November 15, 2018, from New York Times: https://www.nytimes.com/interactive/2018/09/04/technology/facebook-influence-campaigns-quiz.html

Geers, K. (2015). *Cyber war in perspective: Russian aggression against Ukraine*. NATO Cooperative Cyber Defence Centre of Excellence.

Kevin Liptak, T. S. (2015, 6 6). *China might be building vast database of federal worker info, experts say*. Retrieved July 7, 2019, from CNN: https://www.cnn.com/2015/06/04/politics/federal-agency-hacked-personnel-management/index.html

Koerner, B. I. (2016a, 10 23). *Inside the Cyberattack That Shocked the US Government*. Retrieved from Wired: https://www.wired.com/2016/10/inside-cyberattack-shocked-us-government/

Koerner, B. I. (2016b, October 23). *Inside the cyberattack that shocked the US government.* Retrieved September 7, 2018, from Wired: https://www.wired.com/2016/10/inside-cyberattack-shocked-us-government/

Kotler, P. (2017). Marketing management. .

Krebs, B. (2011, 6 11). *$72M scareware ring used Conficker worm.* Retrieved June 27, 2020, from KrebsnSecurity: https://krebsonsecurity.com/2011/06/72m-scareware-ring-used-conficker-worm/

Krebs, B. (2013, 6 13). *Iranian elections bring lull in Bank attacks.* Retrieved September 14, 2019, from KrebsonSecurity: https://krebsonsecurity.com/2013/06/iranian-elections-bring-lull-in-bank-attacks/

Krebs, B. (2014). *Spam nation: The inside story of organized cybercrime - from global epidemic to your front door.* Surcebooks.

Krekel, B. (2009a). *Capability of the People's Republic of China to Conduct Cyber Warfare and Computer Network Exploitation.* Retrieved April 11, 2019, from https://nsarchive2.gwu.edu/NSAEBB/NSAEBB424/docs/Cyber-030.pdf

Krekel, B. (2009b, 10 9). *Capability of the People's Republic of China to Conduct Cyber Warfare and Computer Network Exploitation.* Retrieved April 10, 2019, from The US-China Economic and Security Review Commission: https://nsarchive2.gwu.edu/NSAEBB/NSAEBB424/docs/Cyber-030.pdf

Lambert, N. A. (2017). Brits-Krieg: The strategy of economic warfare. In A. L. G. Perkovich (Ed.), *Understanding Cyber conflict - 14 analogies.* Georgetown.

Lamothe, D. (2017, 12 16). *How the Pentagon's cyber offensive against ISIS could shape the future for elite U.S. forces.* Retrieved August 27, 2018, from Washington post: https://www.washingtonpost.com/news/checkpoint/wp/2017/12/16/how-the-pentagons-cyber-offensive-against-isis-could-shape-the-future-for-elite-u-s-forces/?utm_term=.8cce44e017f9

Landau, S. (2014). Under the radar: NSA's efforts to secure private-sector telecommunications infrastructure. *Journal of National Security Law and Policy, 7.*

Leversage, D. (n.d.). *RiSI.* Retrieved September 11, 2019, from RISI: https://www.risidata.com/Database

Lewis, D. (2015, March 31). *Heartland Payment Systems Suffers Data Breach.* Forbes.

Lewis, J. (2012). *Significant cyber incidents since 2006.* Center for Strategic and International Studies.

Lynn Arnhart, M. K. (2019). *Analytics, operations Research and strategic decision making in the.* Military.

MacEslin, D. (2006). *Methodology for determining EW JMEM.* TECH TALK.

Maeve Duggan, J. B. (2012, 2 14). *The Demographics of Social Media Users — 2012.* Retrieved March 13, 2019, from Pew Research: http://www.pewinternet.org/2013/02/14/the-demographics-of-social-media-users-2012/

Maidment, P. (2009, 3 29). *GhostNet in the machine.* Retrieved April 10, 2019, from Forbes: https://www.forbes.com/2009/03/29/ghostnet-computer-security-internet-technology-ghostnet.html#455f71b7d00e

Major Bradley Greaver, M. L. (2018). CARVER 2.0: integrating the Analytical Hierarchy Process's multi-attribute decision-making weighting scheme for a center of gravity vulnerability analysis for US Special Operations Forces. *Journal of Defense Modeling and Simulation, 15*(1).

Mandiant. (2013). *APT1.* Retrieved April 10, 2019, from Mandiant: https://www.fireeye.com/content/dam/fireeye-www/services/pdfs/mandiant-apt1-report.pdf

Mark Gallagher, M. H. (2013). Cyber joint munitions effectiveness manual (JMEM). *Modeling and Simulation Journal.*

Mark Stokes, J. L. (2011, 11 11). *The Chinese People's Liberation Army Signals Intelligence and Cyber Reconnaissance Infrastructure.* Retrieved April 10, 2019, from Project 2049 Institute: https://project2049.net/2011/11/11/the-chinese-peoples-liberation-army-signals-intelligence-and-cyber-reconnaissance-infrastructure/

Maurer, T. (2018, 7 27). *Cyber proxies and their implications for Liberal democracies*. Retrieved August 17, 2019, from WASHINGTON QUARTERLY : https://carnegieendowment. org/2018/07/27/cyber-proxies-and-their-implications-for-liberal-democracies-pub-76937

Maurer, T. (2016, 12 17). *'Proxies' and cyberspace*. Retrieved August 17, 2019, from Carnegie endowment for international peace: https://carnegieendowment.org/2016/12/17/ proxies-and-cyberspace-pub-66532

Maurer, T. (2014, 11 14). *The future of war: Cyber is expanding the Clausewitzian spectrum of conflict*. Foreign Policy.

Max Boot, M. D. (2013). Political warfare. *Council on Foreign Relations*.

Mazetti, M. (2018, July 13). 12 Russian agents indicted in Mueller investigation.

McGee, M. K. (2017, January 10). *A new in-depth analysis of anthem breach*. Retrieved September 7, 2018, from Bank info security: https://www.bankinfosecurity.com/ new-in-depth-analysis-anthem-breach-a-9627

McGlasson, L. (2009, 1 21). *Heartland payment systems, Forcht Bank discover Data breaches*. Retrieved September 9, 2018, from Bank info security: https://www.bankinfosecurity.com/ heartland-payment-systems-forcht-bank-discover-data-breaches-a-1168

Merriam Webster. (n.d.). *Cyber*. Retrieved from https://www.merriam-webster.com/ dictionary/cyber

Merriam-Webster. (n.d.). *Dictionary*. Retrieved January 30, 2020, from www.merriam-webster. com/dictionary

Messmer, E. (1999, 5 12). *Kosovo cyber-war intensifies: Chinese hackers targeting U.S. sites, government says*. Retrieved June 27, 2020, from CNN: http://www.cnn.com/TECH/comput- ing/9905/12/cyberwar.idg/

Gordon, M. R., H. C. (2017, 4 6). *Dozens of U.S. missiles hit Air Base in Syria*. Retrieved August 21, 2019, from New York times: https://www.nytimes.com/2017/04/06/world/middleeast/us- said-to-weigh-military-responses-to-syrian-chemical-attack.html

Michael Riley, L. D. (2012, 7 26). *Hackers linked to China's Army Seen from EU to DC*. Retrieved April 10, 2019, from Bloombert: https://www.bloomberg.com/news/articles/2012-07-26/ china-hackers-hit-eu-point-man-and-d-c-with-byzantine-candor

Microsoft. (2011). How conficker continues to propagate. *Microsoft Security Intelligence Report*.

Microsoft. (n.d.). *Microsoft Cybersecurity Reference Architecture*. Retrieved June 12, 2020, from Microsoft Cybersecurity Reference Architecture: https://virtualizationandstorage.wordpress. com/2020/02/03/microsoft-cybersecurity-reference-architecture-the-microsoft-cybersecurity- reference-architecture/

Middleton, C. (2018, 6 25). *Cyber attack could cost bank half of its profits, warns IMF*. Retrieved August 22, 2018, from Internet of Business: https://internetofbusiness.com/ fintech-cyber-attack-could-cost-bank-half-of-its-profits-warns-imf/

Mir, A., & Moore, D. (2018, 12 13). Drones, surveillance, and violence: Theory and evidence from a US drone program. *International Studies Quarterly*.

MITRE. (n.d.-a). Retrieved from https://cybox.mitre.org/language/version2.0/

MITRE. (n.d.-b). Retrieved from https://www.mitre.org/publications/technical-papers/ standardizing-cyber-threat-intelligence-information-with-the

MITRE. (n.d.-c). *CARET*. Retrieved September 30, 2018, from CARET: https://car.mitre.org/ caret/#/.

MITRE. (n.d.-d). *Cyber analytics repository exploration tool (CARET)*. Retrieved September 30, 2018, from CAR exploration tool: https://car.mitre.org/caret/#/.

MITRE. (n.d.-e). *Lazarus group*. Retrieved June 11, 2020, from MITRE ATT@CK: https://attack. mitre.org/groups/G0032/

MITRE. (n.d.-f). *MITRE ATT@CK framework*. Retrieved September 20, 2019, from MITRE: https://attack.mitre.org/

MITRE. (n.d.-g). *The Cyber analytics repository (CARET)*. Retrieved July 8, 2019, from MITRE: https://mitre-attack.github.io/caret/#/.

Morgan, J. B. (2015). *The ISIS twitter census defining and describing the population of ISIS supporters on twitter*. Brookings.

Morison, S. E. (1963). *The Two-Ocean war: A short history of the United States navy in the second world war*. Little, Brown.

Mueller, R. (2019). *Report on the investigation into Russian interference in the 2016 presidential election. U.S. Department of Justice*. U.S. Department of Justice.

Nakashima, E. (2012, 9 12). *Iran blamed for cyberattacks on U.S. banks and companies*. Retrieved August 18, 2019, from Washington post: https://www.washingtonpost.com/world/national-security/iran-blamed-for-cyberattacks/2012/09/21/afbe2be4-0412-11e2-9b24-ff730c7f6312_story.html?noredirect=on

Nakashima, E. (2018, 10 23). *Pentagon launches first cyber operation to deter Russian interference in midterm elections*. Retrieved March 12, 2019, from Washington post: https://www.washingtonpost.com/world/national-security/pentagon-launches-first-cyber-operation-to-deter-russian-interference-in-midterm-elections/2018/10/23/12cc6e7e-d6df-11e8-83a2-d1c3da28d6b6_story.html?utm_term=.8c47d573557b

Nakashima, E. (2019, 2 27). *US disrupted internet access of Russian troll factory on day of 2018 midterms*. Retrieved March 12, 2019, from Washington post.

Nakasone, P. M. (2019). A Cyber force for persistent operations. *Joint Forces Quarterly, 92*(1), 10–15.

Youssef, N. A., S. H. (2017, 11 25). *Why Did Team Obama Try to Take Down Its NSA Chief?* Retrieved August 27, 2018, from The Daily Beast: https://www.thedailybeast.com/why-did-team-obama-try-to-take-down-its-nsa-chief

National Cyber Security Center. (2018, 10 11). *Joint report on publicly available hacking tools*. Retrieved March 10, 2019, from National Cyber Security Center: https://www.ncsc.gov.uk/joint-report

National Institute of Standards (NIST). (2018, 4 16). Cybersecurity Framework. Retrieved June 12, 2020, from NIST: https://www.nist.gov/cyberframework

National Research Council. (2010). *Committee on deterring cyberattacks: Informing strategies and developing options for U.S. policy*. NAP.

Newman, L. H. (2017, September 8). *The equifax breach exposes america's identity crisis*. Retrieved September 7, 2018, from Wired: https://www.wired.com/story/the-equifax-breach-exposes-americas-identity-crisis/

Nichols, M. (2019, 8 5). *North Korea took $2 billion in cyberattacks to fund weapons program: U.N. report*. Retrieved August 17, 2019, from Reuters: https://www.reuters.com/article/us-northkorea-cyber-un/north-korea-took-2-billion-in-cyber-attacks-to-fund-weapons-program-u-n-report-idUSKCN1UV1ZX

Nicole Perlroth, Q. H. (2013, 1 8). *Bank hacking was the work of Iranians, officials say*. Retrieved September 21, 2019, from New York times: https://www.nytimes.com/2013/01/09/technology/online-banking-attacks-were-work-of-iran-us-officials-say.html

NSA. (n.d.). *Secure Architecture*. Retrieved June 12, 2020, from NSA: https://apps.nsa.gov/iaarchive/library/ia-guidance/secure-architecture/index.cfm

Nusca, A. (2010, 3 29). *Malware capital of the world is Shaoxing, China*. Retrieved April 10, 2019, from ZDNet: https://www.zdnet.com/article/report-malware-capital-of-the-world-is-shaoxing-china/

Nuttall, C. (1998, 8 19). *Chinese protesters attack Indonesia through Net*. Retrieved June 26, 2020, from BBC: http://news.bbc.co.uk/2/hi/science/nature/154079.stm

OASIS. (n.d.). *Advancing open standards for the information society*. Retrieved February 10, 2018, from OASIS Cyber threat intelligence (CTI) TC: https://www.oasis-open.org/committees/tc_home.php?wg_abbrev=cti

Office of Personnel Management. (2015). *Cybersecurity Resource Center*. Retrieved July 7, 2019, from https://www.opm.gov/cybersecurity/cybersecurity-incidents/

Office of the Assistant Secretary of Defense Networks and Information Integration (OASD/NII). (2010). *Reference Architecture Description*. Retrieved June 11, 2020, from https://dodcio. defense.gov/Portals/0/Documents/DIEA/Ref_Archi_Description_Final_v1_18Jun10.pdf

Office of the Director of National Intelligence. (n.d.). *Cyber Threat Framework*. (DNI, Producer) Retrieved September 20, 2019, from DNI: https://www.dni.gov/index.php/ cyber-threat-framework

Office of the National Manager for NSS. (n.d.). *DoDCAR*. Retrieved March 8, 2019, from https:// csrc.nist.gov/CSRC/media/Presentations/DODCAR-no-class-markings-Pat-Arvidson/images-media/DODCAR_-no%20class%20markings%20-%20Pat%20Arvidson.pdf

O'Flaherty, K. (2019, 5 6). *Israel retaliates to A Cyber-attack with immediate physical action in A world first*. Retrieved May 15, 2019, from https://www.forbes.com/sites/kateofla-hertyuk/2019/05/06/israel-retaliates-to-a-cyber-attack-with-immediate-physical-action-in-a-world-first/#1319bce9f895

Orr, A. (2018, 11 7). *China re-routed US internet traffic for 2.5 years*. Retrieved November 13, 2018, from https://www.macobserver.com/link/china-reroute-internet-traffic/

Osnos, E. (2014). *Age of ambition: Chasing fortune, truth, and faith in the new China*. Farrar, Straus and Giroux.

Ottis, R. (2007). *Analysis of the 2007 Cyber attacks against Estonia from the infor-mation warfare perspective*. Retrieved September 14, 2019, from coopera-tive Cyber defence Centre of Excellence: https://ccdcoe.org/uploads/2018/10/ Ottis2008_AnalysisOf2007FromTheInformationWarfarePerspective.pdf

Overy, R. (2014). *The bombers and the bombed: Allied air war over Europe*. Viking.

Singer, P. W., & E. T. (2018). *LikeWar - the Weaponization of social media*. Houghton Mifflin.

Pape, R. A. (1996). *Bombing to win: Air power and coercion in war*. Cornell University Press.

Parham, J. (2017, 10 18). *Russians posing as black activists on facebook is more than fake news*. Retrieved August 22, 2018, from Wired: https://www.wired.com/story/ russian-black-activist-facebook-accounts/

Paul Ducheine, J. V. (2014). *Fighting Power, Targeting and Cyber Operations*. Retrieved August 4, 2019, from CCDOE - 2014 6th International Conference on Cyber Conflict: https://www. ccdcoe.org/uploads/2018/10/d2r1s9_ducheinehaaster.pdf

Pegues, J. (2018). *Kompromat - how Russia undermined American democracy*. Prometheus.

Peter Feaver, K. G. (2017). "When the Urgency of Time and Circumstance Clearly Does Not Permit ..." - Pre-Delegation in Nuclear and Cyber Scenarios. In A. E. G. Perkovich (Ed.), *Understanding Cyber Conflict - 14 Analogies*. Georgetown.

Philip Bennett, S. C. (1999, May 25). *NATO Warplanes Jolt Yogoslav Power Grid*. Washington Post.

Phillip Porras, H. S. (2009, 2 4). *An analysis of Conficker's logic and rendezvous points*. Retrieved June 29, 2020, from USENIX: https://www.usenix.org/legacy/events/leet09/tech/full_papers/ porras/porras_html/index2.html

Pomerleau, M. (2019, 5 8). *New authorities mean lots of new missions at Cyber command*. (F. Domain, Producer) Retrieved September 14, 2019, from Fifth Domain: https://www.fifthdomain.com/dod/cybercom/2019/05/08/ new-authorities-mean-lots-of-new-missions-at-cyber-command/

Popular Mechanics. (2018, March 13). *How Long Does It Take Hackers To Pull Off a Massive Job Like Equifax?* Retrieved from Popular Mechanics: https://www.popularmechanics.com/ technology/security/a18930168/equifax-hack-time/

Poulsen, K. (2012). *Kingpin: How one hacker took over the billion-Dollar cybercrime under-ground*. Broadway.

Poulsen, K. (2018, 10 25). *'Lone DNC Hacker' Guccifer 2.0 slipped up and revealed he was a Russian intelligence officer*. Retrieved July 8, 2019, from the daily beast: https://www.thedaily-beast.com/exclusive-lone-dnc-hacker-guccifer-20-slipped-up-and-revealed-he-was-a-russian-intelligence-officer

Prosser, M. B. (2006). *Memetics—a growth industry in US military operations (MS thesis)*. Retrieved January 3, 2019, from DTIC: https://apps.dtic.mil/dtic/tr/fulltext/u2/a507172.pdf

Raff, A. (2013, 3 5). *Chinese time bomb.* Retrieved April 10, 2019, from Securlert: http://www. avivraff.com/seculert/test/2013/03/the-chinese-time-bomb.html

Rapid7. (n.d.). *Metasploit.* (R. 7, Producer) Retrieved September 11, 2019, from Metasploit: https://www.metasploit.com/

Clarke, R. A., R. K. (2012). *Cyber war: The next threat to National Security and what to do about it.* Ecco.

Richard Clarke, R. K. (2011). *Cyber war: The next threat to National Security and what to do about it.* Ecco.

Rick Nunes-Vaz, S. L. (2011). A More Rigorous Framework for Security-in-Depth. *Journal of Applied Security Research*, 23.

Rick Nunes-Vaz, S. L. (2014). From strategic security risks to National Capability Priorities. *Security Challenges, 10*(3), 23–49.

Rid, T. (2013). *Cyber war will not take place.* Oxford University Press.

Riley, C. (2019, 7 9). *UK proposes another huge data fine. This time, Marriott is the target.* Retrieved August 21, 2019, from CNN: https://www.cnn.com/2019/07/09/tech/marriott-data-breach-fine/index.html

RISI Data. (2003). *Slammer impact on Ohio nuclear plant.* Retrieved July 7, 2019, from RISI Data: https://www.risidata.com/Database/Detail/slammer-impact-on-ohio-nuclear-plant

RISI. (n.d.). *RISI online incident database.* Retrieved June 24, 2020, from RISI: https://www.risidata.com/Database/

Robb, J. (2007, June). The Coming Urban Terror. *City Journal.*

Lee, R. M., M. J. (2014). *German Steel Mill Cyber Attack. SANS ICS CP/PE (Cyber-to-Physical or Process Effects) case study paper.* SANS.

Romanosky, S. (2016). Examining the costs and causes of Cyber incidents. *Journal of Cybersecurity, 2*(2).

Roque, A. (2020). Assessing the cognitive complexity of Cyber range for peer review environments. *Journal of Defense Modeling and Simulation.*

Roscini, M. (2014). Cyber operations and the use of force in international law. .

Sanger, D. E. (2017). Cyber, drones and secrecy. In A. E. G. Perkovich (Ed.), *Understanding Cyber conflict.* Georgetown.

Sanger, D. E. (2016, 4 24). *U.S. cyberattacks target ISIS in a new line of combat.* Retrieved August 27, 2018, from New York times: https://www.nytimes.com/2016/04/25/us/politics/us-directs-cyberweapons-at-isis-for-first-time.html

SANS. (2016). *Critical security controls.* (SANS, Producer) Retrieved September 24, 2019, from SANS: https://www.sans.org/media/critical-security-controls/critical-controls-poster-2016.pdf

SANS. (n.d.). *SANS institute.* Retrieved September 26, 2019, from https://www.sans.org/

Savage, C. (2015). *Power wars: Inside Obama's post - 9/11 presidency.* Little Brown.

Schmitt, E. (2019, 8 19). *ISIS is regaining strength in Iraq and Syria.* Retrieved June 23, 2020, from New York times.

Schmitt, M. (2011). Cyber operations and the jus in Bello: Key issues. *Naval War College International Law Studies, 2019*(9), 10.

Schmitt, M. N. (2013). *Tallinn manual on the international law applicable to Cyber warfare.* Cambridge University Press.

Schostack, A. (2008). *Experiences threat Modeling at Microsoft.* Retrieved February 19, 2020.

Segura, V. (2009, June 25). *Modeling the economic incentives of DDoS Attacks.* Retrieved from Semantic Scholar: https://pdfs.semanticscholar.org/afdf/d974bc68dc05c48020e-f07a558a61ab94f8a.pdf

Senate Armed Services Committee. (2014). Inquiry in Cyber Instrusions affecting US transportation command contractors. In *113th congress, 2d sess.* Government Printing Office.

Shodan. (n.d.). Shodan. Retrieved November 20, 2020, from https://www.shodan.io/

Simon Pirani, J. S. (2009, 2 15). *The Russo-Ukrainian gas dispute of January 2009: a comprehensive assessment.* Retrieved December 11, 2018, from Oxford Institute for Energy Studies: https://www.oxfordenergy.org/wpcms/wp-content/uploads/2010/11/NG27-TheRussoUkrainia

nGasDisputeofJanuary2009AComprehensiveAssessment-JonathanSternSimonPiraniKatjaYafi
mava-2009.pdf

Singapore. (2016). *Singapore's Cyber Security Strategy*. Retrieved March 10, 2018, from
Singapore's Cyber Security Strategy : https://www.csa.gov.sg/~/media/csa/documents/publi-
cations/singaporecybersecuritystrategy.pdf

Slowik, J. (2019, 8 15). *CRASHOVERRIDE: Reassessing the 2016 Ukraine Electric Power Event
as a Protection-Focused Attack*. Retrieved September 24, 2019, from DRAGOS: https://dragos.
com/wp-content/uploads/CRASHOVERRIDE.pdf

Smith, P. (2013, 9 24). *Internet exchange point design*. Retrieved March 13, 2019, from https://
www.menog.org/presentations/menog-13/192-MENOG13-IXP-Design.pdf

Smith, T. (2001, 10 31). Hacker jailed for revenge sewage attacks - job rejection caused a bit of
a stink. Retrieved July 7, 2019, from the register: https://www.theregister.co.uk/2001/10/31/
hacker_jailed_for_revenge_sewage/

Snowden,   E.   (n.d.).   https://edwardsnowden.com/2015/01/18/the-roc-nsas-epicenter-for-
computer-network-operations/;   https://edwardsnowden.com/2015/01/18/interview-
with-a-sid-hacker-part-I-how-does-tao-do-its-work/;   https://freesnowden.is/2015/01/18/
expanding-endpoint-operations.   Retrieved   from   https://edwardsnowden.com/2015/01/18/
the-roc-nsas-epicenter-for-computer-network-operations/;   https://edwardsnowden.
com/2015/01/18/interview-with-a-sid-hacker-part-I-how-does-tao-do-its-work/;   https://frees-
nowden.is/2015/01/18/expanding-endpoint-operations

Social Media Environment and Internet Replication. (n.d.). Retrieved July 7, 2019, from https://
www.smeir.net/

South Front. (2015, May 25). *Cyberkut hacked the site of Ukrainian Ministry of Finance:
The country has no money*. Retrieved from South Front: https://southfront.org/
cyberkut-hacked-the-site-of-ukrainian-ministry-of-finance-the-country-has-no-money

Spiegel. (2013b, Feb 25). *Digital Spying Burdens German-Chinese Relations*. Retrieved April
10, 2019, from Spiegel: https://www.spiegel.de/international/world/digital-spying-burdens-
german-relations-with-beijing-a-885444.html

Steven Noel, J. L. (2015). *Analyzing Mission impacts of Cyber actions (AMICA)*. NATO.

Stoll, C. (2005). *The Cuckoo's egg: Tracking a spy through the maze of computer espionage*.
Pocket Books.

Strategy and Tactics of Guerilla Warfare. (n.d.). Retrieved September 9, 2018, from Wikipedia:
https://en.wikipedia.org/wiki/Strategy_and_tactics_of_guerrilla_warfare

Symantec. (2012, 8 9). *Complex Cyber Espionage Malware Discovered: Meet W32.Gauss*.
Retrieved   July   7,   2019,   from   Symantec: https://www.symantec.com/connect/blogs/
complex-cyber-espionage-malware-discovered-meet-w32gauss

Symantec. (2019). *Internet Security Threat Report*. Retrieved March 8, 2019, from Symantec:
https://www.symantec.com/security-center/threat-report

Symantec. (n.d.). *Symantec*. Retrieved November 29, 2018, from https://www.symantec.com/

Symantec. (2011, 10 18). *W32.Duqu: The Precursor to the Next Stuxnet*. Retrieved July 7, 2019,
from Symantec: https://www.symantec.com/connect/w32_duqu_precursor_next_stuxnet

Symantec. (2003). *W32.Sobig.F@mm*. Retrieved July 7, 2019, from Symantec: https://www.
symantec.com/security-center/writeup/2003-081909-2118-99

SysML.org. (n.d.). *SysML Partners: Creators of the SysML*. Retrieved June 12, 2020, from SysML.
org: https://sysml.org/sysml-partners/

Syverson, P. (n.d.). *Paul Syverson web page*. Retrieved November 29, 2018, from http://www.
syverson.org/

Tang, R. (2001, 5 1). *China-U.S. cyber war escalates*. Retrieved June 25, 2020, from http://www.
cnn.com/2001/WORLD/asiapcf/east/04/27/china.hackers/

Temple-Raston, D. (2016, 9 12). *Cyber bombs reshape U.S. Battle against terrorism*.
Retrieved   September   3,   2019,   from   NPR: https://www.npr.org/2016/09/12/493654985/
cyber-bombs-reshape-u-s-battle-against-terrorism

Temple-Raston, D. (2019a, 1 11). *Hacks are getting so common that companies are turning to 'Cyber Insurance'*. Retrieved September 3, 2019, from NPR: https://www.npr.org/2019/01/11/684610280/hacks-are-getting-so-common-that-companies-are-turning-to-cyber-insurance

Temple-Raston, D. (2019b, 9 26). *National Public Radio (NPR)*. (NPR, Producer) Retrieved September 29, 2019, from How The U.S. Hacked ISIS: https://www.npr.org/2019/09/26/763545811/how-the-u-s-hacked-isis

Temple-Raston, D. (2019c, 8 14). *Task force takes on Russian election interference*. Retrieved September 3, 2019, from NPR: https://www.npr.org/2019/08/14/751048230/new-nsa-task-force-takes-on-russian-election-interference

Thomas, T. (2016). *Russian strategic thought and Cyber in the armed forces and society: A viewpoint from Kansas. Center for strategic and international studies*. Center for strategic and International Studies.

Timur Chabuk, A. J. (2018, 9 1). *Understanding Russian information operations*. Retrieved September 9, 2018, from signal (AFCEA): https://www.afcea.org/content/understanding-russian-information-operations

Tkacik, J. J. (2008, 2 8). *Trojan dragons: China's Cybrer threat*. Retrieved April 10, 2019, from Heritage Foundation: https://www.heritage.org/asia/report/trojan-dragon-chinas-cyber-threat

Tsyganok, A. (2012). *Informational Warfare – a Geopolitical Reality*. Retrieved from http://en.fondsk.ru/article.php?id-1714

TTP vs Indicator: A simple usage overview. (2018). Retrieved March 11, 2019, from STIX Project: https://stixproject.github.io/documentation/concepts/ttp-vs-indicator/

Twitter. (2020, 6 12). *Disclosing networks of state-linked information operations we've removed*. Retrieved June 12, 2020, from Twitter: https://blog.twitter.com/en_us/topics/company/2020/information-operations-june-2020.html

U.S. Joint Forces Command. (2011, 5 20). Commander's Handbook for Attack the Network. Retrieved August 4, 2019, from https://www.jcs.mil/Portals/36/Documents/Doctrine/pams_hands/atn_hbk.pdf

U.S. Justice Department. (2019, March). *Report on the investigation into Russian interference in the 2016 presidential election*. Retrieved May 9, 2019, from U.S. Justice Department: https://www.justice.gov/storage/report.pdf

U.S. Justice Department. (n.d.). *Federal Register*. Retrieved May 19, 2019, from Federal Register: https://www.federalregister.gov/presidential-documents/executive-orders

U.S.-China Economic and Security Review Commission. (2014). *USCC Annual Report*. Retrieved April 10, 2019, from https://www.uscc.gov/sites/default/files/annual_reports/Complete%20Report.PDF

United Nations. (n.d.). *Charter of the United Nations*. Retrieved October 20, 2019, from http://legal.un.org/repertory/art1.shtml

United States of America. (2018). *United states of America vs. Internet research agency LLC. Indictment*. Department of Justice.

US Army. (n.d.). Joint technical coordinating Group for Munitions Program Office.

US Cyber Consequences Unit (CCU). (2009). Overview by the US-CCU of the Cyber campaign against Georgia in august of 2008. *US Cyber Consequences Unit (CCU)*.

US Department of Defense. (2008). *Cyberspace operations*. US Department of Defense, Joint Staff.

Vaas, L. (2019, 5 13). *Two people indicted for massive anthem health data breach*. Retrieved July 7, 2019, from Naked Security: https://nakedsecurity.sophos.com/2019/05/13/two-chinese-hackers-indicted-for-massive-anthem-breach/

Vera Zakem, M. K. (2018, 4 1). *Exploring the utility of memes for U.S. government influence campaigns*. Retrieved January 3, 2019, from Center for Naval Analyses (CNA): https://www.cna.org/cna_files/pdf/DRM-2018-U-017433-Final.pdf

Verizon. (2018). *2018 Data Breach Investigations Report*. Retrieved October 17, 2018, from Verizon: http://www.verizonenterprise.com/industry/public_sector/docs/2018_dbir_public_sector.pdf

Warner, M. (2017). Intelligence in Cyber - and Cyber in intelligence. In A. E. G. Perkovich (Ed.), *Understanding Cyber conflict - 14 analogies*. Georgetown University Press.

Weaponeering. (n.d.). *History of the Joint Technical Coordinating Group for Munitions Effectiveness*. Retrieved August 19, 2019, from Weaponeering: http://www.weaponeering.com/jtcg_me_history.htm

Whaley, B. (2007). *STRATAGEM - deception and surprise in war*. Artech House.

White House. (n.d.). *Cyber executive orders*. Retrieved March 10, 2019, from homeland security digital library: https://www.hsdl.org/?collection&id=2724

White House. (2018, 9). *National Cyber Security Strategy of the United States of America*. Retrieved March 10, 2019, from National Cyber Security Strategy of the United States of America: https://www.whitehouse.gov/wp-content/uploads/2018/09/National-Cyber-Strategy.pdf

Wikipedia. (n.d.). *Chaos Computer Club*. Retrieved September 9, 2018, from Wikipedia: https://en.wikipedia.org/wiki/Chaos_Computer_Club

Wilsher, K. (2009, February 7). *French fighter planes grounded by computer virus*. The Telegraph.

Winnona DeSombre, D. B. (2018). *Thieves and geeks: Russian and Chinese hacking communities*. Retrieved June 26, 2020, from Recorded Future: https://go.recordedfuture.com/hubfs/reports/cta-2018-1010.pdf

Zachman, J. (n.d.). *Conceptual, logical, physical: It is simple*. Retrieved June 11, 2020, from Zachman International: https://www.zachman.com/ea-articles-reference/58-conceptual-logical-physical-it-is-simple-by-john-a-zachman

Zetter, K. (2014a). *Countdown to zero day - Stuxnet and the launch of the World's first digital weapon*. Crown.

Zetter, K. (2015, 10 15). *Darpa is developing a search engine for the dark web*. Retrieved November 15, 2018, from WIRED: https://www.wired.com/2015/02/darpa-memex-dark-web/

Zetter, K. (2016a, January). Everything we know about Ukraine's power plant hack. *Wired*.

Zetter, K. (2011a, April 26). FBI vs. Corefloot botnet: Round 1 goes to the feds. *Wired*.

Zetter, K. (2016b, 3 3). *Inside the cunning, unprecedented hack of Ukraine's power grid*. Retrieved June 27, 2020, from Wired: https://www.wired.com/2016/03/inside-cunning-unprecedented-hack-ukraines-power-grid/

Zetter, K. (2014b, 12 3). *Sony got hacked hard: what we know and don't know so far*. Retrieved September 9, 2018, from Wired: https://www.wired.com/2014/12/sony-hack-what-we-know/

Zetter, K. (2011b, April 11). *With court order, FBI hijacks 'Coreflood' botnet*. Sends Kill Signal.

Zurasky, M. W. (2017). Methodology to perform Cyber lethality assessment. .

Chapter 12
Cyberspace Analysis and Targeting Conclusions

This book describes the role of analysis and targeting in cyberspace operations. We began, in Chap. 2, with a review of current policy, doctrine and TTPs, which gave us background on the resilience focus of both United States and International executive level cyber policy. In addition, we reviewed published doctrine (e.g., Joint Publication 3–12), along with best practices, employed as tactics, techniques and procedures (TTPs) (e.g., critical security controls [CSCs]), by the Information Assurance (IA) community.

In Chap. 2, we also looked at developing policy and doctrine concerning how cyber fits in the spectrum of diplomatic to kinetic policy options. We used "the line" (Fig. 2.2) to describe how cyber response options are currently in an ill-defined space between traditional diplomatic actions and war. We also looked developing doctrine. This included the Tallinn Manual (Schmitt) and "Droit International Applique aux Operations dans le Cyberspace" (Roguski) as examples of how policy theorists are working to place cyber options within a prescribed policy framework.

While clear overarching policy is still developing, the information assurance (IA) community continued to provide guidance in Chap. 3, "Taxonomy of Cyber Threats," where we reviewed both industry standard risk evaluation techniques (e.g., DREAD, STRIDE and CVSS) and looked at active methods of vulnerability detection (i.e., PASTA). While our focus is defense, vulnerability analysis can also be used to develop tools for targeting (e.g., penetration testing). For example, we looked at Metasploit as a commercial example that uses open-source vulnerability information to continually update its exploitation suite.

In Chap. 4, we reviewed information operations (IO) by looking at a history of active measures by the former Soviet Union, assessing operations for their area or point targeting effects. This included reviewing a Disinformation Process used to describe Soviet era active measures. In addition, we looked at active measures examples from Oleg Kalugin (Kalugin) on how he leveraged journalists to collect information and disseminate messages for the former Soviet Union.

In Chap. 4, we also looked at current examples of Russian "news" organizations (e.g., RT News, Sputnik, and Internet Research Agency [IRA]) and provided

© Springer Nature Switzerland AG 2022
J. M. Couretas, *An Introduction to Cyber Analysis and Targeting*,
https://doi.org/10.1007/978-3-030-88559-5_12

models on how they use the Internet to propagate messages (e.g., 2019 Mueller Report). This included looking at how journalists convert content provided by hackers to publicize private e-mails (e.g., 2016 U.S. Presidential Election). In addition, we looked at how Internet based information operations (IO) can look a lot like a marketing campaign, leveraging Cialdini's six influence approaches (Cialdini) and providing an overall campaign cycle that describes how messages are developed. We also reviewed power laws in terms of how a few key influencers can shape the message for a majority of followers.

Similar to the focus on social networks and open-source intelligence in Chaps. 4 and 5 uses "Cyber ISR and Analysis" to perform detailed collection, providing for "big data" platforms that can be used for analysis and targeting. In Sect. 5.3.2.1.1, we discussed how the ISIS financier, Abu Sayyaf, kept detailed records of the Caliphate, making his data system a key target in a 2015 counter ISIS campaign. Similarly, Chap. 5 also introduced the idea of how multiple data exfiltration might be used to provide a platform for targeting U.S. citizens. For example, the credit (2017 Equifax), health (2017 Anthem) and marital fidelity (2016 FriendFinder) exfiltrations, combined with the 2014 OPM description of Government and contractor personnel holding a security clearance, contains information on health, wealth, and personal morals that could be used for personnel targeting.

In Chap. 5, we also looked at social network analysis (SNA). SNA, originally developed to "attack the network" (i.e., human network) in countering improvised explosive devices (IEDs), is now applied to social media. In one example (Sect. 5.4.3), a private firm, Bellingcat, used open source intelligence (OSINT), to track down the Russian 53rd Anti-Aircraft Brigade as being responsible for the 2014 shooting down of the Dutch MH-17 civilian airliner (Bellingcat Investigation Team).

Using social media to pinpoint a nation state military unit, and call them out by both organization and, sometimes, individual names, shows the resolution available for passive intelligence using cyber means. In addition to passive ISR, we also looked at the STUXNET campaign, as reported by Zetter (Zetter), to better understand how active ISR is conducted via cyber. We reviewed Duqu, Flame, Gauss, and SPE "bots" used to collect information on the Iranian nuclear program, using over 80 estimated locations, around the world, for data storage and command and control. While it is unclear how long the cyber ISR phase of the STUXNET campaign lasted, it is clear that the scale, scope, and stealth of the operation rivalled what an entire nation state human collection operation might have accomplished only a generation ago.

Bots are also highlighted in Chap. 6, where we look at system security in terms of end points, connections, and data stores across the Lockheed Martin cyber kill chain. In addition, Chap. 6 provided the basics on end point security that spanned from every day antivirus (AV) systems to the application of moving target defense (MTD) for a defended network. This included network analysis and netflow as techniques to analyze network communications for possible malware, or botnet, communications. We also reviewed botnets, including their controllers, C2 channels, and implants/bots. Chapter 6 wrapped up with a review of Security Operations

Centers (SOCs), the place where an organization's policies, processes, and technologies are implemented in accordance with leadership security guidance.

While an SOC describes the organization, processes, and technologies used to defend a system of interest, Chap. 7 provided a possible attacker's view of the security systems that we reviewed in Chap. 6. For example, in Chap. 7, we looked at how an attacker might use the CARVER matrix to identify key nodes, and use these nodes to determine how to apply effects (Fig. 7.2) in order to estimate resource requirements via the cyber process evaluator (Fig. 7.4). In addition, the cyber process evaluator helped us estimate imposed time and cost, on a cyber attacker, in navigating defended cyber terrain, for example, security technologies reviewed in Chap. 6.

Chap. 8, cyber systems design, provided a standard architecture-based approach for looking at the cyber system components that we discussed in the previous chapters. For example, each of the components in the defensive system can also be described by a software architecture, with well-developed frameworks available to describe the terms (i.e., reference architecture), functions (i.e., solution architecture), connections (i.e., logical architecture), or implementation (i.e., physical architecture). Describing a cyber system in terms of general artifacts provides the network defender with a method for ensuring that the respective components are up to date, in terms of individual components and overall system security.

In Chap. 8, we also provided a solution architecture, using the system entity structure (SES), an architecture description method, to organize system components into an entity-relational approach. This included using the SES to describe a cyber collection system, based on the 2016 U.S. Presidential election attack, to show how entities (e.g., component technologies) and their relationships benefit from architecture description for design analysis, including quantifying metrics for each section of the system.

In Chap. 9, we used the cyber collection system architecture (i.e., solution architecture) from Chap. 8 and went beyond the time/cost view of a cyber operation (i.e., from Chap. 7) to derive the key performance parameters (KPPs), the measures of performance (MOPs), and the measures of effectiveness (MOEs) for a cyber system. In addition, we developed cyber system KPPs, MOPs, and MOEs in tandem with a derivation of metrics for an autonomous system at the system, engagement, mission, and campaign system modeling levels.

Continuing with the idea of finding measurables, in Chap. 10 we focused on cyber as a technical system, describing the inherent parallel structure of the physical, logical and persona operational processes, each with its own operations sequence and potential vulnerabilities. In addition, we introduced key observable series processes as discrete event systems to describe the memorized processes that a system traverses in carrying out a specified task or set of tasks. Decomposing the system this way, we used Cohen's d as an effect estimator for individual process phase behavior modifications in order to measure the application of a cyber effect. This technical approach for effect estimation was complimented by a Live-Virtual-Constructive (LVC) framework discussion for evaluating human performance via live simulation.

The effect estimation in Chap. 10 is designed to introduce an alternative method, Cohen's d, for evaluating system behavior modifications due to a cyber attack. As mentioned in Chap. 10, a statistically derived phase time anomaly for a system under inspection may not constitute an appreciable effect as described by MOPs or MOEs. However, the phase time anomaly does provide an approach for observing how a system has changed due to a cyber effect.

Due to the challenges in data collection, and the complexity of cyber systems, we also looked at a higher-level approach for estimating adversary resource requirements in Chap. 7. Breaking out a target system in terms of associated people, processes, and technologies provides the targeteer with some gross measures of time and money costs that might be imposed on a cyber attacker via defensive or active means. Tying up these considerations into an overall framework for cyberspace analysis and targeting is provided in Fig. 12.1.

As shown in Fig. 12.1, cyberspace operations, guided by doctrine and policy (Chap. 2), include threat modeling for system defense (Chap. 3) and an understanding of information operations (Chap. 4) to protect against the entire range of

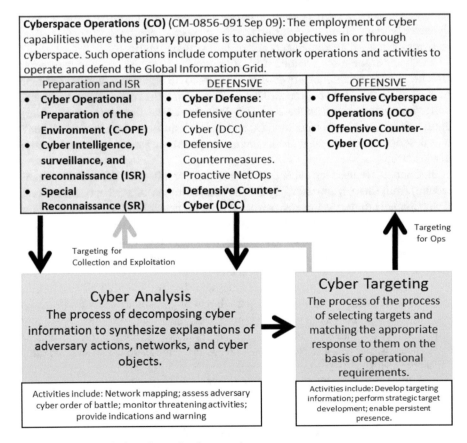

Fig. 12.1 Cyber analysis and targeting framework

information-related capabilities. In addition, ISR (Chap. 5) provides additional understanding for both defensive (Chap. 6) and offensive (Chap. 7) operations. Describing cyber systems (Chap. 8) is facilitated by formulating metrics (Chap. 9) and using modeling and simulation (Chap. 10). The framework in Fig. 12.1 therefore summarizes key elements in cyber analysis and targeting.

Glossary

5G "5G is the fifth generation mobile network. It is a new global wireless standard after 1G, 2G, 3G, and 4G networks. 5G enables a new kind of network that is designed to connect virtually everyone and everything together including machines, objects, and devices."[1] This also includes a greater attack surface.

Active Directory "Active Directory stores information about objects on the ne work and makes this information easy for administrators and users to find and use. Active Directory uses a structured data store as the basis for a logical, hierarchical organization of directory information."[2]

Active Measure "Active measure is a term for the actions of political warfare conducted by the Soviet and Russian security services (Cheka, OGPU, NKVD, KGB, FSB, GRU) to influence the course of world events, in addition to collecting intelligence and producing "politically correct" assessment of it; framing information to change perspective."[3]

Advanced Persistent Threat (APT) An APT is an advanced adversary that attacks and stays on a network. The whole purpose of an APT attack is to gain ongoing access to the system. Hackers achieve this in a series of five stages.[4]

- Stage One: Gain Access.
- Stage Two: Establish a Foothold.
- Stage Three: Deepen Access.
- Stage Four: Move Laterally.
- Stage Five: Look, Learn, and Remain.

---

[1] https://www.qualcomm.com/5g/what-is-5g

[2] https://docs.microsoft.com/en-us/windows-server/identity/ad-ds/get-started/virtual-dc/active-directory-domain-services-overview

[3] en.unionpedia.org

[4] https://www.kaspersky.com/resource-center/definitions/advanced-persistent-threats

© Springer Nature Switzerland AG 2022
J. M. Couretas, *An Introduction to Cyber Analysis and Targeting*,
https://doi.org/10.1007/978-3-030-88559-5

All Source Analysis   "Analyzes threat information from multiple sources, disciplines, and agencies across the Intelligence Community. Synthesizes and places intelligence information in context; draws insights about the possible implications."[5]

Amplifier   In social media message propagation, these are sometimes real, sometimes bots, who primarily retweet and spread material on-line, increasing message frequency so as to increase the general population's belief in a narrative.

A-NIDS   Anomaly Based Network Intrusion Detection System (A-NIDS) observes network characteristics, like protocols used, point-to-point system communications and the network load of individual systems in order to develop patterns and look for anomalies.

Analyzing Mission Impacts of Cyber Attacks (AMICA)   2015 MITRE study on the effects of a cyberattack on an Air Tasking Order (ATO) system

Anonymity Networks   "An anonymity network enables users to access the Web while blocking any tracking or tracing of their identity on the Internet. This type of online anonymity moves Internet traffic through a worldwide network of volunteer servers. Anonymity networks prevent traffic analysis and network surveillance - or at least make it more difficult."[6] (e.g., The Onion Router, ToR)

Anti-Virus System   Cyber security system that uses published malware signatures to detect and deny entry of malicious code onto a protected system

Arab Spring   Civil unrest, channeled through the Internet, eventually resulted in the overturn of governments in North Africa (i.e., Tunisia, Libya, and Egypt) and ended with the beginning of the Syrian civil war.

Architecture   The body of definitions for describing an object to be designed. Architecture usually includes the language, function, and interconnection of the system to be built.

Area Target   A target consisting of an area, rather than an actual point. In cyber, this could be popular opinion (e.g., in an election).

Ashley Madison   Social media site to help married people meet for an extra marital affair.

Astroturf   "An artificially created advocacy community, generally centrally directed, that poses as a spontaneous grassroots movement arising from genuine popular demand."[7]

Attack Framework (MITRE ATT@CK)   "An attack framework and its associated lexicon provide a means for consistently describing cyber threat activity in a manner that enables efficient information sharing."[8]

Attack Surface   "A software system's attack surface is the sum of the different points, or "attack vectors," where an unauthorized user (the "attacker") can try to enter data to, or extract data from, an environment. Keeping an attack surface as small as possible is a basic security measure."[9]

---

[5] https://public.cyber.mil/dcwf-specialty-area/all-source-analysis/

[6] https://www.techopedia.com/definition/25187/anonymity-network

[7] https://thecyberwire.com/glossary/astroturf

[8] resources.sei.cmu.edu

[9] www.cybrary.it

Attack the Network (AtN) AtN can be defined as part of the Counter—Improvised Explosive Device (C-IED) capability. Within C-IED, AtN focuses on the offensive activities necessary to prevent threat networks from planning and implementing attacks, such as the use of IEDs.

Attack Vector "An attack vector is a path or means by which a hacker can gain access to a computer or network server in order to deliver a payload or malicious outcome. Attack vectors enable hackers to exploit system vulnerabilities, including the human element."[10]

Attribution The action of regarding something as being caused by a person or thing.

Authorities Official permission to perform a function or activity.

Availability Effect Control a system's availability for use (e.g., intermittent operations)

Backdoor A backdoor is a typically a covert method of bypassing normal authentication or encryption in a computer or router.

Bellingcat Open-source firm that attributed the Russian military unit responsible for the 2014 downing of the Dutch MH-17 airliner over the Ukraine.

Biological Weapon "Biological weapon, also called germ weapon, includes any of a number of disease-producing agents—such as bacteria, viruses, rickettsiae, fungi, toxins, or other biological agents—that may be used as weapons against humans, animals, or plants."[11]

Bitcoin A type of digital currency in which a record of transactions is maintained and new units of currency are generated by the computational solution of mathematical problems, and which operates independently of a central bank.

Black Propaganda Misinformation that claims to be coming from one side but is actually produced by the opposing side.

Blockade A naval maneuver whereby a point of entry (e.g., port) is blocked from functioning (e.g., allowing people and goods to enter or leave) via naval means. The cyber equivalent is a denial of service (DoS) attack.

Blockchain "A system in which a record, or ledger, of transactions, is maintained across several computers that are linked in a peer-to-peer network in order to increase cyber security."[12]

Bomb Damage Assessment (BDA) "BDA is as the overall estimate of damage composed of the physical damage assessment (PDA) and the functional damage assessment (FDA), as well as target system assessment, resulting from the application of lethal or nonlethal military force."[13]

Botmaster An entity that controls a botnet.

Botnet A network of compromised computers controlled by a botmaster.

---

[10] Internet
  www.wisconsin.edu

[11] www.britannica.com

[12] inechain.com

[13] www.jcs.mil

Bots   An autonomous program on a network (especially the Internet) that can interact with a computer system or users, especially one designed to respond or behave like a player in a game. A malicious program that infects and recruits the host to join the botnet.

Break Out Time   The time that it takes for a cyber attacker to maneuver to the next stage of a network

Campaign Plan   "Campaign plan is a plan to achieve an objective, sometimes over an extended period of time. A campaign plan usually coordinates many activities uses resources that involve multiple organizations. A campaign plan could also have subordinate objectives, with intermediate milestones, broken down by phases."[13]

Capabilities Analysis   Cyber analysis includes the application of conceptual models to understand (1) system vulnerabilities, (2) where an attacker is in an attack cycle, and (3) the effects resulting from a cyberattack.

Capabilities Capabilities   Cyber capabilities are the resources and assets available to an organization that it can draw on or use to resist or project influence through cyberspace.

Capability Maturity Model (CMM)   The CMM is a conceptual model used to guide US defense contractors toward adequate cyber security controls.

CARET   Cyber Analytic Repository Exploration Tool (CARET) provides attacker teams (e.g., APTs), their tools and example targets.

CARVER   "CARVER is an acronym that stands for Criticality, Accessibility, Recuperability, Vulnerability, Effect and Recognizability and is a system to identify and rank specific targets so that attack resources can be efficiently used. CARVER was developed by US Special Forces as a simple, uniformly and somewhat quantifiable means of selecting targets for possible interdiction. CARVER can be used from an offensive (what to attack) or defensive (what to protect) perspective."[13]

Center of Gravity (COG)   "The definition of a CoG is "the source of power that provides moral or physical strength, freedom of action, or will to act." Thus, the center of gravity is usually seen as the 'source of strength'". [13]

Cloud Computing   Centralized processing and storage of computing resources, provided as a service, in order to outsource an organization's computing requirements.

Cohen's d   "Cohen's d is an effect size used to indicate the standardized difference between two means. Cohen's d can be used, for example, to accompany reporting of t-test and ANOVA results. It is also widely used in meta-analysis."[14]

Cold War   The state of political hostility that existed between the Soviet bloc countries and the US-led Western powers from 1945 to 1990.

Command and Control (C2)   "Command and control is the exercise of authority and direction by a properly designated commander over assigned and attached forces in the accomplishment of a mission. Commanders perform command and

---

[14] en.wikiversity.org

control functions through a command and control system. This term is also used for commanding remote assets (e.g., drone, cyber implant) in providing direction and extracting data / information." [13]

Command and Control (C2) Obfuscator   Actors will often want to disguise their location when compromising a target. Cyber actors can use this technique to redirect their packets through multiple nodes to gain greater access to hosts in a network of interest.

Command-and-Control (C&C) server   C&C server is a computer controlled by an attacker or cybercriminal that is used to send commands to systems compromised by malware and receive stolen data from a target network. Many campaigns have been found using cloud-based services, such as webmail and file-sharing services, as C&C servers, to blend in with normal traffic and avoid detection.

Commercial Off the Shelf (COTS)   COTS products are packaged solutions that are then adapted to satisfy the needs of the purchasing organization, rather than the commissioning of custom-made solutions.

Common Vulnerability Enumeration (CVE)   CVE is a dictionary of publicly disclosed cybersecurity vulnerabilities and exposures that is free to search, use, and incorporate into products and services

Common Vulnerability Scoring System (CVSS)   Open framework for communicating the characteristics and impacts of IT vulnerabilities

Communications Intelligence (COMINT)   COMINT is information gathered from the communications of individuals, including telephone conversations, text messages, and various types of online interactions

Competition Continuum   Description of organizational competition spanning from trade to hostile military action, with cyber viewed as currently being in an undefined area, bordering on hostile action.

Computer Network Operations (CNO)   CNO is cyberspace operations, offensive and defensive, with its own separate doctrine in Joint Publication 3–12 (Joint Chiefs of Staff, 2013).

Computer Worm   Independent computer program, often just a few lines of code, that can replicate itself across a network

Conficker   Conficker, also known as Downup, Downadup and Kido, is a computer worm targeting the Microsoft Windows operating system that was first detected in November 2008.

Confidentiality Effect   Externally accessing a system that is believed (e.g., by the owner) to be secure.

Confidentiality, Integrity, Availability (CIA)   Confidentiality, integrity and availability, also known as the CIA triad, is a conceptual model designed to guide policies for information security within an organization.

Constructive Modeling Environment   An environment that provides for the development of models, usually in software, that have both a structured design (e.g., entity relationships) and behavioral representation (e.g., phase diagrams).

Conventional Media   Conventional media includes television, newspapers, radio, and magazines.

Conventional Warfare   Conventional warfare is a form of warfare conducted by using conventional weapons and battlefield tactics between two or more states in open confrontation. The forces on each side are well-defined, and fight using weapons that primarily target the opponent's military.

Coreflood Botnet   The Coreflood botnet (Zetter 2011) was a seven-year-old botnet, comprising over two million infected computers, and was used daily to harvest usernames, passwords, and financial information. 190 gigabytes of data stolen from more than 400,000 victims were found in one server.

Course of Action (COA)   A procedure adopted to deal with a situation

Credential Stealer   Credential stealers are tools used for obtaining credentials from memory. A Credential stealer's main purpose is to allow an actor to collect credentials of other users who are logged in to a targeted machine by accessing them in memory within the Local Security Authority Subsystem Service (LSASS). These credentials can be reused to give access to other machines on a network.

Critical Security Control (CSC)   CSCs are a recommended set of actions for cyber defense that provide specific and actionable ways to stop today's most pervasive and dangerous attacks

Crown Jewels Analysis (CJA)   CJA is a process for identifying those cyber assets that are most critical to the accomplishment of an organization's mission. CJA is also an informal name for Mission-Based Critical Information Technology (IT) Asset Identification.

Cryptocurrency   A digital currency in which encryption techniques are used to regulate the generation of currency units and to verify the transfer of funds, operating independently of a central bank.

Cryptojacking   It is a type of cybercrime that involves infecting computers with malware to use its resources to generate cryptocurrency.

Cyber   A general description of using computer-based systems to perform existing, or futuristic, functions.

Cyber Analysis   Describing the physical, logical, or persona target in terms of the full spectrum of confidentiality, integrity and availability (CIA) effects achievable via cyber means; the process of decomposing cyber information to synthesize explanation of adversary actions, networks, and cyber objects.

Cyber Analytics Repository Exploration Tool (CARET)   The MITRE Cyber Analytics Repository (CAR) is a knowledge base of analytics developed by MITRE based on the MITRE ATT&CK adversary model. The CAR Exploration Tool (CARET) provides an on-line tool, with drop-downs, to evaluate APT attack cycle signatures, tools and targets

Cyber Collection Cycle   The cyber collection cycle describes the identification, exploitation, and extraction of data from a device that includes both instructions and memory. In addition, the cyber collection cycle also includes the processing of this collected data to decision quality information and reporting.

Cyber Collection Pyramid   The cyber collection pyramid progresses from a base of the billions of Internet users through their specific collection technology (peak).

Cyber Domain - Physical, Logical and Cyber Persona Layers Cyberspace is the global domain within the information environment consisting of the interdependent network of information technology infrastructures, including the Internet, telecommunications networks, computer systems, and embedded processors and controllers. Cyberspace can be viewed as three layers (physical, logical, and social) made up of five components (geographic, physical network, logical network, cyber persona, and persona).

Cyber Effect Defined in JP 3–12, cyber effects are denial and manipulation with derivations therefrom.

Cyber Influence Operation It is a process of using online techniques, usually social media, to create support or contention for an issue or person. For example, the 2019 Mueller report states that Russian operatives actively used social media to create divisive issues during the 2016 US Presidential Election.

Cyber Intelligence Cycle Cyber leverages the US Department of Defense's (DoD) intelligence process, which consists of six phases: direction, collection, processing, analysis, dissemination, and feedback (JP 2–0, 2013)

Cyber JMEM The cyber joint munitions effectiveness manual (JMEM) is an approach for projecting the impact of a cyber munition on a proposed target. One goal of such an approach is to ensure that only the desired effects are realized from the use of a cyber munition, that is, minimize collateral damages.

Cyber Kill Chain "Kill chain" is a term originally used by the military to define the steps an enemy uses to attack a target. In 2011, Lockheed Martin released a paper defining a Cyber Kill Chain. Similar in concept to the military's model, it defines the steps used by cyber attackers in today's cyber-based attacks. The theory is that by understanding each of these stages, defenders can better identify and stop attackers at each of the respective stages. The more points at which you can intercept the malicious cyber actors, the better the chance you have to deny them from their objective or force them to make enough noise where you can more easily detect them.

Cyber Operations Lethality & Effectiveness (COLE) The objective of COLE is to estimate the effectiveness of a cyber munition, similar to what has been performed for conventional munitions via joint munition effectiveness manuals (see JMEM)

Cyber Policy It is a course or principle of action adopted or proposed by a government, party, business, or individual. Also, it is a high-level overall plan embracing the general goals and acceptable procedures especially of a governmental body

Cyber Process Cube The cyber process cube uses the people, processes, and technologies that compose a cyber target and breaks them down by their cyber threat framework components (i.e., preparation, engagement, presence, effect), with the observe–orient–decide–act (OODA) conceptual model as an additional dimension to determine the types of operational activities that will occur over the course of a cyberattack.

Cyber Process Evaluator   The Cyber Process Evaluator captures the people, processes, and technologies across the targeting cycle for estimating level of effort required on the part of the attacker

Cyber Proxy   Web actor, usually with an anonymous identity, that advocates for a particular position (e.g., Guccifer 2.0 in the 2016 US Presidential Election).

Cyber Range   Cyber ranges are virtual environments that use actual network equipment, as required. They can range from single stand-alone ranges in a single schoolhouse to Internet replicating ranges that are accessible from around the world.

Cyber Resilience   Cyber resilience refers to an entity's ability to continuously deliver the intended outcome despite adverse cyber events.

Cyber Risk   "Cyber risk" means any risk of financial loss, disruption, or damage to the reputation of an organization from some sort of failure of its information technology systems.

Cyber Risk Bow-Tie   A "bowtie" is a diagram that visualizes the risk you are dealing with in just one, easy to understand the picture. The diagram is shaped like a bow-tie, creating a clear differentiation between proactive and reactive risk management.

Cyber Risk Evaluation Framework   Methodical approach for information technology system's weaknesses and vulnerabilities. These frameworks are usually associated with improving system resilience.

Cyber Security Controls (CSCs)   CSCs distill several decades of experience to provide a set of rules for taking steps to protect a network. In addition, CSCs are prioritized, from most to least effective, providing the user with an added benefit of getting the highest estimated effect with the lowest numbered CSC.

Cyber Security Maturity Model   A cybersecurity maturity model provides a path forward and enables your organization to periodically assess where they are along that path.

Cyber Security Metric   Cyber security metrics identify what to measure and when, in obtaining a clear picture of the ability of a system to maintain a secure posture.

Cyber Security Risks   Operational risks to information and technology assets that affect the confidentiality, availability, or integrity (CIA) of information or information systems.

Cyber Targeting   (1) The process of selecting targets and matching the appropriate response to them on the basis of operational requirements; (2) Cyber operations conducted against cyber identities and cyber objects, resulting in a predefined effect vis-à-vis an actor. If successful, they result in a direct effect against these two cyber elements but, although targeting cyber objects and cyber identities, secondary effects are generated against or in support of persons, objects, and psyche

Cyber Threat Framework (CTF)   The Cyber Threat Framework was developed by the US government to enable consistent characterization and categorization of cyber threat events, and to identify trends or changes in the activities of cyber adversaries

Cyber Threat Intelligence   Cyber threat intelligence is what cyber threat information becomes once it has been collected, evaluated in the context of its source and reliability, and analyzed through rigorous and structured tradecraft techniques by those with substantive expertise and access to all-source information.

Cyber Threat Taxonomy   A cyber threat taxonomy categorizes instances of operational cyber security risks defined as operational risks to information and technology assets that have consequences affecting the confidentiality, availability, or integrity of information or information systems.

Cyber Vulnerability   Cyber vulnerability is a cyber-security term that refers to a flaw in a system that can leave it open to attack. A vulnerability may also refer to any type of weakness in a computer system itself, in a set of procedures, or in anything that leaves information security exposed to a threat.

Cybersecurity Framework (CSF)   NIST standard on defensive Cybersecurity Framework.

Data Model   A data model is an abstract model that organizes elements of data and standardizes how they relate to one another and to properties of the real-world entities.

DCLeaks   DCLeaks is a website that was established in June 2016. Since its creation, it has been responsible for publishing leaks of emails belonging to multiple prominent figures in the US government and military. Cybersecurity research firms say the site is a front for the Russian cyber-espionage group Fancy Bear.

Defensive Cyber Operations   Passive and active cyberspace operations intended to preserve the ability to use friendly cyberspace capabilities and protect data, networks, net-centric capabilities, and other designated systems.

Defense in Depth   Layered defense approach, with policy actions at the outer layer and the potential for kinetic actions at the inner layer.

Denial   Key JP 3–12 effect where the operator denies use of an adversary system.

Denial of Service   Computer system is denied the ability to perform legitimate services due to the actions of a malicious cyber actor.

Denial and Deception   Denial and deception (D&D) is a theoretical framework for conceiving and analyzing military intelligence techniques pertaining to secrecy and deception in order to secure a system.

Department of Defense Architecture Framework (DoDAF)   Standard framework for describing information technology operational, system, and technical views.

Department of Defense Cybersecurity Analysis and Review (DODCAR)   The DoDCAR is sponsored by the DOD deputy CIO for cybersecurity, the NSA deputy manager for National Security Systems, and the DISA director. DoDCAR is a threat-based, analysis-driven, repeatable process to synchronize and balance cybersecurity investments, minimize redundancies, eliminate inefficiencies, and improve all-around mission performance. DODCAR performs threat-based cybersecurity architecture assessments to ensure DOD leadership has the insight and knowledge to make well-informed, prioritized cybersecurity investment decisions that enable dependable mission execution of the unclassified and secret environments

Deter   Discourage (someone) from doing something by instilling doubt or fear of the consequences.

Diplomatic Cables   A confidential message exchanged between a diplomatic mission, like an embassy or a consulate, and the foreign ministry of its parent country. A diplomatic cable is a type of dispatch.

Disinformation   False information that is intended to mislead, especially propaganda issued by a government organization to a rival power or the media.

Distributed Denial of Service (DDoS)   The intentional paralyzing of a computer network by flooding it with data sent simultaneously from many individual computers.

Doctrine   Doctrine describes lessons learned and best practices, or teachings in the field.

Domain Generation Algorithm (DGA)   An algorithm to automatically generate domain names that are pseudo-random. Domain-flux botnets often use such an algorithm to generate their domain names in order to obfuscate communications.

Domain Name System (DNS)   A hierarchical, distributed naming system used to map from domain names to their corresponding Internet Protocol (IP) addresses and for other mappings.

Domain Name System (DNS) failure graph   A bipartite graph that represents the mapping between domain names and the Internet Protocol (IP) addresses of host interfaces that generated a query. This graph is constructed from the information extracted from failed DNS queries.

Domain Name System (DNS) lookup graph   A bipartite graph represents the mapping between fully qualified domain names and the Internet Protocol (IP) addresses mapped to them; this graph is constructed from the information extracted from successful DNS queries.

Domain-Flux   A technique based upon changing the domain name very frequently with (usually) algorithmically generated domain names in order to obfuscate communications.

DREAD   Technique invented at Microsoft to help security engineers build security into systems. DREAD stands for damage, reproducibility, exploitability, affected users and discoverability.

Droit International Applique aux Operations dans le Cyberspace   This document, still being translated at the time of this book's publication, is reportedly similar to the Tallinn Manual 2.0 in many key ways, but is more expansive in its definition of an attack in cyberspace.

Drone   See UAS.

Dropper   A dropper is a kind of Trojan that has been designed to "install" some sort of malware (virus, backdoor, etc.) to a target system.

Dumb Bomb   Air launched (usually) munition with no guidance or control system, that is, guided purely by gravity.

Duqu   Malware, associated with STUXNET; Duqu got its name from the prefix "~DQ" it gives to the names of files it creates. Duqu is a remote access Trojan

(RAT) that steals data from computers it infects. Duqu has been targeted at industrial equipment manufacturers, illegally collecting information about the manufacturer's systems and other proprietary data (Zetter 2014).

Effect Size   Effect size is a simple way of quantifying the difference between two groups that has many advantages over the use of tests of statistical significance alone. Effect size emphasizes the size of the difference rather than confounding this with sample size.

Election Tampering   Electoral fraud is illegal interference with the process of an election.

Electronic Warfare   Electronic warfare (EW) is any action involving the use of the electromagnetic spectrum (EM spectrum) or directed energy to control the spectrum, attack an enemy, or impede enemy assaults.

Element of Surprise   The unexpected or surprising character of something.

Equifax   Credit rating firm whose files were compromised via cyber means in 2017.

Executive Order   A rule or order issued by the president to an executive branch of the government and having the force of law.

Exemplar   A definitive example of a phenomena that provides a pattern.

Exfiltration Attack   Data exfiltration is the unauthorized copying, transfer, or retrieval of data from a computer or server. Data exfiltration is a malicious activity performed through various different techniques, typically by cybercriminals over the Internet or other network.

Factor Analysis for Information Risk (FAIR)   It is a portfolio-based approach for understanding a given enterprise's risk. It is based on an international standard (Jack Freund 2014).

Fake News   Deliberate disinformation designed to confuse or slander a targeted issue or individual, respectively.

Fast-Flux   A technique for changing the Internet Protocol (IP) addresses associated with a domain at high frequency in order to obfuscate communications.

Financial Intelligence   Financial intelligence (FININT) is the gathering of information about the financial affairs of entities of interest, to understand their nature and capabilities, and predict their intentions. Generally, the term applies in the context of law enforcement and related activities.

Flame   Flame, also known as Flamer, sKyWIper, and Skywiper, is modular computer malware discovered in 2012 that attacks computers running the Microsoft Windows operating system.] The program is being used for targeted cyber espionage in Middle Eastern countries (Zetter 2014).

Gauss   Bot believed to perform cyber intelligence in preparation for the STUXNET attack.

General Data Protection Regulation (GDPR)   Consisting of seven principles, the GDPR provides guidelines for the collection and dissemination of person information for individuals living in the European Union (EU).

Gray Propaganda   Propaganda that is never identified.

Great Game   Nineteenth century geo-political competition between Great Britain and Russia that spanned from diplomatic to military actions.

GRU   GRU is in the military intelligence service of the Russian Federation (formerly the Red Army of the Soviet Union). "GRU" is the English version of a Russian acronym that means Main Intelligence Directorate. The GRU is Russia's largest foreign intelligence agency.

Guccifer 2.0   "Guccifer 2.0" was a persona that hacked into the Democratic National Committee (DNC) computer network and then leaked its documents to the media and the website WikiLeaks. It was later found out that Guccifer2.0 was actually a 10–12 operative cell working for the GRU, or Russian Intelligence.

Hacker   A person who uses computers to gain unauthorized access to data.

Hactivist   A person who gains unauthorized access to computer files or networks in order to further social or political ends.

Hamilton 68   The Hamilton 68 dashboard, launched as part of the Alliance for Securing Democracy, provides a near real-time look at Russian propaganda and disinformation efforts online.

Honey File   File with beacon inserted so that the defender can identify the attacker at a future time. A Honey File is similar to the use of marked bills in bank robberies.

Honeypot   Honeypots mimic the behavior of potentially vulnerable applications in order to engage with, collect on, and analyze potentially malicious files and programs. In addition, honeypots may also be proactive, designed to browse the web or read spam emails like a regular user and open the received content. In addition, honeypots are categorized in terms of their interaction. A high interaction honeypot, for example, has all the features and applications that a human user would have. The goal is to capture the maximum amount of data on the adversary. Low interaction honeypots, on the other hand, have a reduced set of capabilities and focus on capturing only targeted adversarial behavior.

Honker Union   Honker or red hacker is a group known for hacktivism, mainly present in Mainland China. Literally the name means "Red Guest", as compared to the usual Chinese transliteration of hacker.

Host based Intrusion Detection System (HIDS)   Computer network intrusion detection system that looks for anomalous behavior to detect unauthorized operations on a system.

Human Intelligence   Intelligence gathered by means of interpersonal contact, as opposed to the more technical intelligence gathering disciplines such as signals intelligence (SIGINT), imagery intelligence (IMINT) and measurement and signature intelligence (MASINT).

Improvised Explosive Device (IED)   A bomb constructed and deployed in ways other than in conventional military action. It may be constructed of conventional military explosives, such as an artillery shell, attached to a detonating mechanism. IEDs are commonly used in roadside bombs and became a popular means of attacking Coalition Forces during Operation Iraqi Freedom (OIF) and Operation Enduring Freedom (OEF).

Incapacitate   Prevent something from functioning in a normal way.

Industrial Control System (ICS)   ICS is a collective term used to describe different types of control systems and associated instrumentation, which include the devices, systems, networks, and controls used to operate and/or automate industrial processes.

Information Assurance (IA) Information assurance is the process of securing computer systems for their confidentiality, integrity, and availability (CIA) to perform their specified functions.

Information Environment An information environment includes the people, processes, and technical systems that collect, process, and disseminate information.

Information Operations (IO) Information operations, also known as influence operations, include the collection of tactical information about an adversary as well as the dissemination of propaganda in pursuit of a competitive advantage over an opponent.

Information Operations (IO) Effect Change perception by re-framing of an issue in targeting a particular audience in order to change the perceived outcome (e.g., information operation to make a victor believe that a success was actually a failure).

Information Technology Technical systems (especially computers and telecommunications) for storing, retrieving, and sending information.

Information-Related Capability (IRC) A tool, technique, or activity employed within a dimension of the information environment that can be used to create effects and operationally desirable conditions.

Integrity Effect Change reliability of outcomes from a system; or cause the system to behave outside its designed scope of operations.

Intelligence The ability to acquire and apply knowledge and skills.

Intelligence, Surveillance and Reconnaissance (ISR) ISR functions are principal elements of US defense capabilities and include a wide variety of systems for acquiring and processing information needed by national security decision makers and military commanders.

Internet Research Agency (IRA) A Russian news organization implicated in disseminating "fake news" via social media posts during the 2016 US Presidential election.

Internet of Things (IoT) The Internet of things (IoT) is the extension of Internet connectivity into physical devices and everyday objects. Embedded with electronics, Internet connectivity, and other forms of hardware (such as sensors), these devices can communicate and interact with others over the Internet, and they can be remotely monitored and controlled.

IPv6 Internet Protocol version 6 (IPv6) is the most recent version of the Internet Protocol (IP), the communications protocol that provides an identification and location system for computers on networks and routes traffic across the Internet. IPv6 was developed by the Internet Engineering Task Force (IETF) to deal with the long-anticipated problem of IPv4 address exhaustion. The total number of possible IPv6 addresses is more than 7.9_10^{28} times as many as IPv4, which uses 32-bit addresses and provides approximately 4.3 billion addresses.

Irregular Warfare   Irregular warfare is defined in US joint doctrine as a violent struggle among state and non-state actors for legitimacy and influence over the relevant populations.

ISIL / ISIS   The Islamic State of Iraq and the Levant (ISIL), also known as the Islamic State of Iraq and Syria (ISIS), officially known as the Islamic State (IS), is a Salafi jihadist militant group and former unrecognized proto-state that follows a fundamentalist, Salafi doctrine of Sunni Islam.

ISO 31000   International standard for risk evaluation.

Joint Concept for Operating in the Information Environment (JCOIE) (2018)   The 2018 Joint Concept for Operating in the Information Environment addresses the incorporation of information considerations into all aspects of operations (Joint Chiefs of Staff, 2018).

Joint Integrating Concept (2009)   The Strategic Communications Joint Integrating Concept (Joint Chiefs of Staff, 2009) provided an overarching view of how communications are used throughout a military campaign cycle.

Joint Munitions Effectiveness Manual (JMEM)   The Joint Munitions Effectiveness Manual (JMEM) provides typical conventional weapon effectiveness against various elements. The JMEM is used in the capabilities analysis of the Joint Targeting Cycle to evaluate existing military capabilities, usually weapon systems, against approved targets.

Joint Publication (JP) 3–12 Cyber   Cyber operations joint doctrine.

Joint Publication (JP) 3–13 Cyber   Information operations joint doctrine.

KGB   Main security organization in the former Soviet Union.

Key Performance Parameter (KPP)   KPP, at the system / engineering level, prescribe the design thresholds (e.g., size, weight, power) that a system needs to meet its operational objectives.

Lateral Movement Framework   Lateral movement tools are designed to allow an actor (or penetration tester) to move around a network after gaining initial access.

Law of Armed Conflict (LOAC)   Branch of international law that states have generally agreed to. While there are ongoing updates, the four key principles are distinction, military necessity, unnecessary suffering, and proportionality.

Letter of Marque   A license to fit out an armed vessel and use it in the capture of enemy merchant shipping and to commit acts which would otherwise have constituted piracy.

(The) Line   Imaginary line on the spectrum from diplomacy to war that provides the border where an actor transition from diplomatic to hostile kinetic action.

Line of Communication (LOC)   A line of communication is the route that connects an operating military unit with its supply base. Supplies and reinforcements are transported along the line of communication. Therefore, a secure and open line of communication is vital for any military force to continue to operate effectively.

Live, Virtual, and Constructive (LVC)   LVC is a broadly used taxonomy for classifying models and simulations.

Lockheed Martin Attack Cycle   Step-by-step map of how a cyber attacker maneuvers from initial reconnaissance of a target to actions on objectives.

Log File Analysis   Network attack forensics performed by looking at log files, where an attacker will likely leave evidence for unauthorized system activity.

Main Stream Media (MSM)   MSM is a term and abbreviation used to refer collectively to the various large mass news media that influences a large number of people, and both reflect and shape prevailing currents of thought.

Malicious Cyber Actor (MCA)   A threat actor or malicious actor is a person or entity that is responsible for an event or incident that impacts, or has the potential to impact, the safety or security of another entity.

Managed Service Security Provider (MSSP)   An MSSP provides outsourced monitoring and management of security devices and systems

Manipulation   Key JP 3–12 effect where operator manipulates operations of adversary system

Maturity Modeling   A maturity model is a tool that helps people assess the current effectiveness of a person or group and supports figuring out what capabilities they need to acquire next in order to improve their performance.

Measure of Effectiveness (MOE)   An MOE is used to show how a system meets overall objectives. For example, in the modeling and simulation hierarchy (i.e., system / engagement / mission / campaign), MOEs will describe organization level performance at the mission/campaign level for overall scenarios. MOEs can also be used for individual systems, quantifying performance along an agreed upon dimension of performance, or even a probability.[15]

Measure of Performance (MOP)   An MOP describes a system's performance in terms of both the physical dimensions of the system (e.g., size, weight, power) and the human / operator skill. For example, in the military modeling and simulation hierarchy (i.e., system / engagement / mission / campaign), MOPs will describe the ability of the system to perform at the engagement / mission level.

Meme   Idea, including a picture or phrase, that passes between individuals, often through social media.

Metasploit   An tool that can perform automated identification and possible exploitation of system vulnerabilities, Metasploit (Rapid7) uses open-source data to continually improve Metasploit's ability to find and exploit network vulnerabilities.

Military Deception (MILDEC)   Military Deception includes taking action to mislead an adversary; elements of which will lead to mission accomplishment (Department of Defense, 2012).

Military Decision Making Process (MDMP)   The MDMP is a S Army seven-step process for military decision-making in both tactical and garrison environments.

Military Order   A military command or order is a binding instruction given by a senior rank to a junior rank in a military context.

---

[15] http://acqnotes.com/acqnote/tasks/measures-of-effectivenessrequirements

Mimikatz Privilege escalation tool, originally used for white hacking, but when released into the wild became a standard tool for malicious cyber actors.

Misinformation Misinformation is incorrect information that may be spread due to plausible rumors, or partial truths, that obfuscate actual facts. For example, conspiracy theories spread via social media make it a challenge to know which trending messages are true.

Modeling and Simulation Hierarchy Used for the construction of models, the M&S hierarchy divides the spaces from system (bottom), to engagement, then mission, finally to campaign as a structure for visualizing where models and their capability fit for application to a particular conflict domain.

Moonlight Maze Russian intelligence operation that exfiltrated US military documents via the Internet in the late 1990s.

Mosul Capture (2014) (Twitter Offensive) On 6 June, 2014, ISIL attacked Mosul from the northwest and quickly entered the western part of the city. The ISIL forces numbered approximately 1500, while there were at least 15 times more Iraqi forces. This attack was accompanied by near real-time tweeting, by ISIL, of movements; twitter reporting was believed to have contributed to the Iraqi force's decline in morale.

Moving Target Defense (MTD) MTD is the concept of controlling change across multiple system dimensions in order to increase uncertainty and apparent complexity for attackers, reduce their window of opportunity, and increase the costs of their probing and attack efforts. MTD assumes that perfect security is unattainable.

National Security Presidential Memorandum (NSPM) 13 Presidential Memorandum with a stated goal of "persistent engagement" (Nakasone 2019) that provides operators with a set of pre-delegated authorities to defend forward.

NETFLOW Method of monitoring network health by collecting and analyzing network traffic.

Network Intrusion Detection System (NIDS) Computer security system that monitors for malicious activity or policy violations.

Network System Under Test (NSUT) System under test (SUT) refers to a system that is being tested for correct operation.

Non-lethal Effect Of or pertaining to a weapon or effect not intended to cause death or permanent injuries to personnel.

Offensive Cyber Operations Offensive cyber capabilities are defined as operations in cyberspace to manipulate, deny, disrupt, degrade, or destroy targeted computers, information systems, or networks.

Office of Personnel Management (OPM) US government organization that housed personal information of individuals holding security clearances, which was compromised via cyber in 2015.

OODA Loop The observe–orient–decide–act (OODA) loop is the cycle, developed by military strategist and United States Air Force Colonel John Boyd. Boyd applied the concept to the combat operations process, often at the operational level during military campaigns.

Open Source Analysis   Data collected from publicly available sources to be used in an intelligence context. In the intelligence community, the term "open" refers to overt, publicly available sources (as opposed to covert or clandestine sources).

Operation Ababil (2012)   US banks were attacked with Distributed Denial of Service (DDoS) for 9 months in 2012 (Krebs 2013)

Operation Aurora (2009)   Directed collection effort (Carlin, 2018), where Chinese agents leveraged existing US law enforcement taps on Internet accounts (e.g., Google and Microsoft) in order to see who US authorities were interested in and the type of information that they were collecting.

Operation Orchard (2007)   2007 military operation where the Israeli Air Force used cyber to blind Syrian anti-aircraft radars prior to bombing a potential nuclear development site.

Operations Assessment   Evaluation of working effectiveness and suitability of a system through test methods aimed at (1) identification of defects, gaps, areas of risk; (2) measurement of the adequacy of the output; and (3) assessment of the reliability of the operations.

Operations Maturity Model   In the context of an individual organization, operations maturity is a measure of how well run the organization's IT operations are, e.g., the extent to which standard processes, monitoring tools, and resource consolidation characterize the organization's systems management approach.

Operations Research   The application of mathematical techniques to describe competitive scenarios, usually military, using pre-defined entities and over a short to intermediate term time horizon (e.g., days to months).

Operations Security (OPSEC)   OPSEC includes identifying critical information and ensuring procedures to keep that information secure.

Order of Battle   The order of battle of an armed force participating in a military operation or campaign shows the hierarchical organization, command structure, strength, disposition of personnel, and equipment of units and formations of the armed force.

OSINT   OSINT is Open Source INTelligence Information in the public domain or accessible from public sources.

Panama Papers   The Panama Papers are 11.5 million leaked documents that detail financial and attorney–client information for more than 214,488 offshore entities. The documents contain personal financial information about wealthy individuals and public officials that had previously been kept private.

Parallel System   A system is said to be a Parallel System where multiple processors have direct access to shared memory that forms a common address space.

Patch Management   Patch management is the process that helps acquire, test, and install multiple patches (code changes) on existing applications and software tools on a computer, enabling systems to stay updated on existing patches and determining which patches are the appropriate ones.

Patriotic Hacker   Hacker that sees himself operating on behalf of his country.

Peer-to-Peer   Denoting or relating to computer networks in which each computer can act as a server for the others, allowing shared access to files and peripherals without the need for a central server.

People's Liberation Army (PLA)   Unified organization of China's land, sea, and air forces. It is one of the largest military forces in the world. The PLA traces its roots to the 1927 Nanchang Uprising of the communists against the Nationalists.

Persona   Role that one adopts for an on-line activity.

Playbook   Borrowed from sports, a playbook provides a team (e.g., cyber defense) with a set of actions to take based on adversary behavior.

Point Target   A target of such small dimension that it requires the accurate placement of ordnance in order to neutralize or destroy it.

Policy   A course or principle of action adopted or proposed by a government, party, business, or individual.

Power Law   The power law describes a situation where a small number of influencers affect the thinking and opinion of a large number of influencees.

Power Shell   Task automation system consisting of command line shell and scripting language (Microsoft)

Pre Delegation of Authorities   A formal process by which an organization gives another entity the authority to perform certain functions on its behalf. Although the organization may delegate the authority to perform a function, it may not delegate responsibility for ensuring that the function is performed appropriately.

Precision Guided Munitions (PGMs)   A weapon that uses a seeker and guidance system to close in on a target with a defined location or energy emanation. Also called PGM.

Privateer   An armed ship with governmental authorities to fight and harass enemy ships—sometimes operating under the authority of a Letter of Marque.

Procedures   Procedures are the steps to get something done.

Process for Attack Simulation and Threat Analysis (PASTA)   The PASTA is a risk-centric methodology. It provides a seven-step process for aligning business objectives and technical requirements, taking into account compliance issues and business analysis. The intent of the method is to provide a dynamic threat identification, enumeration, and scoring process. In addition, this methodology is intended to provide an attacker-centric view of the application and infrastructure from which defenders can develop an asset-centric mitigation strategy.

Propaganda   A term made famous by Bernays (Bernays E. L., 1918), propaganda is used to project a narrative to influence a population as a center of gravity in a messaging campaign (Lasswell, 1927). Corporations, governments, and private individuals might use propaganda to influence a population concerning a product, program, or political issue, respectively. The end-state for a propaganda campaign is to steer opinion to the desired position by the entity generating the message.

Psychological Operations (PSYOP)   PSYOP includes the planning and use of information (propaganda) to influence and shape behaviors of governments / organizations / groups / individuals. Strategic PSYOP includes using information to influence target audiences. At the operational level, PSYOP are used for mission support. The term PSYOP was changed to military information support operations (MISO) in recent doctrine.

Pwn   Pwn is slang term derived from the verb own, meaning to appropriate or to conquer to gain ownership. The term implies domination or humiliation of a rival, used primarily in the Internet-based video game culture to taunt an opponent who has just been soundly defeated.

Ransomware   Cyber attack where the victim's data is held hostage until a ransom is paid.

Remote Access Tools (RATs)   A Remote Access Tool (RAT) is a program, which, once installed on a victim's machine, allows remote administrative control. In a malicious context, RATs can provide the ability for an actor to upload and download files, execute commands, log keystrokes, and/or record a user's screen.

Resilience   The capacity to recover quickly from difficulties; toughness.

Risk Bow-Tie   General technique that looks like a "bow-tie," with the attacker, controls and candidate mitigations on the left side, the event in the middle, and the remediations on the right side.

RT News   RT (formerly Russia Today) is a Russian international television network funded by the Russian government. RT is a brand of ""TV-Novosti"", an "autonomous non-profit organization", founded by the Russian news agency, RIA Novosti, on 6 April 2005.

SANS20   The SANS Critical Security Controls (CSCs), 20 best practices, in order of priority, is one example of practical knowledge, distilled from known TTPs, to provide defensive cyber personnel with step-by-step approaches for securing cyber infrastructure from possible attack.

SPE   Sometimes referred to as "mini-Flame," SPE is surveillance software believed to be used to collect information for the STUXNET industrial controller attack.

Schmitt Criteria   Six Criteria to Establish State Responsibility for a cyber attack (M. Schmitt 2011).

Security Operations Center (SOC)   A SOC can take multiple forms (e.g., dedicated, virtual, and outsourced) to implement an organization's security policies.

Security Technical Implementation Guide (STIG)   Based on DoD policy and security controls, these are implementation guides geared to specific products and versions, containing all requirements that have been flagged as applicable for the product that has been selected on a DoD baseline.

Shodan   Web searching tool to look for open industrial control ports on the Internet.

Sinkhole   A way of redirecting malicious Internet traffic so that it can be captured and analyzed by experts and/or law enforcement officials.

Social Media   Websites and applications that enable users to create and share content or to participate in social networking.

Social Network Analysis (SNA)   Social network analysis (SNA) is the process of investigating social structures through the use of networks and graph theory. SNA characterizes networked structures in terms of nodes (individual actors, people, or things within the network) and the ties, edges, or links (relationships or interactions) that connect them.

Sortie   For air operations, a sortie indicates the time to take off, perform of activity (e.g., collect information and drop a bomb), and return to base.

Soviet Union   A former communist country in Eastern Europe and northern Asia; established in 1922; included Russia and 14 other soviet socialist republics (Ukraine and Byelorussia and others); officially dissolved 31 December 1991. Synonyms: Russia, USSR, Union of Soviet Socialist Republics.

Spear Phish   Spear phishing is an email or electronic communications scam targeted towards a specific individual, organization or business. Although often intended to steal data for malicious purposes, cybercriminals may also intend to install malware on a targeted user's computer.

Sputnik   Formerly The Voice of Russia, and RIA Novosti, is a news agency, news website platform, and radio broadcast service established by the Russian government-owned news agency Rossiya Segodnya.

Star Wars   US defense program during the 1980s with the goal of providing a shield for the United States from intercontinental ballistic missiles (ICBMs).

State Diagram   A state diagram is a type of diagram used in computer science and related fields to describe the behavior of systems. State diagrams require that the system described is composed of a finite number of states.

STRIDE   Security engineering technique that helps developers with easy to remember memonic—spoofing, tampering, repudiation, information disclosure, denial of service, and elevation of privilege.

STUXNET   STUXNET is a malicious computer worm, first uncovered in 2010, thought to have been in development since at least 2005. Stuxnet targets SCADA systems and is believed to be responsible for causing substantial damage to Iran's nuclear program (Zetter 2014).

Supervisory Control and Data Acquisition (SCADA) System   Overall control system to coordinate the activities of the industrial processes in a continuous or discrete operations or manufacturing center. SCADA systems are also used to manage the respective industrial controllers for individual machines.

SysML   The systems markup language (SysML) was developed by the Object Management Group (OMG) to provide a technique for general system description

Systems Engineering Hierarchy   A structuring of models that includes levels of campaign, mission, engagement, and system. Goal is to match the right model to the right level for analysis.

Table Top Exercise   Discussion-based sessions where team members meet in an informal, classroom setting to discuss their roles during an emergency and their responses to a particular emergency situation. A facilitator guides participants through a discussion of one or more scenarios.

Tactics   Describes the way the actor chooses to operate over a course of action (COA). For example, tactics are associated with the achievement of short/medium term goal(s) via one out of many possible ways that involve human factors as the subject and means of these tactics.

Tactics, Techniques, and Procedures (TTPs)   The term tactics, techniques, and procedures (TTP) describes an approach for analyzing an actor's operation or can be used as a means of profiling behavior. The word tactics is meant to outline the way the actor chooses to operate over a course of action (COA). For example, tactics are somehow associated with achievement of short/medium term goal(s) via one out of many possible ways involving human factors as the subject and means of these tactics. Techniques are more related to the subject and its technicalities and specific details that imply or suggest a specific way to get something done; for example, installing / uninstalling a backdoor, performing a scan. Another definition is that tactics are means to implement strategies. Techniques are means to implement tasks. Procedures are then the standard, detailed steps that prescribe how to perform specific tasks.

Tallinn Manual   The first policy-level document addressing nation state response to a cyber attack that provides an academic, non-binding, assessment of how international law applies to cyber warfare. In addition, The Tallinn Manual has been proposed to NATO, and, although not adopted, is currently used as a reference

Target   A person, object, or place selected as the aim of an operation.

Target Element   A target element is a specific part of a target whose compromise or denial will result in an effect.

Target System Analysis (TSA)   An all-source examination of potential target systems to determine relevance to stated objectives, military importance, and priority of attack.

Taxonomy   A scheme of classification.

Techniques   Techniques are related to a target, its technicalities, and specific details that imply or suggest a specific way to get something done; for example, installing/uninstalling a backdoor, performing a scan.

Time Sensitive Targeting   A time-sensitive target with an extremely limited window of vulnerability or opportunity, the attack of which is critical to ensure successful completion of the Joint Force Commander's operations.

Trojan   In computing, a Trojan horse, or Trojan, is any malware that misleads users of its true intent.

Troll / Troll Farm   An organization whose employees or members (trolls) attempt to create conflict and disruption in an online community by posting deliberately inflammatory or provocative comments.

Unmanned Autonomous System (UAS)   High-tech, intelligent machines capable of traveling by air, land, or sea without a human crew on board.

U Boat   Anglicized term for World War II German Unterseeboot, or submarine. U Boats were used to slow Allied shipping in the North Atlantic during the first few years of World War II by sinking allied ships.

Unified Markup Language (UML)   The Unified Markup Language (UML) provides a method for describing computer programs for their design.

URL-flux   A technique using a username generation algorithm (so the URLs are associated with different user profiles) to disguise the command and control traffic in botnet

Virtual Private Network (VPN)   A VPN provides a secure connection between network connected nodes.

Wargame   A military exercise carried out to test or improve tactical expertise.

Watering Hole   A watering hole is an infected website where a targeted group is known to frequent.

Weaponeering   The process of determining the quantity of a particular type of weapon required to achieve a specific level of target damage by considering the effects of target vulnerability, warhead damage mechanism, delivery errors, damage criterion, and weapon reliability.

Weapons Platform   A weapons platform is generally any structure or system on which a weapon can be mounted. For example, a fighter jet is a weapons platform for missiles, bombs, or cannons.

Web Defacement   An attack on a website that changes the visual appearance of a website or a web page. This is typically the work of a someone who breaks into a web server and replaces the hosted website with one of their own.

Web Shell   Web shells are malicious scripts that are uploaded to a target host after an initial compromise and grant an actor remote access into a network. Once this access is established, web shells can facilitate lateral movement within a network.

White Propaganda   Openly revealing its source, white propaganda relies on persuasion and public relations techniques (e.g., RT News, Sputnik (Rutenberg, 2017)).

WikiLeaks   WikiLeaks is an independent, non-profit, online media organization that publishes submissions of otherwise unavailable documents from anonymous sources.

Zero Day   Zero day actually refers to two things—a zero-day vulnerability or a zero-day exploit. Zero-day vulnerability refers to a security hole in software—such as browser software or operating system software—that is yet unknown to the software maker or to antivirus vendors.

Cyber   Warfare   Lexicon   https://publicintelligence.net/the-vocabulary-of-cyber-war/

Index

A
Acceptable course of action (COA), 20
Active cyber ISR
~DQ (*see* Duqu (~DQ))
computer network, 108
cyber integrity attack, 107
Duqu, 110
flame, 110, 112–113
implementation, 108
mapping, 108
Stuxnet, 108, 109
Active defense, 26
Actor verification, 20
Advanced persistent threat (APT), 30, 51, 103, 124, 226
Adversary courses of action (COAs), 2
Adversary's scheme of maneuver, 105
Advertising, 61
Air tasking order system (ATO) production system, 223
Aircraft sorties, 202
Al Qaeda, 95
Al Qaeda in Iraq (AQI), 97
Amazon's data centers, 135
AMICA simulation, 223
Analytical framework, 7
Analyzing information operations framework, 9
Analyzing Mission Impacts due to Cyber Attacks (AMICA), 177, 223
and COATS, 223
MOPs, 223
Anomaly-based NIDS (A-NIDS) observes network characteristics, 132
Antivirus (AV) systems, 125, 280
APT10 to Western officials and researchers, 135

Arab Spring, 106
Architecture description language (ADL), 177, 178
Area targeting, 67, 76
Area *vs.* point targeting, 76
Asset target interactions (ATIs), 157
Astroturf, 68
Attack cycle, 122, 124, 160–162
Attacker, 136
and cyber kill chain, 122–124
Attack process methodology, 6
Attack surface, 124, 133, 137, 212
Australian essential eight, 151
Authorities
challenge, 21
delayed authorities, 22–23
functional organizations, 21
pre-delegated, 23–24
questions, 22
time constant, 23
Automated Computer Network Defense (ARMOUR), 51
Automated search/evaluation, 104

B
Ballistic missiles, 23
Battle damage assessment (BDA), 205
and biological weapons, 210
COLE, 207–209
commander uses, 206
cyber attacks, 206
cyber effect, 209, 210
defined, 205
JMEMs, 206, 207

© Springer Nature Switzerland AG 2022
J. M. Couretas, *An Introduction to Cyber Analysis and Targeting*,
https://doi.org/10.1007/978-3-030-88559-5

Printed in the United States
by Baker & Taylor Publisher Services